Sixth International Congress "Water, Waste and Environment" (EDE6-2017)

Africa, State of Play and Future Prospects

E3S Web of Conferences Volume 37 (2018)

Beni Mellal, Morocco
12 – 14 December 2017

Editors:

Ahmed Boukdir
Mustapha El Mabrouki

ISBN: 978-1-5108-6392-7

Printed from e-media with permission by:

Curran Associates, Inc.
57 Morehouse Lane
Red Hook, NY 12571

Some format issues inherent in the e-media version may also appear in this print version.

Printed by Curran Associates, Inc. (2018)

For additional information, please contact EDP Sciences – Web of Conferences at the address below.

EDP Sciences – Web of Conferences
17, Avenue du Hoggar
Parc d'Activité de Courtabœuf
BP 112
F-91944 Les Ulis Cedex A
France

Phone: +33 (0) 1 69 18 75 75
Fax: +33 (0) 1 69 28 84 91

contact-edps@webofconferences.org

Additional copies of this publication are available from:

Curran Associates, Inc.
57 Morehouse Lane
Red Hook, NY 12571 USA
Phone: 845-758-0400
Fax: 845-758-2633
Email: curran@proceedings.com
Web: www.proceedings.com

TABLE OF CONTENTS

FUNCTIONING OF THE PRIMARY AQUIFER RELATING TO THE MAIDER BASIN, MOROCCO: CASE OF THE ORDOVICIAN AQUIFER. 1
Ben-Said E., Boukdir A., Mahboub A., Younsi A., Zitouni A., Alili L., Ikhmerdi H.

HYDROGEOLOGICAL FUNCTIONING OF THE TABLECLOTH OF THE MIDELT FURROW (HIGH MOULOUYA, MOROCCO) 11
Ikhmerdi H., Boukdir A., Kossir A., Alili L., Ben-Said E.

ASSESSMENT OF THE EFFECTS REJECTIONS OF FEED FISH ON WATER RESOURCES.: (OUEDOUMERRBIA, MOROCCO) 20
Ouaissa Khadija, Kritihi Assia, Oumessoud Youness, Maychal Abdelaziz, Hasnaoui Mustapha

KARST GROUNDWATER VULNERABILITY MAPPING TO THE POLLUTION: CASE OF DIR SPRINGS LOCATED BETWEEN EL KSIBA AND OUAOUMANA (HIGH ATLAS, MOROCCO) 29
Alili L., Boukdir A., Maslouhi M. R., Ikhmerdi H.

ASSESSMENT OF GROUNDWATER QUALITY IN THE AL- BURAYHI AND HEDRAN SUB-BASIN, TAIZ, YEMEN – A GIS APPROACH 40
Naser Ramzy, El Bakkali Mohammed, Darwesh Nabil, El Kharrim Khadija, Belghyti Driss

DIAGNOSIS OF THE GHISS NEKOR AQUIFER IN ORDER TO ELABORATE THE AQUIFER CONTRACT 46
Baite Wissal, Boukdir A., Zitouni A., Dahbi S. D., Mesmoudi H., Elissami A., Sabri E., Ikhmerdi H.

THEORETICAL STUDY OF THE DISSOLUTION KINETICS OF GALENA AND CERUSSITE IN AN ABANDONED MINING AREA (ZAIDA MINE, MOROCCO) 58
El Alaoui Lamiae, Dekayir Abdelilah

IMPACT OF WASTEWATER FROM THE RURAL COMMUNE OF JMAA MOULBLAD ON THE PHYSICO-CHEMICAL QUALITY OF THE WATERS OF THE GROU RIVER (RABAT REGION, MOROCCO) 64
Arifi Karim, Tahri Latifa, El Abid Abdallah, Hefiane Fatima Zahra, Elblidi Souad, Yahyaoui Ahmed, Fekhaoui Mohammed

STUDY OF THE TECHNICAL PERFORMANCE OF LOCALIZED IRRIGATION AND ITS ENVIRONMENTAL AND AGROECONOMIC IMPACT IN THE FIRST AREAS OF COLLECTIVE RECONVERSION AT THE IRRIGATED PERIMETER OF THE TADLA - BENI MOUSSA PERIMETER OF THE WEST – MOROCCO 73
Mouradi Abdellah, Ait Yacine Zehor, El Harti Abderrazak

PUBLIC SERVICES FOR DISTRIBUTION OF DRINKING WATER AND LIQUID SANITATION IN URBAN ZONES IN MOROCCO RELEVANCE OF INTRODUCTION THE PERFORMANCE INDICATORS FOR PRESERVATION WATER RESOURCES. 81
Habib Akka, Abdelhamid Bouzidi, Said Housni

NEW TUBULAR CERAMIC MEMBRANES FROM NATURAL MOROCCAN CLAY FOR MICROFILTRATION APPLICATION 93
Ait Taleb A., El Baraka N., Saffaj N., Laknifli A., Mamouni R., Fatni A., El Hammadi A., El Qacimi N.

ADSORPTION OF CR (III) FROM AQUEOUS SOLUTION BY TWO FORMS OF A SUPERABSORBANT POLYMER : PARAMETRIC STUDY AND EFFECT OF ACTIVATION MODE 102
Ouass A., Essaadaoui Y., Kadiri L., Lebkiri I., Lafreme C., Cherkaoui M., Lebkiri A., Rifi E. H.

ADSORPTION OF AN ANIONIC DYE ON A NOVEL LOW-COST MESOPOROUS ADSORBENT: KINETIC, THERMODYNAMIC AND ISOTHERM STUDIES 110
Msaad Asmaa, Belbahloul Mounir, Zouhri Abdeljalil

ASSESSMENT OF LEACHATES FROM UNCONTROLLED LANDFILL: TANGIER CASE STUDY 119
Elmaghnougi I., Afilal Tribak A., Maatouk M.

CHARACTERIZATION OF PYROLYSIS PRODUCTS DERIVED FROM THREE BIOLOGICAL WASTES AND THEIR EFFECT ON PLANT GROWTH AND SOIL WATER RETENTION 129
Bouqbis Laila, Werner Koyro Hans, Kammann Claudia, Zohra Ainlhout Lalla Fatima, Boukhalef Laila, Cherif Harrouni Moulay

KINETIC STUDIES OF ADSORPTION OF CU (II) FROM AQUEOUS SOLUTION BY CORIANDER SEEDS (CORIANDRUM SATIVUM) 138
Kadiri L., Lebkiri A., Rifi E. H., Ouass A., Essaadaoui Y., Lebkiri I., Hamad H.

OPTIMIZATION AND CHARACTERIZATION OF GELATIN AND CHITOSAN EXTRACTED FROM FISH AND SHRIMP WASTE 145
M Ait Boulahsen, H Chairi, A Laglaoui, A Arakrak, S Zantar, M Bakkali, M Hassani
PHYSICOCHEMICAL AND BACTERIOLOGICAL CHARACTERIZATION OF RAW SEWAGE FROM THE MEKNES REGION MOROCCO 153
Hassan Ouallal, Hamzaoui Najia El, Mohammed Sbiti, Mohammed Machkor, Lahcen Messaoudi, Mohamed Azrour
INTEGRATING THE ENVIRONMENT IN LOCAL STRATEGIC PLANNING : GUIDELINES (CASE OF MOROCCO) 162
Benbrahim Hafsa
FROM THE PROMOTION OF BIODIVERSITY TO THE RECOVERY OF ORGANIC WASTE 171
Jammoukh Mustapha, Mansouri Khalifa, Salhi Bachir
CLIMATE CHANGE IMPACT ON RAINFALL: HOW WILL THREATEN WHEAT YIELD? 180
Tafoughalti K., El Faleh E. M., Moujahid Y., Ouargaga F.
ANALYSIS OF ENVIRONMENTAL IMPACTS OF RENEWABLE ENERGY ON THE MOROCCAN ELECTRICITY SECTOR: A SYSTEM DYNAMICS APPROACH 187
Chentouf M., Allouch M.
THE CLIMATE CHANGES IN THE SUB-BASIN OF THE OUM ER RBIA CENTRAL AND THE IMPACT ON THE SURFACE WATERS. 198
Echakraoui Zhour, Boukdir Ahmed., Aderoju Olaide., Hassan Ben – Saïd El, Zitouni Abdelhamid., El Maslouhi Rachid, Guerner Dias António
RADIATIVE FORCING OF THE DESERT AEROSOL AT OUARZAZATE (MOROCCO) 212
Tahiri Abdelouahid, Diouri Mohamed
HYDROLOGICAL AND HYDROCLIMATIC REGIMES IN THE OUERGHA WATERSHED 222
Msatef Karim, Benaabidate Lahcen, Bouignane Aziz
MODELLING SOIL SALINITY IN OUED EL ABID WATERSHED, MOROCCO 233
Mouatassime Sabri El, Boukdir Ahmed, Karaoui Ismail, Arioua Abdelkrim, Messlouhi Rachid, El Amrani Idrissi Abdelkhalek
QUANTIFICATION OF SOIL LOSS IN VARIOUS LITHOLOGICAL AREAS OF THE WESTERN MIDDLE ATLAS CENTRAL: APPLICATION TO THE RAS-ELMA, TAMELALET AND SEBAB WATERSHED (TIGRIGRA WATERSHED, MOROCCO) 242
Achiban Hassan, Taous Ali, El-Khantoury Ismail, El Mderssa Mohamed, Amechrouq Ali
ESTIMATION OF PARTICLE MATERIAL AND DISSOLVED FLOWS DURING FLOODS IN THE INAOUENE WATERSHED. (NORTHEAST OF MOROCCO) 250
Sibari Hayat, Haida Souad, Foutlane Mohamed
FLOOD RISK MANAGEMENT IN THE SOUSS WATERSHED 257
Bouaakkaz Brahim, El Abidine El Morjani Zine, Bouchaou Lhoussaine, Elhimri Hamza
APPLICATION OF ELECTRICAL GEOPHYSICS TO THE RELEASE OF WATER RESOURCES, CASE OF AIN LEUH (MOROCCO) 268
Zitouni A., Boukdir A., El Fjiji H., Baite W., Ekouele Mbaki V. R., Ben Said H., Echakraoui Z., Elissami A., El Maslouhi M. R.
CONTRIBUTION OF GRAVITY DATA AND SENTINEL-1 IMAGE FOR STRUCTURAL MAPPING. CASE OF BENI MELLAL ATLAS AND BENI MOUSSA PLAIN (MOROCCO). 285
Boutirame Ikram, Boukdir Ahmed, Akhssas Ahmed, Boutirame Fatima, Manar Ahmed, Aghzzaf Brahim
APPLICATION OF THE ELECTRICAL (GEOPHYSICAL) METHOD FOR THE IDENTIFICATION OF THE SEALED BEVEL IN THE COASTAL REGION OF WALIDIA PROVINCE OF EL JADIDA – MOROCCO 307
Elfjiji H., Boukdir A., Zitouni A.
SOLAR SYSTEM DESIGN FOR WATER PUMPING 321
Abdelkader Hadidi, Mohammed Yaichi
MODELING AND SIMULATION OF THE SHADING EFFECT ON THE PERFORMANCE OF A PHOTOVOLTAIC MODULE IN THE PRESENCE OF THE BYPASS DIODE. 330
Zebiri Mohamed, Mediouni Mohamed, Idadoub Hicham
Author Index

UNIVERSITY OF SULTAN MOULAY SLIMANE

FACULTY OF SCIENCE AND TECHNOLOGY

ABHOER

ACDPE

The 6th Edition of International Congress

Water Waste & Environment

EDE6-2017

under the theme :

« Africa, State of play and future prospects »

during 23, 24 and 25 November 2017
at the FST of Béni Mellal - Morocco

Sponsors

Coordinator : Mobile : 0661 07 07 47 – Email : ahboukdir@yahoo.fr

PROGRAMME DU CONGRES

Jour	Horaire	Amphi 5	Amphi de conférences	Salle C13 (Bloc C)	Salle C14 (Bloc C)
Mardi 12 décembre 2017	8h00 A 13h00	Accueil et inscription des participants et pauses café Cérémonie officielle d'ouverture à l'Amphi de conférences Mots d'ouverture Conférences plénières			
	13h00 A 14h30	DEJEUNER (1) Lieu : Forum FST			
	14h30 A 18h30	**Conférence 1 : Modélisation de la nappe de Boudnib Evaluation des ressources en eau et modèle de Gestion - Abderahman MAHBOUB - ABHGZR**	**Conférence 1 : Valorisation des eaux usées GHALI EL KHIYATI Mohamed – GIZ**	**Conférence 1 : Nouvelle approche pour l'evaluation de l'impact des changements climatique sur les ressources en eau dans le bassin d'oum Er R'bia - Raja ELKSABI - ABHOER/NOVEC-**	**Conférence 1 : Catalogue de bonnes pratiques de gestion des eaux pluviales Jihane NAIM -GIZ-**
		Thème **1**: (20 Exposés) **Modérateur :** ZEROUALI Abdelaziz -Directeur ABHBC- **Rapporteurs :** FIKRI Ahmed MESLOUHI Mohamed Rachid	Thème **2 (St2.2)**: (18 Exposés) **Modérateur :** WONG-WAH-CHUNG Pascal -Aix-Marseille Université- **Rapporteurs :** ZOUHRI Abdeljalil MILI El Mostafa	Thème **3**: (16 Exposés) **Modérateur :** HADDY Mohamed -IRAT-Kénitra- **Rapporteurs :** HILALI Mohamed ELBOUHALI AHMED	Thème **4**: (14 Exposés) **Modérateur :** BENJIRA Said/ DREF AKRAJAI Lahoucin/-Ex-Directeur ABHOER- **Rapporteurs :** YOUNSI Abdelkader AKHSSAS Ahmed
Mercredi 13 décembre 2017	8h15 A 13h00	**Conférence 2 : Planification intégrée de l'alimentation en eau potable au Maroc -ELISSAMI Abdeslam -ONEE-RABAT-**	**Conférence 2 : Réutilisation des eaux usées épurées au Maroc, expérience de l'ONEE, El HIDAOUI Abdelghni – ONEE-RABAT**	**Conférence 2 : Evaluation des changements climatiques au niveau du bassin d'Oum Er R'bia -Fatima DRIOUCH –DMN – Casablanca-**	**Conférence 2 : Appui institutionnel et opérationnel aux trois ABH Loukous, Moulouya et Sous Massa-Projet Morocco-Belge MOR 1204211 -Dahbi Salah /Mohamed BOUEZMARNI – DRPE/ CTB, /ABHL/ABHM,/ABHSM-**
		Thème **1** : (19 Exposés) **Modérateur :** AGHAZZAF Brahim -Directeur ABHOER- **Rapporteurs :** BOUCHAOU Lhoussaine HSSAISOUNE Mohammed	Thème 2 **(St2.1)**: (20 Exposés) **Modérateur :** EL HOUSSAINI Amine - Vice-Président Général Régional- **Rapporteurs :** HILALI Mohamed EL ALEM Noureddine	Thème **3**: (15 Exposés) **Modérateur :** MAHBOUB Abderrahman -ABHGZR- **Rapporteurs :** NAJINE Abdessamad ELBOUHALI AHMED	Thème **4**: (14 Exposés) **Modérateur :** BENJIRA Said/AKRAJAI Lahoucin -DREF/Ex-Directeur ABHOER- **Rapporteurs :** AKHSSAS Ahmed YOUNSI Abdelkader
	13h00 A 14h30	DEJEUNER (2) Lieu : Forum FST			
	14h30 A 18h30	**Conférence 3 : Contrat de nappe du complexe aquifère du Tadla - Ahmed OUAKILI -ANZAR-** Thème **1** : (19 Exposés) **Modérateur :** DAHBI Salahddine -ABHL- **Rapporteurs :** AKHSSAS Ahmed MILI El Mostafa	Thème **2 (St2.1)**: (19 Exposés) **Modérateur :** EL HIDAOUI Abdelghni -ONEE-RABAT- **Rapporteurs :** FIKRI Ahmed YOUNSI Abdelkader	Thème **5**: (12 Exposés) **Modérateur :** BAYOUMI Mustapha -ABHOER- Lhoussaine BOUCHAOU -Université Ibn Zohr- **Rapporteurs :** NAJINE Abdessamad HSSAISOUNE Mohammed	Thème **2** : **(St2.4)**: (11 Exposés) **Modérateur :** AKRAJAI Lahoucin -Ex-Directeur ABHOER- ELISSAMI Abdesslam/ ONEE Rabat **Rapporteurs :** AMIR Soumia HILALI Mohamed
Jeudi 14 décembre 2017	8h15 A 13h00	**Conférence 1 : La Smart City : Une solution innovante pour faire face aux défis de l'eau et de l'environnement ? Pr Isam SHAHROUR (Université de Lile - Lile, France)**	**Conférence 3 : Stratégie nationale du développement durable Hassan BAHHAR -Direction Régionale de l'Environnement-**	Activité Extra programme: Cours sur les bassins versants Par Velibor SPALEVIC UNIVERSITY DE MONTENEGRO, NIKSIC, MONTENEGRO	**Conférence OCP Mr Ramdani Khouribga - Maroc**
		Thème **6** : (19 Exposés) **Modérateur :** EL FOULADI Abderrahman - Université de Montréal - **Rapporteurs :** NAJINE Abdessamad BARAKAT Ahmed	Thème 2 **(St2.3)**: (16 Exposés) **Modérateur :** ELISSAMI Abdesslam -ONEE Rabat- **Rapporteurs :** ELBOUHALI AHMED AKHSSAS Ahmed		Thème 2 **(St2.6)**: (18 Exposés) **Modérateur :** WONG-WAH-CHUNG Pascal -Aix-Marseille Université- **Rapporteurs :** AMIR Soumia FIKRI Ahmed
	13h00 A 14h30	DEJEUNER (3) Lieu : Forum FST			
	14h30 A 16h30	Thème 6 : (10 Exposés) **Modérateur :** SHAHROUR Isam -UNIVERSITE DE LILLE- Mustapha MABROUKI -FST-BENI MELLAL- **Rapporteurs :** NAJINE Abdessamad BARAKAT Ahmed	Thème 2 **(St2.5)**: (10 Exposés) **Modérateur :** Pr Pascal WONG-WAH-CHUNG -Aix-Marseille Université- ELISSAMI Abdesslam -ONEE Rabat- **Rapporteurs :** MILI El Mostafa AKHSSAS Ahmed	Activité Extra programme: Filme OCP	******
	16h30 A 17h30	Pause-café			
	17h30 A 19h00	Table ronde, Remise des prix et clôture des journées (Amphi de conférences)			

UNIVERSITY OF SULTAN MOULAY SLIMANE
FACULTY OF SCIENCE AND TECHNOLOGY

l'ACDPE
&
l'ABHOER

in partnership with

organize

The 6th Edition of International Congress « Water, Waste and Environment » EDE6

under the theme:

« Africa,
State of play and future prospects»

during 12, 13 and 14 December 2017
at the FST of Béni Mellal - Morocco

Sponsors

PRESENTATION

Since twelve years, EDE has been a biennial platform for international scientific discussion on Water , waste and environment . it represents a good opportunity to exchange ideas, to strengthen existing and to create new academic networks, and to foster dialogue between the academia, public institutions, the private sector and civil society organizations on the recent global and regional trends in the environmental sector. Multidisciplinary results reported during EDE will contribute to the dissemination of knowledge and good practices to all actors and socio economics operators as well as the general public about the importance of environment , one of the most important strategic areas of many national research strategies.

The rationalization of water use and waste management have become an essential component of modern civilization. At present, environmental concerns become an inevitable challenge in the management of water, waste and the environment to ensure the sustainable development of the territory.

Coordinator :

Pr. Ahmed BOUKDIR
Ahboukdir@yahoo.fr
Tél : +212 661 07 07 47
ede6.2017@gmail.com

TOPICS

- **Theme 1 :** water resources protection– Aquifer contracts
  - Current state of water resources;
  - Aquifer contracts;
  - Integrated water resources management;
  - Artificial groundwater recharge;
  - Karstic aquifer Modelling;
  - Vulnerability to pollution;
  - Protection and safeguard perimeter (Sources, Dams…);
- **Theme 2 :** Waste Management and Environment
  - Sanitation and waste management;
  - Recovery and reuse;
  - Management and legislative aspects;
  - Leachate and landfill gas;
  - Environmental protection;
  - Environmental impact
- **Theme 3 :** Climate Change
  - Evaluation of climate change impacts;
  - Climate change and et water ressources management;
  - Climate change and sustainable development;
  - Climate change and renewable energy;
- **Theme 4 :** Watershed Management
  - Sustainable management of watersheds;
  - Management of natural resources and landscape conservation;
  - Water and soil conservation;
  - Natural risks;
- **Theme 5 :** Geophysics
  - Geophysical Methods;
  - Refraction seismics;
  - Applied geophysics to environment;
- **Theme 6 :** Renewable Energy
  - Green energy sources (wind,solar);
  - Geothermal energy;
  - Energy storage;

ORGANIZING COMMITTEE

Ahmed BOUKDIR (Coordonnateur : Faculté des Sciences et Technique de Béni Mellal)
Abdelhamid ZITOUNI (Faculté des Sciences et Technique « FST » de Béni Mellal)
Lahcen ALILI (Faculté des Sciences et Technique de Béni Mellal)
Hanaa BELHOU (Secrétaire du Congrès)
Zhour ECHAKRAOUI (Faculté des Sciences et Technique de Béni Mellal)
Elhassan BEN-SAID (Faculté des Sciences et Technique de Béni Mellal)
Hassan IKHMERDI (Faculté des Sciences et Technique de Béni Mellal)
El mouatassime SABRI (Faculté des Sciences et Technique de Béni Mellal)
Nadia LMOUDN (l'Association de Citoyenneté, du Développement et de Protection de l'Environnement « ACDPE » de la Région du Tadla Etendue Béni Mellal
Meryem CHAHBI (ACDPE-Béni Mellal)
Khaoula SAOUD (ACDPE-Béni Mellal)
El mahdi SAAF(ACDPE-Béni Mellal)
Mustapha JYAT (ACDPE-Béni Mellal)
Wissal BAITE (Faculté des Sciences et Technique de Béni Mellal)
Vivien R. EKOUELE MBAKI (Faculté des Sciences et Technique de Béni Mellal)
Hamou EL FJIJI (Faculté des Sciences et Technique de Béni Mellal)
Ikram BOUTIRAME (Faculté des Sciences et Technique de Béni Mellal)
Brahim AGHAZZAF (Agence du Bassin Hydraulique de l'Oum Er-Rbia)
El mostapha RAKIB (Université Sultan Moulay Slimane, Béni Mellal)
Mustapha MABROUKI (Faculté des Sciences et Technique de Béni Mellal)
Habiba BOUCHAKOUR (Faculté des Sciences et Technique de Béni Mellal)
Abdelkader YOUNSI (Faculté des Sciences et Technique d'El Jadida)
Mustapha BAYOUMI (Agence du Bassin Hydraulique de l'Oum Er-Rbia)
Mohamed Rachid MESLOUHI (Agence du Bassin Hydraulique de l'Oum Er-Rbia)
Ahmed AKHSSAS (Ecole Mohamadia des Ingenieurs - Rabat)
Latifa OUADIF (Ecole Mohamadia des Ingenieurs - Rabat)
Abdessalam ELISSAMI (Office National d'Electricité et de l'Eau - Branche Eau (ONEE-BO)-Rabat)
Ilyas BENTECHFINE (Faculté des Sciences et Technique de Béni Mellal)
Mounia OUBOUSELHAM (Agence du Bassin Hydraulique de l'Oum Er-Rbia)
Ahmed SAYH (Faculté des Sciences et Technique de Béni Mellal)
Aziz BALGHITI (Faculté des Sciences et Technique de Béni Mellal)

SCIENTIFIC COMMITTEE

Ahmed BOUKDIR (Université Sultan Moulay Slimane, Béni Mellal, Maroc);
Mustapha EL MABROUKI (Université Sultan Moulay Slimane, Béni Mellal , Maroc);
Mohammed HSSAISOUNE(Université Sultan Moulay Slimane, Béni Mellal , Maroc);
Abdelkader YOUNSI (Université Chouaib Doukkali, El Jadida, Maroc);
Ahmed AKHSSAS (Université Mohamed V, Rabat, Maroc);
Roger GUERIN (Université pierre et marie curie, Paris, France);
Nadia SAIYOURI (Université de bordeaux, France) ;
Luliu BOBOS (Université de Porto, Portugal) ;
António GUERNER DIAS (Université de Porto, Portugal) ;
Abdessamad NAJINE (Université Sultan Moulay Slimane, Béni Mellal , Maroc);
Ahmed BARAKAT (Université Sultan Moulay Slimane, Béni Mellal, Maroc);
Lhoussaine BOUCHAOU (Université Ibn Zohr, Agadir, Maroc);
Mohamed HILALI (Université Moulay Ismail, Er-Rachidia, Maroc);
Mustapha MILI (Université Mohamed Ismail, Meknès, Maroc);
Noureddine FTOUHI (Université Cadi Ayyad, Marrakech, Maroc);
Soumia AMIR (Université Sultan Moulay Slimane, Beni Mellal, Maroc);
Latifa OUADIF (Université Mohamed V, Rabat, Maroc);
Abdeljalil ZOUHRI (Université Hassan I, Settat, Maroc) ;
Nour-eddine EL ALEM (Université Ibn Zohr Agadir, Maroc);
MERZOUKI Mohammed (Université Sidi Mohammed Ben Abdellah, Fès, Maroc);
Mohammed SBAA (Université Mohammed Premier, Oujda, Maroc);
Lahcen BENABIDATE (Université Sidi Mohammed Ben Abdellah, Fès, Maroc);
Mina AMHARREF (Université Abdelmalek Essaadi, Tanger, Maroc);
Mohamed Kefi (Water Researches and Technologies Center (CERTE), Ecopark Borj Cedria, Tunisie);
Driss KHATTACH (Université Mohammed Premier, Oujda, Maroc);
Ahmed FEKRI (Université Hassan II, Casablanca, Maroc);
Ahmed EL BOUHALI (Institut Royal d'Administration Kénitra, Maroc).

COMMITTEE OF HONOR

Le Secrétaire Général du Secrétariat d'Etat Chargé de l'Eau-Rabat;
Le Secrétaire Général du Secrétariat chargée du développement durable-Rabat;
Le Secrétaire d'Etat Chargé de l'Enseignement Supérieur de la Recherche Scientifique –Rabat;
Le Wali de la Région Beni Mellal Khenifra;
Le Président du Conseil régional de la région Beni Mellal Khenifra;
Le Président de l'Université Sultan Moulay Slimane (USMS);
Le Directeur de l'Agence du Bassin Hydraulique de l'Oum Er R'Bia (ABHOER)-Beni Mellal;
Le Président de l'Association « ACDPE »;
Le Doyen de la Faculté des Sciences et Techniques de Béni Mellal (FST-BM);
Le Directeur de l'Agence allemande pour la Coopération internationale(GIZ)-Allemagne;
Le Directeur Régional de la Direction de la Recherche et de la Planification de l'Eau-Rabat;
Le Président Directeur Général de CIMAT-Casablanca;
Le Directeur de l'Office Chérifien des Phosphates-Khouribga (OCP);
Le Directeur de l'Agence du Bassin Hydraulique du Bouregreg et du Chaouia (ABHBC)-Ben Slimane;
Le Directeur de l'Agence du Bassin Hydraulique du Loukkos (ABHL)-Tétouan;
Le Directeur de l'Agence du Bassin Hydraulique de la Moulouya (ABHM)-Oujda;
Le Directeur de l'Agence du Bassin Hydraulique du Guir Ziz Rhris (ABHGZR)- Errachidia;
Le Directeur de l'Agence du Bassin Hydraulique du Sous Massa Draa (ABHSMD)-Agadir;
Le Directeur Régional de l'Office National d'Electricité et de l'Eau - Branche Eau (ONEE-BO)-Rabat;
Le Directeur Régional de l'Office National d'Electricité et de l'Eau – Branche Electricité (ONEE-BE)-Béni Mellal;
Le Directeur Provincial de l'Office National d'Electricité et de l'Eau – Branche Electricité (ONEE-BE)-Béni Mellal;
Le Secrétaire Général de Fondation de La Banque Populaire Casablanca;
Le Directeur de l'Agence Belge de Développement (CTB)-Rabat;
Le Directeur de la Direction Régionale des Eaux et Forêts - Béni Mellal;
L'Ambassadeur de France au Maroc-Rabat;
Le Directeur de l'Académie Régionale de l'Education et de la Formation-Béni Mellal ;
Le Directeur du Centre National pour la Recherche Scientifique et Technique-Rabat;

edp sciences

Web of Conferences

Statement of Peer review

In submitting conference proceedings to *Web of Conferences*, the editors of the proceedings certify to the Publisher that

1. They adhere to its **Policy on Publishing Integrity** in order to safeguard good scientific practice in publishing.
2. All articles have been subjected to peer review administered by the proceedings editors.
3. Reviews have been conducted by expert referees, who have been requested to provide unbiased and constructive comments aimed, whenever possible, at improving the work.
4. Proceedings editors have taken all reasonable steps to ensure the quality of the materials they publish and their decision to accept or reject a paper for publication has been based only on the merits of the work and the relevance to the journal.

Title, date and place of the conference

The Sixth International Congress "Water, Waste and Environment" EDE6-2017
12-14 December 2017
Faculty of Science and Technology – Beni Mellal, Morocco

Proceedings editor(s):

Pr. Ahmed BOUKDIR

Pr. Mustapha EL MABROUKI

Le Coordonnateur

EDE6 2017

Pr. Ahmed BOUKDIR

Date and editor's signature

24-04-2018

Functioning of the Primary Aquifer Relating to the Maider Basin, Morocco: Case of the Ordovician aquifer .

E. Ben-said*[1], *A.* Boukdir[1], *A.* Mahboub[2], *A.* Younsi[3] *A.* Zitouni[1], *L.*Alili[1], *H.*Ikhmerdi[1].

[1] Faculty of Sciences and Techniques, University Sultan Moulay Slimane, Béni Mellal, Morocco.
[2] Guir-ziz-rheris Hydraulic Basin Agency, Errachidia, Morocco.
[3] Faculty of Sciences /University Choaib Doukali , Eljadida, Morocco.

Abstract. The basin of Maider is limited northly by the vast ensemble Oriental Saghro-Ougnate, from the east by the Tafilalet plain, from the west by the oriental Jbel Bani, finally from the south and south-east by the Cretaceous Hamada of Kem-Kem. During last decades, groundwater in the basin of Maider, is confronting degradation in both cases: Quantitative and qualitative, as a result of the drought, the overexploitation and the salinization. The aim of this action research is to understand the current state of water resources in the area of stady. At the end of this work, we can get the following conclusions: the general flow of the ordovician aquifer is always directed from the north to the south-east of the basin by following the principal axes of the wadis:Taghbalt ,Hssiya and Fezzou. The recharge of the aquifer is primarily done, either by the underground flow, or by the surface runoff of torrential waters from the upstream of Jbel Saghro. The piezometric anomaly noticed at the level of Ait Saàdane, explained by overexploitation linked to the needs of irrigation water. The physicochemical approach for the Maider basin identifies two essential factors of the salinisation of groundwater: the dissolution of the aquifer which is rich in minerals with high temperature on the one hand, and the decrease of the piezometric surface due to the overexploitation and drought on the other hand.
Key words: Maider, groundwater, drought, overexploitation, salinization, flux, physicochemical.

* Email of corresponding : hbensaid40@gmail.com

1 Introduction:

In Morocco, particularly in the anti-atlas domain, water is one of the major problems because it represents a limiting factor key to sustainable development, as it improves the quality of life and peace, bearing hard circumstances as arid climate and lack of groundwater reserves.

During last decades, the groundwater in the Maider basin object of this study is increasingly exploited as a result of sharp increase of water needs, (traditional irrigation), of drought and of the climate semi-desert characterizing this region. within this framework, a good understanding of the functioning of the aquifers of the study area and the use of an appropriate management of water resources, are needed.

The quality of groundwater in the basin, which is moreover degraded, is a true challenge not less important than the scarcity of this vital resource.

This action research, is to apprehend the hydrogeological operation of the Paleozoic aquifer relating the basin Maider in function of the piezometric characteristics, hydrodynamic and hydrochemical of environment; special attention to the knowledge of the water of the Ordovician sandstone which can be a perennial resource of water in the basin.

This approach is carried out thanks to an inventory of water points comprising a hundred of boreholes and wells located in the basin, through measurement of the piézométrie, physico-chemical parameters and the call to the hydrodynamic data points of water during the pumping test.

2 Description of the area of study:

The area of study as a whole located in the Anti-Atlas, constitutes at the same time southern part and eastern part of this area; it is limited northly by the broad set Saghro oriental-Ougnate, from east by the plain of the Tafilalt, and from the west by the Jbel Bani oriental, finally from the south and south-east by the Cretaceous Hamada of Kern-Kem [1].

On this watershed, precipitations are very low, the average annual rainfall observed at the meteorological stations of Tazarine and Alnif, on the period 1982-2007, is respectively of 93 mm and 86mm; most of the rain (>50 %) is concentrated between October and February, these precipitations vary from 200 mm on Jbel Saghro to 87,5mm on the plains of Nkoub and Alnif, and achieved less than 50mm on the plain of the Maider basin. The topography of the basin is varied, with altitudes between 2692 m to the upstream (NNW) and 613 m in the outlet (Daya-el- Maïder SE). The socio-economic activity, in the basin of Maider where the inhabitants number is 418450 (according to the general census of Morocco in 2014) is essentially agricultural and pastoral [1] nevertheless the mining activity occupies economic importance in the region [2]. The density of the population close to 34 hab/km2, knows an annual growth of 0.58 % for the period (2004-2014).

The rocks are essentially sedimentary and metamorphic formations of the anti- atlas domain, dated Paleozoic whose lithologically can be presented as of the shale, sandstone and limestone [3]. Concerning water resources, the potential aquifers, although the reserves are of little important, are Ordovician age essentially sandstone and in Devonian limestone [4].

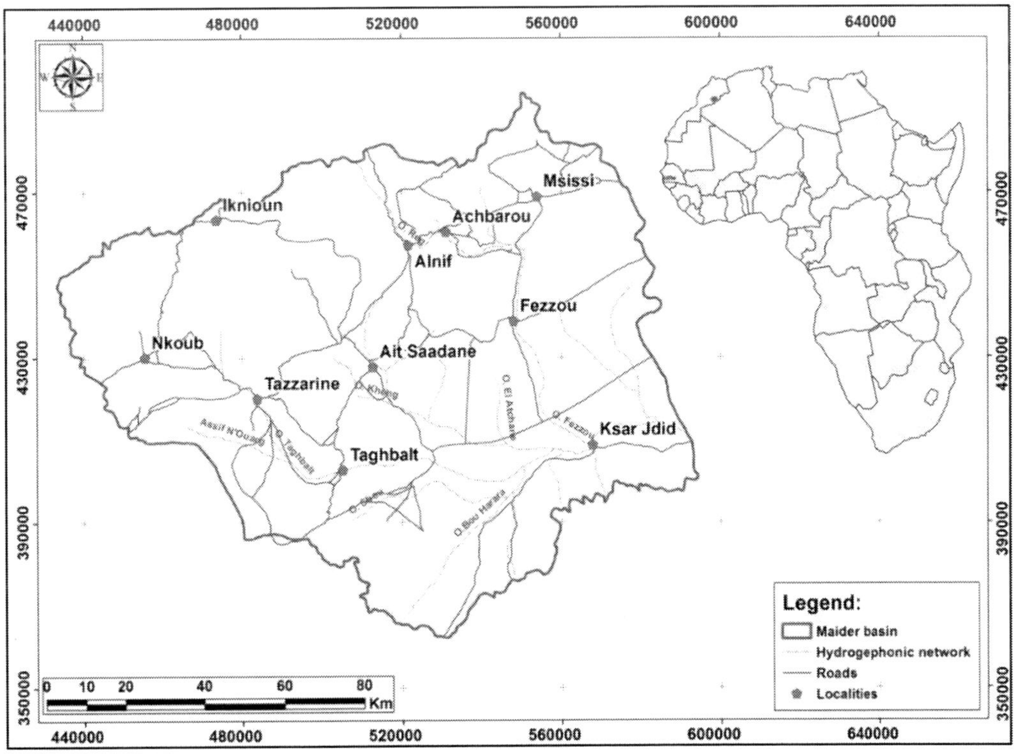

Fig. 1: Geographical situation of the Maider basin

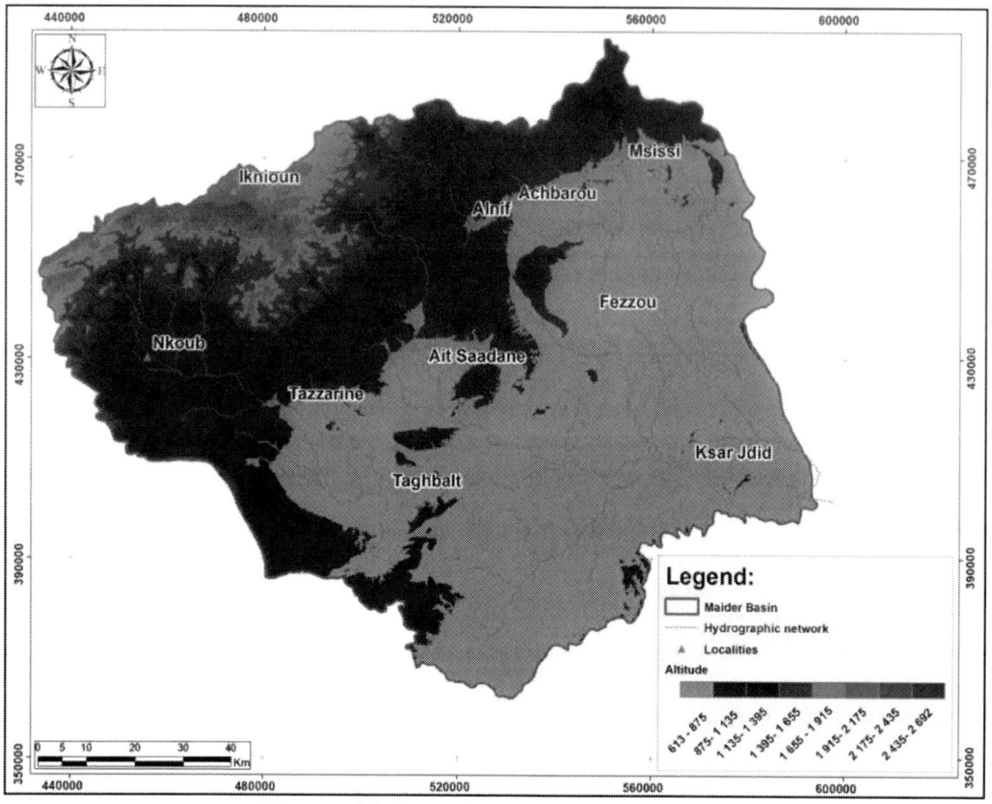

Fig. 2: Digital Terrain Model of Maider basin.

3 Data and Methods:

The methodological approach proposed in this study is based on the collection, treatment and the analysis of the entire piezometric and hydrodynamic data relative to the basin studied. The acquisition of data constitutes the main step in the realization and the functioning of this study. A part of the data used have been acquired from the campaigns of the measures and the technical data sheets of many wells drilled in the basin, in the framework of the partnership with the Guir-Ziz-Rheris Hydraulic Basin Agency.

The hydrochemical data are incorporated the physicochemical parameters: temperature, pH, electrical conductivity (EC, A' 25 8C) on a hundred wells. The follow-up of the piézometry is also been carried out on the same network of water points for the three drowned levels that the basin contains: Ordovician, Devonian and Cambrian.

The Hydrodynamic data are got from 117 water points distributed between the borehole and the wells; it is in particular parameter of debit and transmissivity. After treatment of the serie of gotten data, this undergoes a representation of Krigeage method on the whole basin.

Furthermore, the use of a platform of Geographic Information System (GIS tools), occupies a great importance in the achievement of this work. Indeed the elaboration of piezometric maps and other cartographic documents, which will be shown below, are useful for the confection of the study.

4 Results and Discussion :

4.1 Piézometry and structure of flow:

In order to highlight the current state of water resources in the Maider basin, we have established the piezometric map for the Ordovician aquifer on the basis of the statements of the year 2017.

The analysis of this piezometric map (Fig 3) shows that the general flow of the ordovician aquifer is always directed from the north to the south-east of the basin by following the principal axes of the wadis: Taghbalt, Hssiya and Fezzou. Considering that the area of study do annually receives more than 90m of precipitations, it is clearly shown that the recharge of the aquifer is primarily done, either by the underground flow, or by the Surface runoff of torrential waters through Jbel Saghro. According to this map we can differentiate several zones by their hydraulic gradient:

- The zone center and southeast of the basin are characterized by a very low hydraulic gradient which varies between 0.7 to 0.3%. These variations of the Gradient reflect an easy flow linked to topography plain of the area;
- The Northwest Zone of the basin is characterized by a low hydraulic gradient of the order of 1 per cent. In this area, the tightening of the isopiestic lines is linked to poor hydrodynamic characteristics of the aquifer;

The North-East area (north of the axis Alnif-Mssici) of the basin where the hydraulic gradient is important to the order of 4% for the current state (2017), is distinguished by high values that can be attributed to poor permeabilities and the tilt of the substratum at this level.

Noting the existence of an anomaly in piezometric level of Ait Saadane in the center of the basin, materialized by the curve 750m, that can be explained by the overexploitation of the aquifer because this area has experienced in recent years a pronounced extension of irrigated areas.

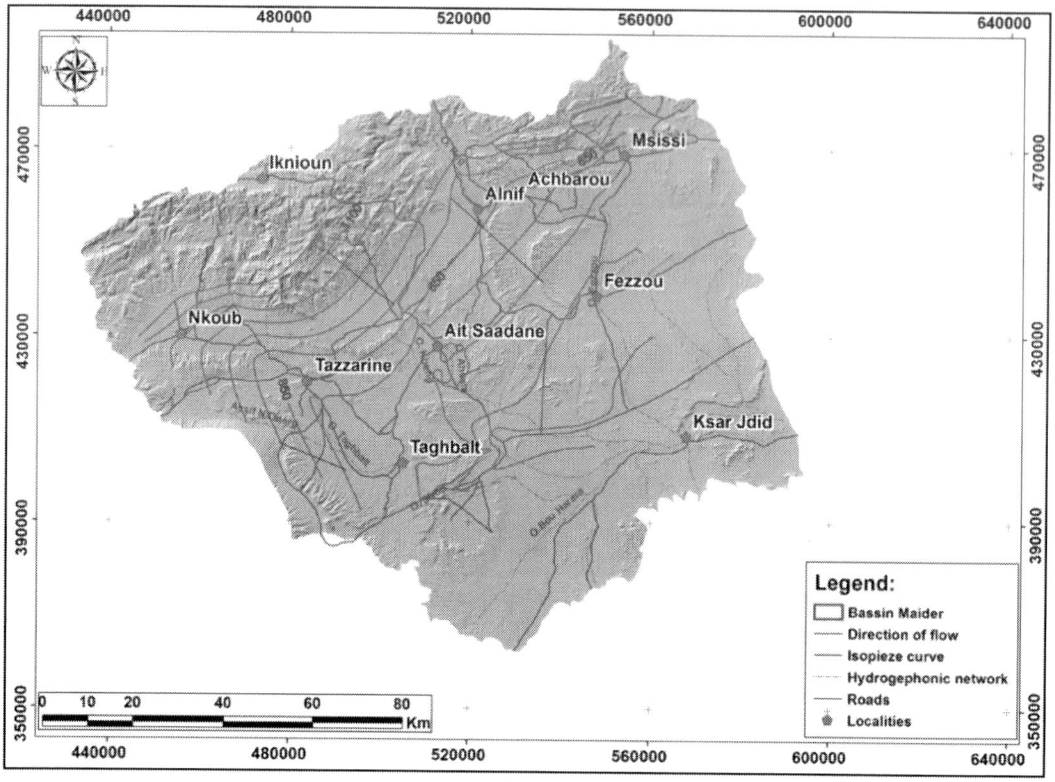

Fig. 3 : The piezometric map of primary aquifer in the Maider basin for year 2017

4.2 Productivity:

In totality, the Paleozoic aquifer of the Maider Basin, due to its low porosity (...), does not offer good transmissivities which constitute a hydrodynamic parameter in determining the evaluation of productivity.

From a database of 117 water points (wells and boreholes), well distributed over the study area, the values map of the transmissivity (Figure 3) of the Maider region was realized. It shows, of course, the lowest transmissivities at the west of the basin, while the maximum values are confined at the east, mainly at the level of Alnif upstream.

Concerning flow rates, we tried to recover as much as possible a good coverage of data, so as to have an idea about the state of circulation of groundwater in the basin. FIG. 4 below, which shows the flow distribution, according to values recorded in 132 water points around the basin, obviously shows that the flow rates are largely very low and do not exceed 2l/s for 80% of the points, while those with relatively high flow rates do not exceed twenty.

 https://doi.org/10.1051/e3sconf/20183701001

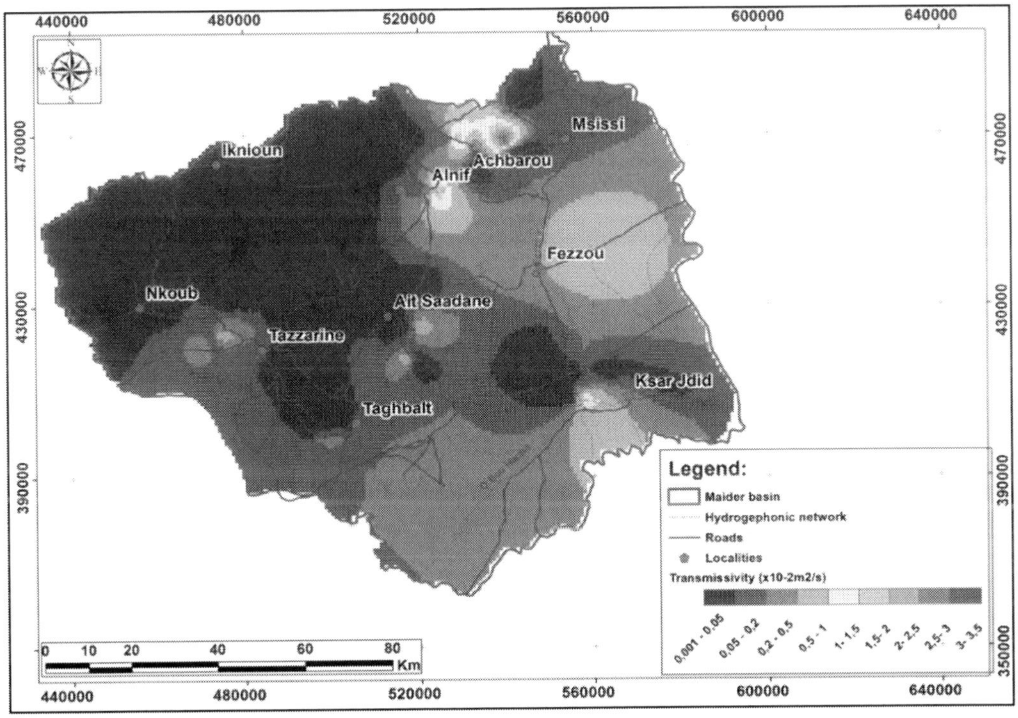

Fig. 4 : the transmissivity Map of primary aquifer in the Maider basin

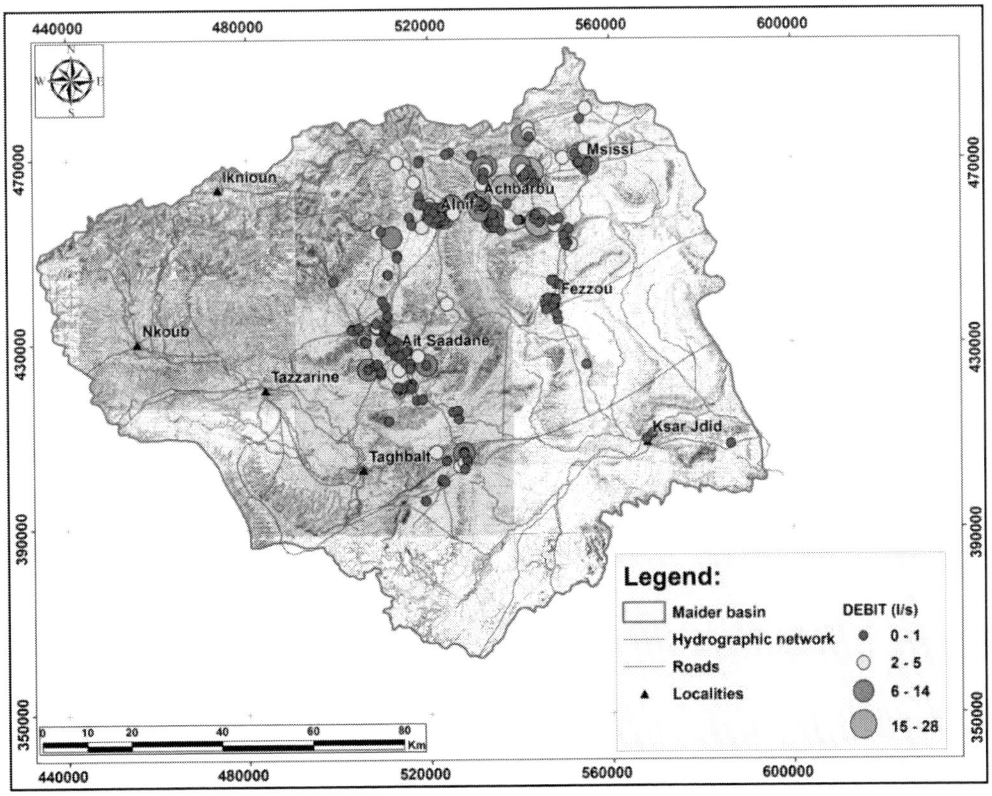

Fig 5 : the distribution map of flow rates in the Maider basin.

 https://doi.org/10.1051/e3sconf/20183701001

4.3 Physicochemical Quality :

To have an idea of the overall quality of groundwater in the basin Maider, we were limited to electrical conductivity. This last is the inverse of the electrical resistivity; it increases with the content of dissolved salts (mineralization). Groundwaters in the basin studied have high values of conductivity in general; which exceed greatly the guide value set by the European standards (100µS.cm-1). They are, generally, included between 349 µS.cm-1 in the vicinity of Taghbalt, and 5200 µS.cm-1 at the south-east of the basin. The map of electrical conductivity (Figure 6) highlights on three groups of different qualities:

- In the eastern portion of the basin (Corridor N'kob-Alnif and basin Tazarine-Taghbalt), the mineralization of groundwater is relatively good and the electrical conductivity measured does not exceed 1000 µS.cm-1.
- In areas that constitute the center and the major part of the basin (the basin of Ait Saadane and the western part of the Fezzou basin), the mineralization of the groundwater is mediocre and the electrical conductivity is between 1000 µS. cm-1 to more than 3500 µS.cm-1.
- In the southern corridor of the basin Tismoumine-Majrane, the mineralization of groundwater which reaches 5400 µS.cm-1is much worse, with a tendency towards increase.

The strong mineralization of the water in the basin is generally explained by the lithological nature of the saline formation and by the changes of the hydrological regime, due to the decrease of the water assimilation process caused by drought (significant reduction at the level of the water table).

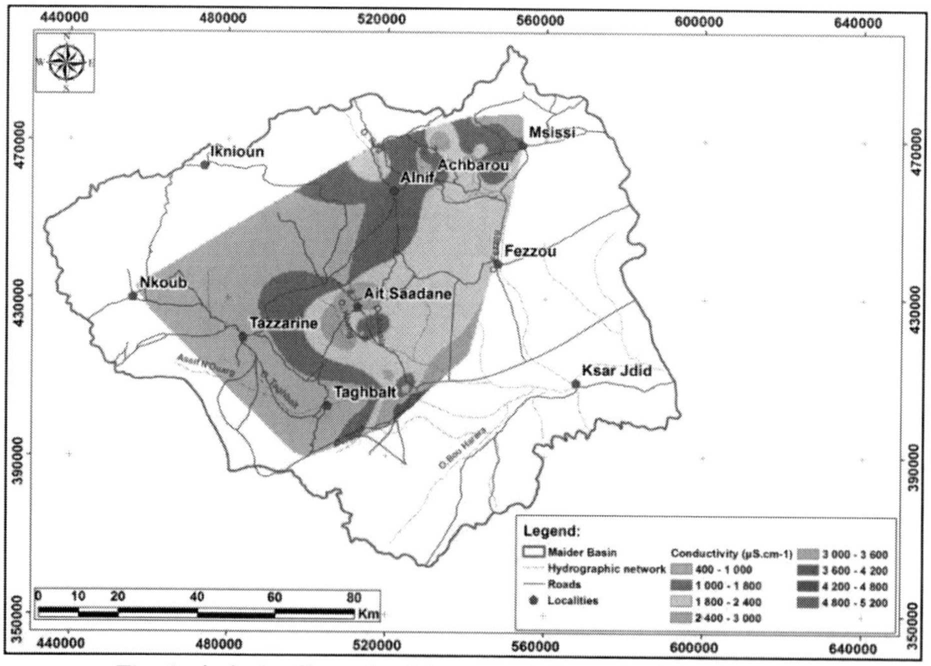

Fig. 6 : the hydraulic conductivity map of the Ordovician Aquifer.

The temperature degrees of the waters of the ordovician groundwater of Maider basin are between 20.8 and 29.70°C, with an average of 24.7 °C. They present a lower difference at the Aquifer level; this is due to the fact that the measures have been distributed in a broad interval of time and that the sampling has been carried out at varying depths.

In what concerning the pH of the groundwater in this basin, it varies between 5.8 and 8.3, with an average of 7.2.

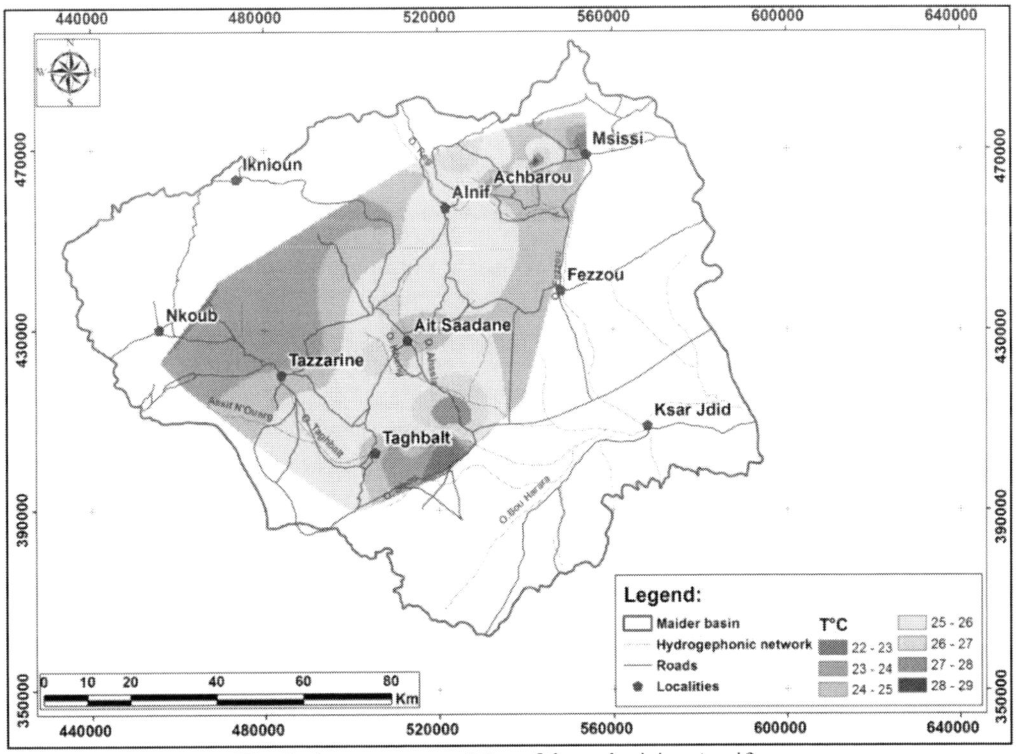

Fig. 7: the temperature Map of the ordovician Aquifer.

5 Conclusion:

In this work, we can describe the primary aquifer of the Maider basin and propose some prospects that can be seen pertinent.

- The general flow of the ordovician aquifer is always directed from the north to the south-east of the basin by following the principal axes of the wadis: Taghbalt ,Hssiya and Fezzou.
- View that the studied basin do annually receives about 90m of precipitations, it is clearly shown that the recharge aquifer is primarily done, either by the underground flow, or by the surface runoff of torrential waters from the upstream of JBEL SAGHRO.

- The piezometric anomaly noticed at the level of Ait Saadane (in the center of the basin) that has materialized by the curve 750 m, is explained by an overexploitation of the water table from the fact that this area knows in last years an important extension at the level of irrigated surfaces.

The physicochemical approach for the Maider basin identifies two essential factors for groundwater salinisation: The dissolution of the aquifer formation which is rich in minerals, with high temperature on the one hand, and the decrease of the piezometric surface due to the overexploitation and drought on the other hand.

Although the scarcity and the salinisation of groundwater appear as irreversible phenomena, it is possible to limit their deterioration and their geographical extension. The control of intensive pumping at the whole basin, the use of saving water techniques in irrigation, the artificial recharge of the aquifer, and the demineralization techniques, all these are the main actions to be taken to reduce the risks of the water resources degradation in both: qualitatively and quantitatively sides.

Acknowledgments:

Our great thanks to the administration and agency staff of the Guir-Ziz-Rhéris Hydraulic Basin of Errachidia to put at our disposal the data used in this article, as well for their assistance and their constructive suggestions. We also thank all those who have contributed in the fullfilment of this work.

References :

[1] Combe M., Ruhard J.P., & Yacoub M., (1977) - Présentation du bassin du Maider. In : Ressources eau du Maroc : Domaine atlasique et sud-atlasique, Notes et mémoires du service géologiqque du Maroc, **N°231** pp. 416-429.

[2] A. Mouttaqi, E.C. Rjimati, L. Maacha, A. Michard, A. Soulaimani & H. Ibouh, 2011. LES PRINCIPALES MINES DU MAROC. In: Nouveaux Guides géologiques et miniers du Maroc, **vol. 9** (Michard, A., Saddiqi, O., Chalouan, A., Rjimati, E.C., Mouttaqi, A., Eds.). Notes & Mém. Serv. géol. Maroc, **564**, 25-71.

[3] Michard, A.,Hoepffner, A., Soulaimani, A. & Baidder, L. 2008. The Variscan belt. In Continental Evolution: The Geology of Morocco. Structure, Stratigraphy, and Tectonics of the Africa-Atlantic-Mediterranean Triple Junction (eds A. Michard, A. Chalouan & O. Saddiqi), pp. 65–132. Lecture Notes in Earth Sciences, **116**. Springer-Verlag.

[4] Agence du Bassin Hydraulique du Guir-Ziz-Rhéris (ABH GZR) 2011. Actualisation du Plan Directeur d'Aménagement Intégré des Ressources en Eau des bassins de Guir, Rhéris, Ziz et Maïder (provinces d'Errachidia, Ouarzazate, Zagora et Figuig). Rapport définitif de la mission **1**, 91-120.

Hydrogeological functioning of the tablecloth of the Midelt Furrow (High Moulouya, MOROCCO)

H. IKHMERDI [1,*1], *A.* BOUKDIR[1], *A.* KOSSIR[2], *L.* ALILI[1] and E. BEN-SAID[1]
(1) FST Béni Mellal/Université Sultane Moulay Slimane, Béni Mellal, Maroc
(2)Agence de bassin hydraulique de la Moulouya, Oujda, Maroc

Abstract: The superficial tablecloth of furrow of Midelt belongs to the bowl of High Moulouya which stretches out from the west eastward between the High Atlas in the South and the Medium Atlas west and in the Northeast. The methodology used includes the synthesis of geological data, piezometry, hydrodynamics, hydroclimatology and water quality. This study provides the following results: The flow mode of the water table is general SW to NE on the left bank of the Moulouya river and on the right bank, the flow is from the NW to the SE. The piezometric ratings vary from 1460 to 1780 m. The hydraulic gradient is the order of 2% on average. The transmissivity is usually about 10-3 m^2/s. the punctual flows can reach 50 l / s (case of the drilling N ° IRE 879/38 realized in the alluviums of the Outat). The flow provided by the sources from conglomerates and lake limestones of the Plio-Villafranchien is 50 l / s. The unit of the Mio-Plio-Quaternary aquifer is fed from the infiltrations of rains, by the wadis which cross the banks of the conglomerates and by the landing of the tablecloths of Lias, Dogger and Cretaceous this feeding is however weak in because of the discontinuity of the formations and the poor permeability of the different levels. From a qualitative point of view the groundwater analysis of the aquifer shows that their overall quality is average to good.

Keywords: Hydrogeological Functioning, Midelt Mio-Plio-Quaternary Furrow Tablecloth, upper Moulouya.

Email * ikhmerdi@gmail.com

1. Introduction

The Midelt trench surface water table constitutes a water resource of the foreground at the level of the upper Moulouya basin which is part of the atlasic domain[1] ,which is characterized by a continental climate, dry and hot in summer, cold and little humid in winter. Average annual precipitation is highly variable and closely related to elevation and exposure. They oscillate 235 mm at Midelt[2][3].

At the level of the study area, water resources are more and more heavily used for agricultural activities, and the drop in the piezometric level is becoming a cause for concern, particularly in the part of the alluvial aquifer, well known in this sector. This study is based on the analysis of the piezometry and the feeding mode, in relation with the evolution of the samples. Geological, hydrogeological and hydrogeochemical synthesis are used to improve knowledge of this aquifer. The determination of the hydrogeological context of the Midelt trench water table is based on the treatment of the drilling work carried out in the area, surveys of the water points, geological field surveys and climatic data.

2. Study area

The superficial sheet of Midelt furrow is part of the Upper Moulouya aquifer (Figure 1) which stretches from west to east between the High Atlas which forms the southern boundary, and the Middle Atlas to the west and northeast. The mole of Aouli and that of Bou Mia separate two depressed corridors, that of Aghbalou N'Serdane-Midelt (Midelt Trench tablecloth) in the South, of direction E-W and that of Itzer in the North of direction SW-NE.

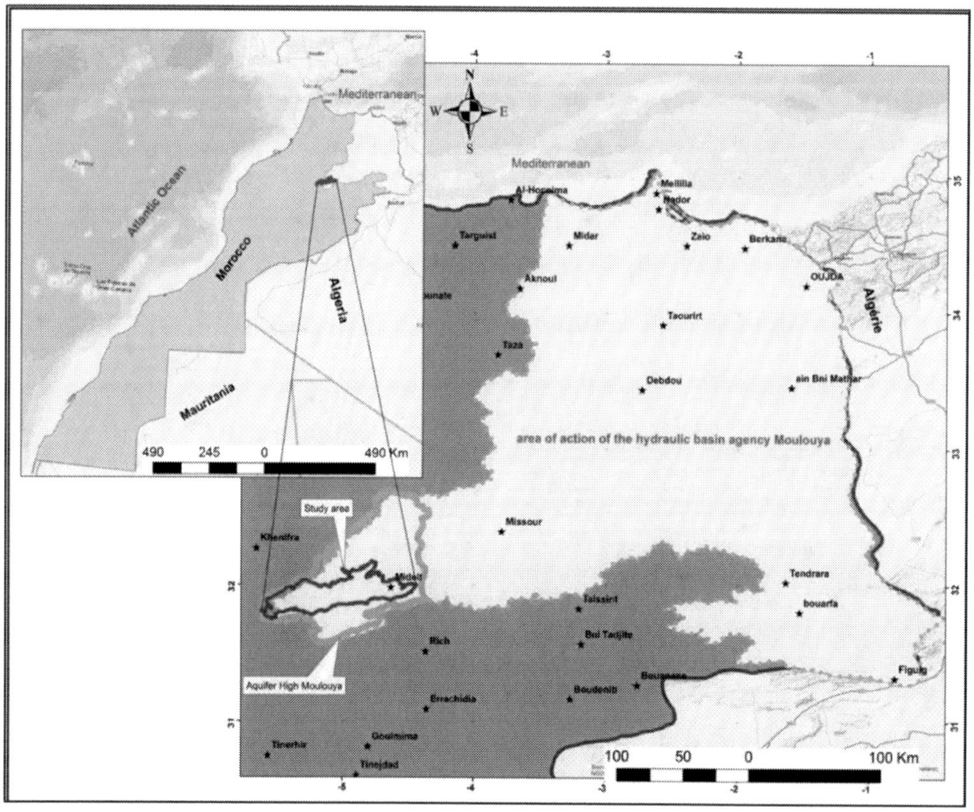

Fig. 1 . Geographical location of the study area

 https://doi.org/10.1051/e3sconf/20183701002

3. Methodology

To understand the hydrogeological functioning of the Midelt trench water table, the proposed methodological approach is based on the collection, processing and analysis of all the data related to the following scheme:

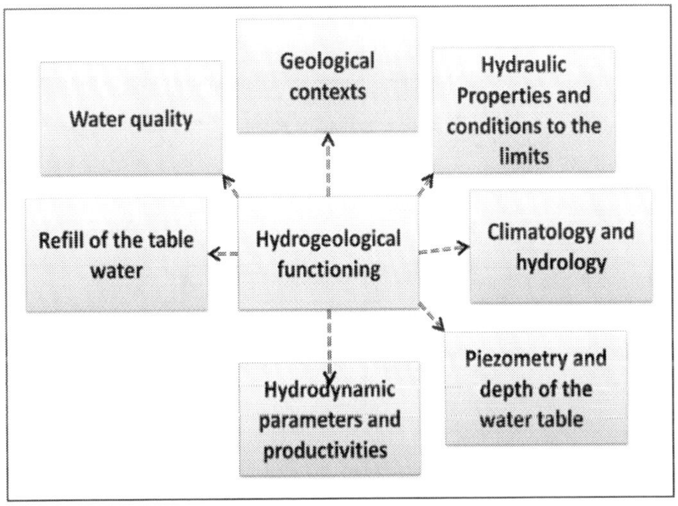

Fig. 2. Hydrogeological operating diagram

4. Results and Discussions

4.1. Aquifer reservoir lithology

The aquifer reservoir of the Moi-Plio-Quaternary is generally enclosed in Marls, conglomerates, sandstones, sands and lacustrine limestones. In the furrow of Midelt, the quaternary terraces of the valleys may contain large alluvial beds; The alluvial deposits of Oued Outat, Ansegmir, Oudghes, the wadis of the left bank of Upper Moulouya, especially the region of Aghbalou N'Serdane. The thickness of lithological formations can range from 5 to more than 350 m [1], [4].

To the south of the Midelt Trench nappe and precisely the boundary between the Midelt plain and the High Atlas, the Edge Fault presents an E-W[1] direction, this fault corresponds to the north atlasic accident which limits the entire Midelt plain to the south.

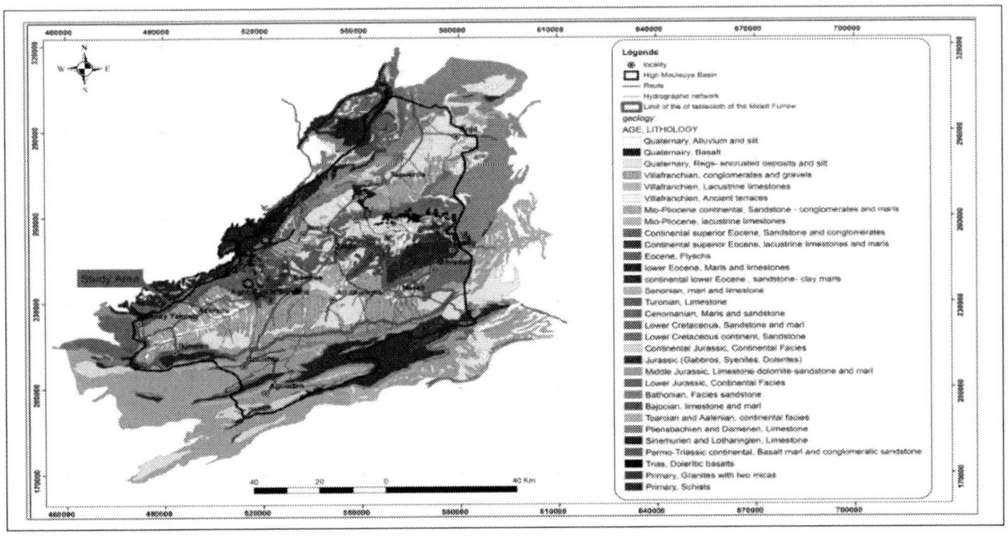

Fig. 3. Geological map of Haute Moulouya (extract from the geological map of Rabat at 1/500000)

 https://doi.org/10.1051/e3sconf/20183701002

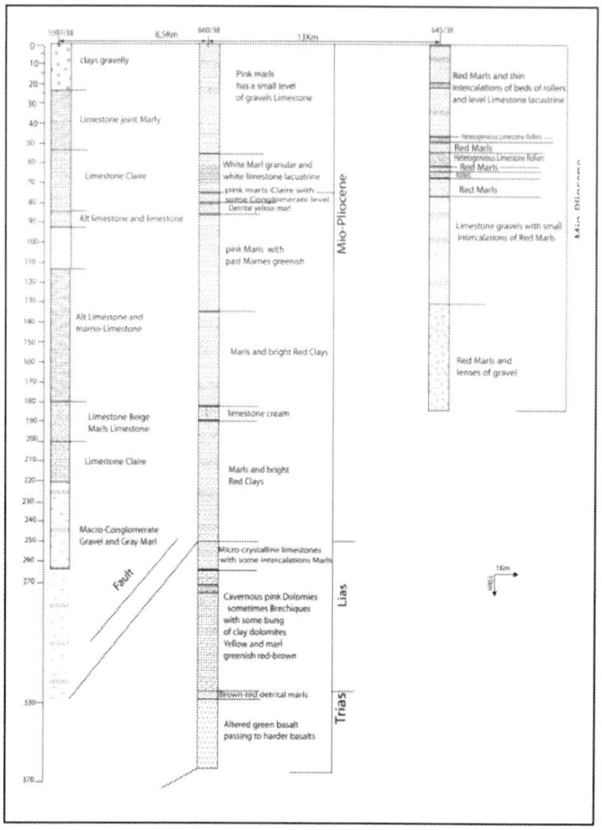

Fig. 4 . Lithological sections of some deep holes drilled in the Moi-Plio-quaternaire Midelt aquifer

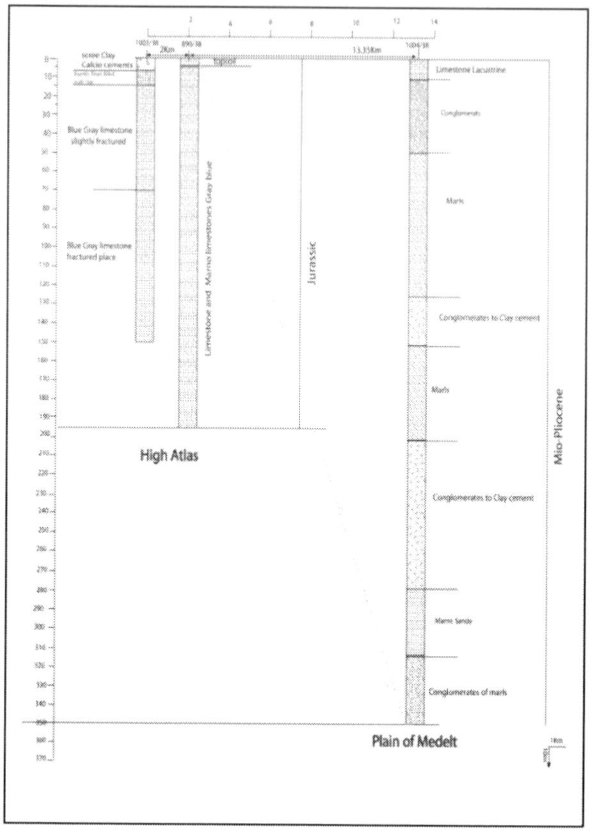

Fig. 5. Lithological sections of some boreholes drilled in the passage zone of the north-high Atlasic accident

Table 1. Characterization of boreholes used in lithological sections

N°IRE	X	Y	Z	NATURE	ZLOCALIT_	P	AQUIFERE	N_P_SOIL
645/38	509000	229150	1640	FR	Arbal Ou N'serdane	186,00	CONGLOMERATES+LIMESTONES	8,94
660/38	521620	227120	1770	FR	Ouest de Midlet	526,00	LACUSTRE LIMESTONE	64,73
890/38	515350	210150	1900	FR	TOUNFITE	194,00	LIMESTONE IN MARNO-LIMESTONE	6,63
1003/38	514550	208350	1940	FE	TOUNFIT	150,00	LIMESTONE BLUE FRACTURE	12,76
1004/38	523000	221100	1800	FR	AARID	350,00	-	87,11
1097/83	500500	229900	1765	FE	Azerzou (ONEP)	415,00		1,00

From a geological point of view, the study area is characterized by the junction between two domains separated by the north-high Atlasic accident (Fig 4). From a geomorphological point of view, the area is marked by a very rugged topography in the South and slightly flat in the North.

The tectonics of the Tablecloth is characterized by the influence of the atlasic edge (the High Atlas), formed of anticlines WSW-ENE axis faults and straddling to the North [1][3]. According to (Fig.3.4.5) the Substratum of the slick can correspond either to Jurassic or Cretaceous formations, or to the Miocene marly soils.

A geophysics campaign (1966) made it possible to specify the deep structure of the eastern part of the slick. It highlighted a major fault in the E-W direction at the foot of the High Atlas and collapsing the Lias (Fig 6). The roof of the Lias then forms a vast asymmetrical syncline under the plain, in the same direction as the atlasic accident, and which stops N on the parallel of Midelt[5][3].

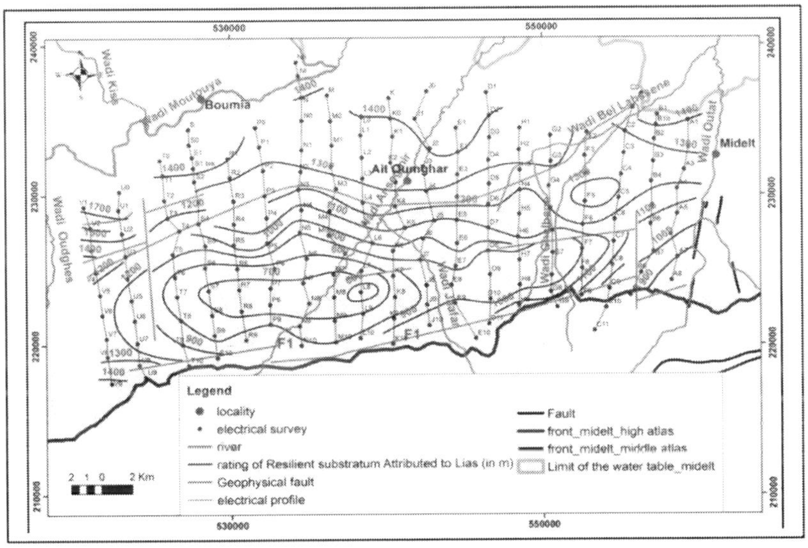

Fig. 6 . Resilient bedrock ratings attributed to Lias limestones (CAG, 1966)

4.2. Piezometry and depth of the slick

In the Midelt Trench, the general flow of the entire aquifer unit is from the NW to the SE on the left bank of Moulouya Wadi. On the right bank, the flow is from SW to NE. The hydraulic gradient is about 2% on average. The piezometric dimensions vary from 1780 to 1460 m.

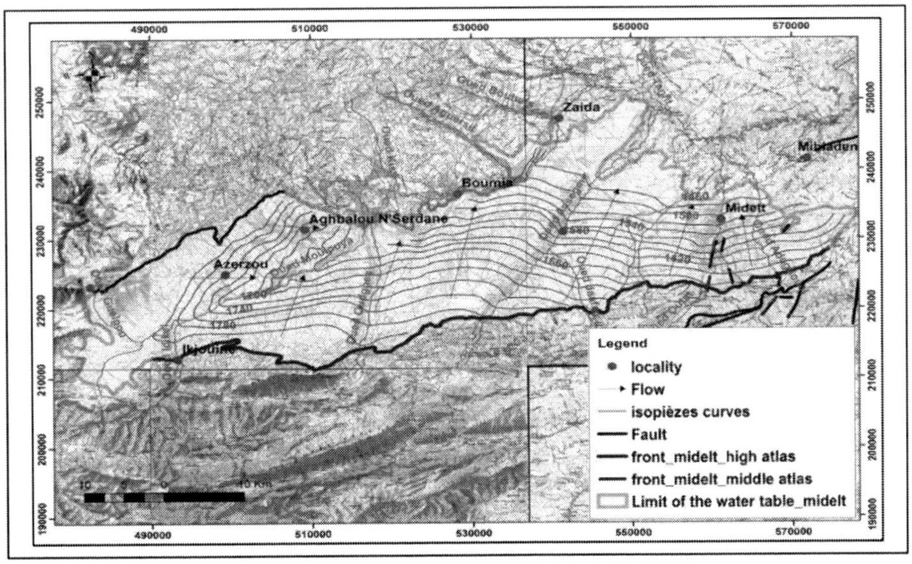

Fig.7. Piezometric map of the Midelt Trench nappe

Piezometric maps which are drawn up for the main alluvial slicks of the Upper Moulouya (Fig 7), show that the flow is generally in the direction parallel to the direction of the wadis which crossed these slicks. The supply of these slicks is very important by the waters of the wadis entering upstream and by the infiltration of precipitations.

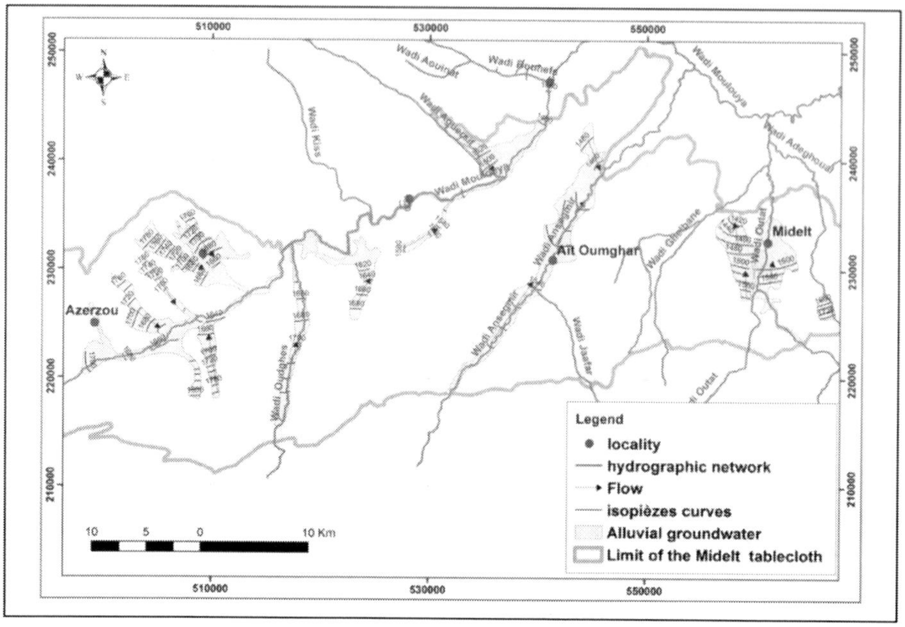

Fig. 8. Piezometric maps of the alluvial slicks of the Midelt plain

4.3. Hydrodynamic parameters and productivity

Quaternary alluvial deposits, largely exploited by wells, are discontinuous but have good hydrodynamic and productivity characteristics. The transmissivities are in the order of 10^{-3} m2/s, the punctual flows can reach 50 l/s (case of drilling N°IRE 879/38 carried out in the alluvium of the Outat).

About ten sources from the Quaternary basaltic formations are located near Bou Mia. They showed low point flows. The available flows are old and do not exceed 5 l/s, the case of the source N°IRE 728/30 with a flow of 5 l/s, the source N°IRE 702/30 with a flow of 2.5 l/s and other sources had very low flows (N°IRE 727/30 and 730/30). The flow provided by springs from lake conglomerates and limestones in Plio-Villafranchien is around 50 l/s.

The conglomeratic levels of the Oligo-Miocene are also exploited by numerous wells and some sources of spillage in contact with marl. The flows are generally low (less than 1 l/s), and can give a total flow of nearly 200 l/s. The few boreholes that exploit the Oligo-Miocene aquifer show its poor hydrodynamic characteristics; transmissivities vary between 10^{-5} and 10^{-6} m2/s. The flow rates are generally low and oscillate between 1 and 4 l / s.

In the Aghbalou N'Serdane hydrogeological basin and the Ansegmir zone, the Mio-Plio-Quaternaire slick may have an artesian character [7]; some boreholes (N°IRE 645/38 and 646/38) are locally artesian due to aquifer loading by marl. Drill hole N°IRE 645/38, located south of Aghbalou N'Serdane, crossed marly levels in intercalation with limestone pebbles. It showed an artesian inflow at 60 m with a low flow not exceeding 0.5 l/s, continuous drilling to a depth of 18 m and total artesian flow had reached nearly 2 l/s.

4.4. Tablecloth supply

The Mio-Plio-Quaternary aquifer unit is fed from rain infiltrations, by the wadis that intersect the banks of the conglomerates and by the butt of the Lias, Dogger and Cretaceous aquifers. However, feeding is low due to discontinuity of formation and poor permeability of the different levels **[3]**.

The alluvial slick of the Ansegmir wadi, well fed thanks to the outcropping in the major bed of thick pebble formations, is exploited by numerous wells dug in the silty terraces.

4. 5. Conditions to the limits

The natural supply of the reservoir is ensured by the lateral contributions coming from the outcrops of the Middle Atlas and the High Atlas overhanging the Midelt Trench. An inflow boundary is drawn along the atlas massifs bordering the Midelt Trench.

4.6. Quality

The water points chosen to monitor the quality of the groundwater were chosen so as to have an overall picture of the water table of the Midelt furrow.

Indeed, the measurement of conductivity makes it possible to appreciate the quantity of salts dissolved in water, thus of its mineralization. The values recorded are 1035, 1040, 804 µS/cm, the minimum recorded at well N°IRE 1135/38 and the maximum recorded at well N°IRE 1134/38. The electrical conductivity depends on the loads of endogenous and exogenous organic matter, which generates salts after decomposition and mineralization [6], it also varies according to the geological substrate crossed. At water peaks 1134/38, 1135/38, located near urban centres (Midelt and Aghbalo nsserdan), the EC is generally 1040 pS/cm. This argues in favour of groundwater contamination by infiltration and percolation of leachate water generated by waste [8].

The analyses of the physico-chemical parameters show two groups of chemical elements. The first has dominant levels of Na+, Ca2+, Mg2+, Cl- NO3-, HCO3- and SO4- followed by NH4+, Fe2+,Mn2+, NO2- and CO3-. The richness of water in elements Na+, Cl, Ca2+ and SO4 2- is linked to the dissolution of minerals likely to be formed by evaporation of salt-laden water, by dissolution of evaporite formations and dissolution of carbonates [10]. The elements Ca2+, Mg2+ and HCO3 - on the other hand are linked to the dissolution of carbonates. In a general way, trace elements import information on the origin of waters, their mixtures and contaminations as well as lithological facies [11]. The factors that determine the spatial evolution of trace element contents, of which we can cite: the presence of deep faults, the intensity and load of rainfall, conditions at the Atlas boundary (High Atlas and Middle Atlas) and also depends on the geological substrate through which the water flows.

Table 2 . Physico-chemical parameters of groundwater in the Midelt Trench surface water table

N° IRE	pH :	CE : µS/cm	NH4+ : (mg/l)	Na+ : (mg/l)	K+ : (mg/l)	Ca2+ : (mg/l)	Mg2+ : (mg/l)	Fe 2+ : (mg/l)	Mn2+ : (mg/l)	Cl- : (mg/l)	NO2- : (mg/l)	NO3- :(mg/l)	HCO3- :(mg/l)	CO3- :(mg/l)	SO4- :(mg/l)
1135/38	7,20	1035	<0,007	2,32	2,33	205	50,0	0,112	<0,014	14,8	0,083	9,20	228	0,0	524
1133/38	8,40	1040	0,019	85,4	1,12	106	42,0	0,237	<0,014	138	0,091	22,0	265	0,00	164
1134/38	7,40	804	<0,007	30,5	7,29	111	43,0	0,068	<0,014	67,0	0,097	38,0	421	0,0	61,0

According to the piper diagram, the wells have a chlorinated and sulphated calcium and magnesium geochemical facies (1133/38) and also bicarbonate and calcium and magnesium (1134/38) (Fig 9).

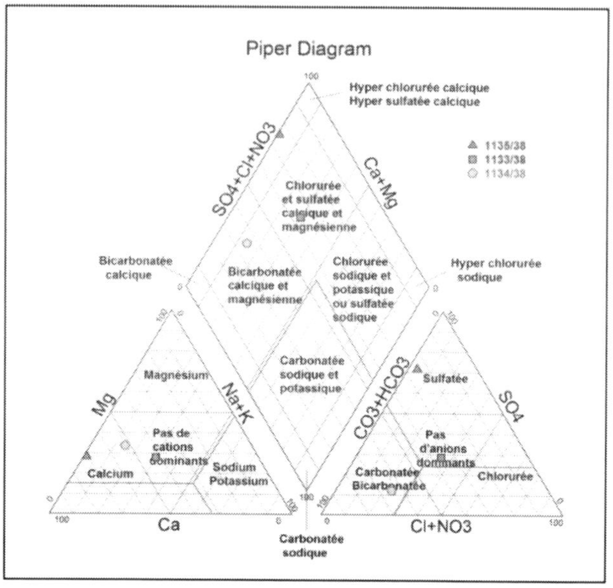

Fig.9. Piper Diagram

5. Conclusion

The interest that hydrogeological data (geology, piezometric, hydrodynamic parameters, etc.) may have for the geometrical and structural reconstruction of the Midelt trench. From a hydrogeological point of view, the geometry, the overall structure of the plain and the lithological nature of the formations have a more or less direct influence on the chemistry and renewal of the water table, as also the recharge to an influence on the vertical and lateral groundwater flows of the Mio-Plo-quaternary aquifer. Piezometric analysis of the water table indicates a decrease in recharge and overexploitation of the underground resource, leading to a significant drop in piezometric levels. The physico-chemical quality of groundwater is generally average to good compared to Moroccan water standards and criteria (Moroccan Standards, 2002).

Acknowledgements

We thank the management of the Moulouya Hydraulic Basin Agency for making the data used in this article available to us. We also thank all those who contributed in any way to the development of this work.

Bibliographical references

[1] A. Michard, Eléments de géologie marocaine (1976).

[2] A. LAABIDI, L. GOURARI, A. HMAIDI, M. AARAB, M. GRETAA. Caractérisation géomorphologique, lithostratigraphique et sédimentologique des dépôts quaternaires de l'Oued Bou Salloum (Haute Moulouya, Maroc), International Journal of Engineering Research and Development e-ISSN: 2278-067X, p-ISSN: 2278-800X, www.ijerd.com Volume **10**, Issue 4, PP.07-19, (2014).

[3] Combe, M., & Simonot, M. "La Haute Moulouya, le sillon d'Itzer et le massif de Boumia-Aouli. Ressources en eaux du Maroc", Notes et Mémoires du Service de Géologie du Maroc, 213, pp. 193-201, (1975).

[4] A. Amrani. Apport des données couplées d'imagerie satellitaire ETM+ de Landsat 7 et model numérique de terrain appliques a l'étude morphogénétique de la Haute et Moyenne Moulouya (Maroc). Thèse de doctorat es Sci., Univ. Mohamed V-AGDAL, Fac. Sci., Rabat, 216 p (2007).

[5] ABHM. CAG. Étude par prospection électrique dans la haute Moulouya, (1966).

[6] S. Bouchelaghem, S. Benzara, W. Méradi, and R. Souhir. Etude comparative de la qualité chimique et microbiologique des eaux de l'oued El kebir (Région d'El Tarf) International Journal of Innovation and Scientific Research, ISSN 2351-8014 Vol. 3 No, pp. 71-74 (2014).

[7] ABHM. Étude est l'élaboration des cartes hydrogéologiques des nappes de la Haute Moulouya, de la Moyenne Moulouya et du bassin de Guercif, **Mission I :** Elaboration des cartes hydrogéologiques de gestion des nappes de la Haute Moulouya et du sillon Itzer-Enjil ; 115P, (2006).

[8] A. BOUKDIR. Fonctionnement Hydrogéologique de L'aquifère du Turonien du Bassin du Tadla-plateau des Phosphates-tassaout Aval (Maroc). Approche Pluridisciplinaire et Etude D'impact – Thèse Doctorat, Université Mohammed v, Ecole Mohammadia des Ingénieurs Rabat, 190pp 116-139, (2007).

[9] K. Baba-Hamed, A. Dribat, A. Bouanani, Hydrogeological Caracterization and Modeling of the Aquifer of Oued in Amguel. (Tamanrasset, Central Hoggar -Ssouth Algeria) Journal of Fundamental and Applied Sciences, ISSN 1112-9867, 19P, (2018).

[10] M.L. BELGHITI, A. CHAHLAOUI, D. BENGOUMI, R.EL MOUSTAINE. Etude de la Qualite Physico-chimique et Bacteriologique des Eaux Souterraines de la Nappe Plio-quaternaire Dans la Région de Meknès (Maroc), Larhyss Journal, ISSN 1112-3680, (2013).

[11] M. MAKHOUKH, M.SBAA, A.BERRAHOU, M.VANCLOOSTER. Contribution à l'étude de l'impact d'un site minier abandonné dans la haute Moulouya sur la qualité de l'Oued Moulouya, Maroc, Afrique SCIENCE 07(3), 33 – 48 (2011).

[12] N.M. Norme Marocaine de qualité des eaux. Bull. Off. n° 5062, (2002).

Assessment of the effects rejections of feed fish on water resources. (Ouedoumerrbia, Morocco)

Khadija Ouaissa[1,2], *Assia*Kritihi [1,2], *Youness*Oumessoud[2], *Abdelaziz*Maychal[2]and *Mustapha*Hasnaoui[1]

[1].Environnemental engineering Teams. Department of Biology, Facylty of Sciences and Techniques Beni-Mellal. University Sultan Moulay Slimane. M' Ghila, B.O 523. 23 000 Beni-Mellal, Morocco
[2]. Fish Farming AinAghbal, Azrou–Morocco.

Summary. In order to compare the effects of three types of extruded food (A,B and C) on the growth of rainbow trout, an experimental test was conducted on June 15, 2015 at a rainbow trout farming station near river of Oumerrabi .Morocco.
The comparison of three foods of different composition and energy is performed in isoenergetic conditions.
Six basins were used for this comparative test. These basins are fed with fresh water according to the open circuit with a renewal of twice an hour. The initial feeding conditions were the same for the three food types and the initial density of 1, 58 kg/m^3 (kg by volume) and an initial flow rate of 1, 04 m^3/h.
Fish are fed by ratios two to three times a day depending on the magnification stage.
The sampling frequency is fortnightly, where we measure the zootechnical performance of fish and collect water samples for physico-chemical analyses in order to assess the quality of the water leaving in the basins before their discharge into the river of Oum Er-Rbia.
The comparative trial of three fish foods (A, B, and C) revealed that diet B is the better formulation reflected by the zootechnical performances and low phosphate release than diet A and C.

Keywords: Food; biochemical; Performance; Releases; Environment; OumErRbia; Morocco.

Khadija Ouaissa1 :E-mail: khadijaouaissa89@gmail.com

1. Introduction

The aquaculture products are becoming increasingly important in human nutrition. In the context of sustainable control of the salmonids production, it is essential to optimize feed and maintain a high quality finished product to meet customer requirements and environmental requirements.

Aquaculture is a sector of food production that has changed in recent years. Farmed fish is currently nearly 50% of the fish world [1] .This explosive growth has led the aquaculture industry to significant challenges. The most important of them is the reduction of the negative impacts of aquaculture on the environment practices. These effects are still many. However, eutrophication of streams effluent receiving aquaculture companies is the aspect most concern [2]. These environmental concerns have led researchers and producers to explore possible solutions to move aquaculture more sustainable development.

Proteins in the diet of farmed fish are mainly brought by fishmeal as an essential element in the artificial diet. With a view to preserving the environment and the non-dependence of sources of raw materials such as fishmeal. Food for trout is becoming increasingly rich in fish meal because its extra energy intake .Fishmeal is the main ingredient in the food composition of rainbow trout, it covers a large percentage in the formulation of carnivore food fish, and this ingredient is rich in phosphate that which participates in the skeletal formation of fish.

The negative environmental impacts of aquaculture are very numerous, but the most danger is the eutrophication of effluents on aquaculture farms [3], [4], [5] and [6]. The main pollutants involved in this phenomenon are phosphorus (P).

In the context of a sustainable control of trout farming , the reserve of water resources and protection of the environment. The objective of this study is to compare three types of food extruded different biochemical compositions on the zootechnical performance of rainbow trout to determine foods that have the best trout growth performance, digestibility in order to limit and optimize fish in emissions more than engage in sustainable aquaculture that respects the environment and that respect environmental standards.

For economic and ecological reasons, the comparative test of three food (A, B and C) realized in the station of Oum Er-Rbiaa magnification allowed to retain the B diet presented a better performance zootechnique and low phosphorus rejection [7].

2. Materials and methods

2.1. Experimental study

The experiment was conducted on June 15, 2015 at the fish farm of Oumerrabia, located about 70 km from Azrou (Morocco).

The six basins are fed by spring water at 14°C. The water, that supplements the tanks, was renewed every half hour (48 times /days) with high level of oxygen (more than 90 % saturation).

4574 Trout, of average weight of 40 g, were sampled from the same batch of eggs and divided randomly into six tanks. The test was conducted in monoculture and the fish were fed manually with a determined daily ration.

Every 15 days, 90 fish (according to Student's law) are captured from each tank, being fasted for 24 hours and then anesthetized to measure the size and the weight of each fish, weight gain (WG), feed conversion ratio (FCR), Specific growth ratio (SGR), and factor condition (K).

The distributed foods were weighted to estimate the consumption by the fish between two successive Sampling

2.2. Experimental food

2.2.1. Composition of the three foods tested

The biochemical analyses (proteins, lipids, moisture and ash) of the extruded foods used in this experiment were performed in duplicate using the standard [8] and concerned:

-Moisture doses the desiccation method at 105 °c.

-Crude proteins (% N * 6.25) are determined by the Kjeldahl method (self-analyser Kjel-FOSS)-

-Lipids by the hot method (type Soxhlet).

The extraction of lipids was carried out by a mixture of chloroform: methanol (2:1 V/V) According to the method of [9].

-Ash after incineration of the samples in the muffle oven à550 °c for 12 hours,

-The carbohydrate content, assimilated to the nitrogen Extractifnon (ENA), by a calculation of difference from the values found for the other constituents of the sample.

-Total phosphorus per continuous flow analysis.

-Raw energy measured by a calorimetry bomb. Apparatus measures the energy (in Joules) released in the form of heat by the total combustion of the sample.

The following table bellow shows the proximal composition of the extruded food

Table 1. Biochemical composition of the three foods tested according to supplier datasheets (different food diameters)

Type of Foods	A			B			C		
Food diameter	3mm	4,5mm	7mm	3mm	4,5mm	7mm	3mm	4,5mm	7mm
Protein (%)	40	39	39	45	43	41	46	45	40
Lipid (%	23	27	27	20	22	24	20	24	26
NFE (%)	21	18,2	18,2	20	20,7	20,5	15	14	13
Cellulose (%)	2,25	1,86	1,8	1,9	1,9	2,3	1	1	1
Ash (%)	7,20	6,77	6	8,1	7,4	7,2	10	10	10
Phosphorus (%)	1,05	0,98	0,9	1	1	1	1,5	1,5	1,40
DE (Mj/kg)	19,2	20,2	20,2	20	20,3	20,6	19	19,90	20,50
DP/DE (g /MJ)	19	17,5	17,5	-	-	-	22,30	22,30	17,60
Vitamin A (UI.kg)	10000	10000	10000	10000	10000	10000	10000	10000	10000
Vitamin D3 (UI.kg)	1750	1750	1750	1000	1000	1000	1850	1850	1850
Vitamin E (UI.kg)	200	200	200	200	200	200	200	200	200
Vitamin C (UI.kg)	150	150	150	-	-	-	180	180	180

2.3. Analyses of fish effluents

To estimate the outflows of phosphorus (HPO_4^{3-}) in outlet of the basins, water samples are taken twice per month consecutively.
The samples are then transported, in a cooler at 4 °C, to the analysis laboratory of the Oum-Er-Rbia hydraulic basin agency in Beni-Mellal.
The water samples are then transported in a cooler at 4 °c to the laboratory of analysis of the hydraulic basin agency of Oumerrabia (ABHOER) of Beni-Mellal. Orthophosphate was measured by spectrophotometry after mineralization followed by acid hydrolysis (NFT 90-013) [10].

3. Results

The following table presents the results of the zootechnical performance of the three foods (A, B and C) in the three stages of rearing fed by the different food diameters (3mm, 4,5mm and 7mm) obtained during this test Experimental.

Table 2. The zootechnical parameters studied for the three foods tested (A, B and C).

Settings	A			B			C		
	3mm	4,5mm	7mm	3mm	4,5mm	7mm	3mm	4,5mm	7mm
Initial weight (g)	41,05	143,28	495	41,15	147,22	501,1	44,65	132,68	481,9
Final Weight (g)	143,28	495	803,14	147,22	501,1	1043,9	132,68	481,9	942,1
Number of days	60	90	72	75	75	72	60	90	72
weight Gain (g)	101,33	351,72	308,14	106,07	353,88	542,8	88,03	449,22	460,2
Specific growth rate% (SGR)	2,08	2,76	0,66	1,7	1,62	1,02	1,81	1,43	0,93
Individual daily Growth (IDG)	1,7	3,9	4,27	1,41	4,71	7,53	1,46	3,88	6,39
Survival rate (%)	98	99,59	99,82	95,5	99,82	99,82	99,4	99,76	99,82
Conversion Index	0,8	1	1,1	0,8	0,9	0,9	0,9	0,8	0,8
Condition factor	1,2	1,22	1,27	1,27	1,45	1,61	1,08	1,28	1,03

During the experimental period, the test was carried out under the same conditions for the three tested foods A, B and C.

Table 2 shows the growth performance of rainbow trout during the experimentation period, food B is still the most efficient (weight gain, specific growth rate and daily individual growth, Conversion index and condition factor) which gives results of greater growth and better performance compared to the other two foods tested.

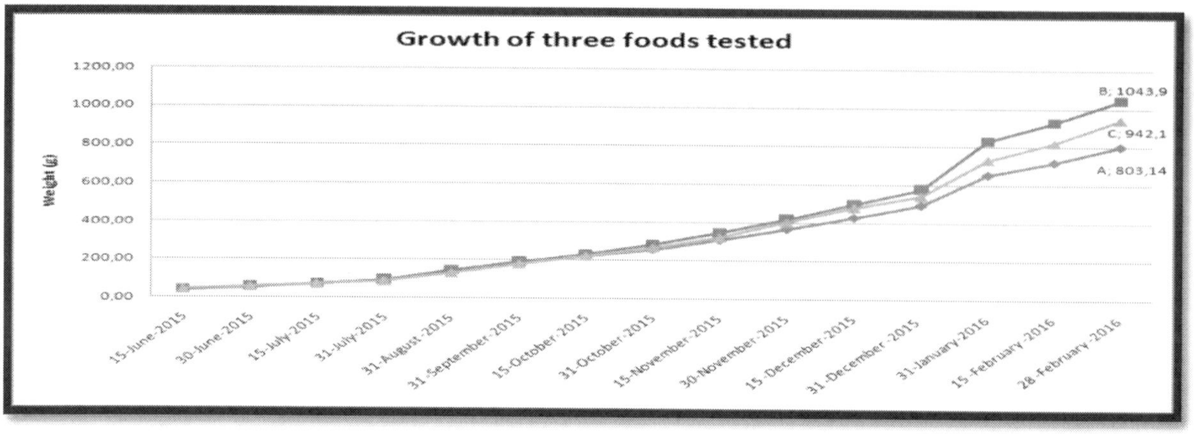

Fig. 1. Growth of three types foods (A, B and C)

During the experimental period, the test was carried out under the same conditions of breeding for the three tested foods A, B and C (temperature, source and flow of water).

Table 2 shows the growth performance of rainbow trout during the experimental period. B food is still the most efficient (weight gain, specific growth rate and daily individual growth). Greater growth and better performance compared to the other two foods tested (A and C).

Weight growth has a major impact on production. It reflects the effectiveness of the food in mass production. For this reason, a regular trout weight measurement was carried out throughout the experimental test.

The graphical representation of the evolution of the average weight of rainbow trout fed by three types of iso-energy food but of different biochemical composition is shown in table 1.

This indicates that B diet is the best food at the end of growth with a growth of 1043.9 g compared to foods A and C (803.14g) and (942.1g) despite a lower lipid content to the other two foods A and C.

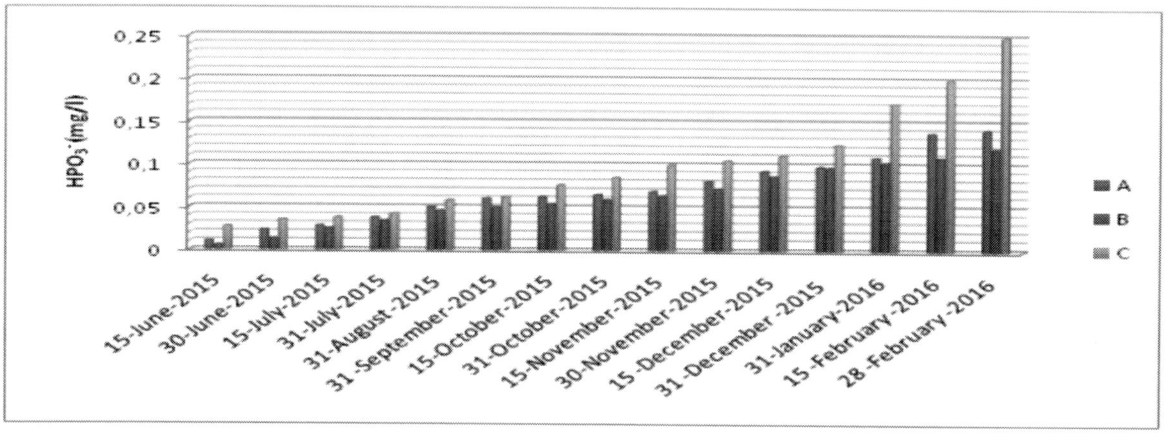

Fig. 2 . Rate of phosphorus discharges

(Fig.2) present the results of phosphate analyses of water from fish effluents in the basins of the three extruded foods tested.

For the phosphorus rejection of trout, the results show that C fed fish release more phosphate releases than food A and B, the amount of releases increases as a function of fish biomass and the food distributed.

Food B releases are always lower than the two foods A and C.

4. Discussion

The trout were kept under optimal conditions throughout the trial period, as the values obtained for temperature, dissolved oxygen and pH met the standards for breeding these fish [11], [12].

In high trophic level fish it is the proteins and dietary lipids that allow them to cover their energy needs because of limited capacities used efficacement the energy of food carbohydrates [13].

According to this comparative study of three foods, the results show that feed B ensures a significant growth of trout compared to food A and C in effect of 1043.9gr, compared to 883.1gr for food A and 942.1gr for food C.

The extruded food B provided better zootechnical performance during the three breeding stages (40-100g; 100-500g; 500-1000 g) compared with food A and C.

It was found that growth also was increased by the increase in lipid levels in diet C, the final weight of fish fed by food B is greater than C food knowing that they have almost the same protein content (B = 39% and C = 40%) in the biochemical composition but with a different lipid level, these results obtained are similar to the results obtained by [14].

Many studies have showed that a high-energy diet acts on the growth and food efficiency of fish food proteins [15] . However, supplementation with fat rather than carbohydrates, as a non-protein source of energy, is generally more effective because lipids are easily metabolized, especially by fish carnivores like rainbow trout [16].

Our results confirm the existence of a load efficiency, where a better growth recorded for food b which translates to 1043.9 g, compared to food C (760 g) and food A (726.4 g), knowing that food B has a low content fat in the stage of magnification which has a content of (24%) against food A and C which have high lipid content (27%) and (26%) This is contradictory to the results found in [16].

The food conversion index is the ratio of food consumed on weight gain. It must be as low as possible, reflecting a good efficacy of the food in question and its proper use by the animal.

The conversion index recorded in this study represented by (CI_B = 0.9) are comparable to those obtained by [17] and those reported by [18]whose experimental studies relate to the effect of two extruded foods on the performance of rainbow trout and are indicative of good growth and food conversion efficiency of 0.88.

Survival rates are clear that we have not recorded major problems in terms of mortality; the survival rate exceeds 95% for the three diets (A, B and C) and in the three breeding stages (40-100g. 100 g-500g and 500-1kg).

The few dead individuals counted during the experiment did not appear to be related to diet but to external factors. Therefore, we consider that our results are in the accepted standard.

We found that the K-factor is very important in diet B of {K_B = 1.61}, which has an excellent state of health, which explains that the fish undergoing this diet grew better at the levels of the weight and length ratio; While for an extruded C diet, we reflect that the fish had a larger performance size than the weight performance and this can only be explained by the richness of C food by phosphorus.

We found that the K factor is very important in the B regimen of {KB = 1.61}, which has an excellent health condition, which explains that the fish that are in this diet increased better at the weight and length ratios; While for an extruded C diet, we reflect that fish had a larger performance size than weight performance and this can only be explained by the richness of C food by phosphorus hence an important development of skeleton of fish bones

All the phosphorus needed for fish comes from their diet. The phosphorus produced will therefore be a function of the quantity present in the food. Phosphorus can be of animal origin (bone meal, fish meal), vegetable or mineral. The digestibility of phosphorus is very variable depending on the species and the origin of phosphorus. The phosphorus of a good quality fish meal is 60% digestible by rainbow trout [19]. On the other hand, phosphorus of plant origin is in the form of calcium and magnesium salts of phytic acid. This phosphorus is very little used by fish because they do not have the phytase enzyme necessary to digest them [20].

Excess phosphorus in fish feeds results in higher levels of excreted phosphate, which is the main cause of eutrophication in aquatic environments and consequently the deterioration of water quality [21]

The reduction of phosphate excretion by fish has become imperative [22] in general, fish with the stomach, such as trout, assimilate more phosphorus than fish without the stomach, such as carp [23].

The (P) requirement available for growth is 0.37 and 0.57% for better mineral accretion. While the food C comprises 1.5 % in the food used and on the other hand it rejects more phosphorus and contributed to growth in size rather than weight.

The results obtained in this study are consistent with those obtained by [24] on rainbow trout (OnchorynkissMykiss), whose test revealed a high concentration of phosphorus releases by food, which is the same food used in our comparative test.

The phosphorus found in fish effluents are presenting in soluble or dissolved form. Phosphate releases are below the environmental standard at a rate of 0.12 mg / l for a weight of 1 kg of fish, this may explain the efficiency of the production process of the feed extruded by excellent gelatinization of raw material vegetable and also by the choice of raw materials that has low phosphate levels. On the other the food tested a rate of 1% phosphate incorporated into the food, some countries developed in the field of food manufacturing have adopted measures limiting phosphorus levels in fish feed by (<1%).(Kaushik,2000)

5. Conclusion

The results of this study show that the extruded food B characterized by its energy content, buoyancy, high digestibility and non-protein energy level, extruded food B trout provides the best zootechnical performancein terms of growth , conversion index. And also by a reduction of phosphate releases by controlling the formulation of food.Considering the food as the main cost of production in intensive fish farming with the use of the diets a better conversion and the contribution to the Sustainable aquaculture.

Acknowledgements

We thank Administration, Managers and staff of the fish farm Trout of the Atlas in Azrou for the financing and the realization of this study.

References:

1. FAO , Rapport de la FAO sur la pêche et l'aquaculture (2010).
2. G. Ouellet, Caractérisation des eaux usées de stations piscicoles québécoises , MAPAQ, Direction de l'innovation et des technologies, 35 (1998).
3. D L .Corell , The role of phosphorus in the eutrophication of receiving waters: a review. Journal of Environmental Quality, 27: 261-266 (1998).
4. G .Ouellet , Les rejets des stations piscicoles et leurs impacts environnementaux, Québec, MAP AQ,42 (1999).
5. G .Persson , Eutrophication resulting from salmonid fish culture in fresh and salt waters: Scandinavian experiences. In Nutritional Strategies and Aquaculture Waste. Proceedings of the First International Symposium on Nutritional Strategies in Management of Aquaculture. University ofGuelph, Ontario, p. 163-185 (1991).
6. G.V. Vandenberg, Encapsulation de la phytase microbienne: Influences sur la disponibilité de nutriments chez la truite arc-en-ciel. Thèse de doctorat, Université Laval, Québec, 277 (2001).
7. K. Ouaissa , A. Kritihi , Y. Oumessoud , A. Maychal et Hasnaoui M, "Effets d'un aliment extrudé sur les performances de croissance de la truite arc-en-ciel (Oncorhynchusmykisswalbaum, 1792) et sur la qualité de l'eau de l'oued Oum Er-Rbia (station Ain Aghbal, Azrou-Maroc). "J. Wat. Env. Sci. Vol. 1, (S.I. COP22), 132-139.(2017).
8. AOAC (Association of Official Analytical Chemists), Official methods of analysis of the Association of Official Analytical Chemists, " 15th edition. AOAC, Arlington, Virgina, USA.(1990).
9. J. Folch , M. Lees et Sloane-stanleyg H. S, A simple method for the isolation and purification of total lipids from animal tissues, " J. Biol. Chem. 226, 497-509 (1957).
10. AFNOR ,Recueil de normes françaises des eaux : méthodes d'essais. Paris (1983).
11. G. A. Wedemeyer ,Physiology of fish in intensive culture systems.chapmanhall,LONDON.UK (1996).
12. S. J .Kaushik, Effect of digestible energy on nitrogen and energy balance in rainbowtrout. Aquaculture, 50, 89-10 (1985)
13. F Medale et J C. Guillaume, Nutrition énergétique. In : Guillaume .J, Kaushik S, Bergot P, Métailler R, Eds. Nutrition des poisons et des crustacés. Paris (France): INRA, pp. 87–111. (1985).
14. P. Luquet , Etude du développement chez la truite. Evolution de la teneur en acides nucléiques dans diverses fractions corporelles. Ann. giol. animal. Bioch. Biophys., 11, 657-668 (1971).
15. NRC, «Nutrient Requirements of Fish. National Academy Press,Washington, D.C., 114p.1993.
16. S. Chaiyapechara , M.T. Casten , R.W.Hardy and DONG F. M , Fish performance, fillet characteristics,and health assessment index of rainbowtrout (Oncorhynchusmykiss) fed diets containing adequate and high concentrations of ipid and vitamin E, " Aquaculture, 219: 715-738 (2003).

17. J.E. Erika , A.H. David , P.F. Anthony , Effect of iso-energetic diets with different protein lipid content on the growth performance and heat increment of rainbow trout. Aquaculture 272 .723–736p (2007) .https://doi:10.1016/j.aquaculture.2007.09.006
18. C. Brauge ,M. Medale , G. Corraze . Effect of dietary carbohydrate levels on growth, body composition and glycaemia in rainbow trout, Oncorhynchus mykiss, reared in seawater. Aquaculture 123, 109–120. (1994) https://prodinra.inra.fr/record/115983
19. A. Dosdat , L'excrétion chez les poissons téléostéens, " La pisciculture française, n. 109,18-29 (1992).
20. S.P. Lall ,Digestibility, metabolism and excretion of dietary phosphorus in fish. Dans : C.B. Cowey et C.Y. Cho (eds.). Nutritional Strategies and Aquaculture Waste. Proceedings of the First International Symposium on Nutritional Strategies in Management of Aquaculture Waste. University of Guelph, Guelph, Ontario, 21-36 (1991).
21. M. Rodehutscord, Z. Gregus, E. Pfeffer , Effect of phosphorus intake on faecal and non- faecal phosphorus excretion in rainbowtrout (Oncorhynchusmykiss) and the consequences for comparative phosphorus availability studies. Aquaculture, Amsterdam, 188:383-398. (2000).
22. J.P. Blancheton , Cahiers Agricultures. Volume 18, Numéro 2-3, 227-34 (2009).
23. A. Hernandez, S. Satoh , V. Kiron , T. Watanabe , Phosphorus retention efficiency in rain bowtroutfed diets with lowfish meal and alternative protein ingredients .FISHERIES SCIENCE 70 ,580–586. (2004).
24. M. Abba, D. Belghyti, M. Benabid . Effet de deux aliments pressés et extrudés sur les performances de croissance de la truite arc en ciel (Oncorhynchus mykiss) et leurs impacts sur environnementaux, ScienceLib,3,4706 (2011)
25. S. Kaushik, Feed formulation, diet development and feed technology CIHEAM" ,p.43-51, (2000)

Karst groundwater vulnerability mapping to the pollution: Case of Dir springs located between EL KSIBA and Ouaoumana (High Atlas, Morocco)

L. ALILI*[1], *A.*BOUKDIR[1], *M. R.* MASLOUHI[2] et *H.* IKHMERDI[1]
[1]Department of Earth Sciences, Faculty of Science and Technology, Sultan Moulay Slimane University, Beni Mellal, Morocco
[2] Oum Er R'bia Hydraulic Basin Agency, Beni Mellal, Morocco

Abstract. The study area is located in the north of the province of Beni Mellal, it covers the Piedmont of the high Atlas between El Ksiba and Ouaoumana. It is characterized by a poorly developed hydrographic network and the presence of very important karstic forms. These forms condition the rapid infiltration to the karstic springs are the subject of this study. In this work we presented a method of mapping the vulnerability to pollution of Karstic springs located between El Ksiba and Ouaoumana. To do this, we have introduced a vulnerability index called F which takes into account four parameters (EPIK): Development of the Epikarst, importance of the protective cover, infiltration conditions and development of the Karst network. The overlay of the thematic maps of these parameters through a GIS software (ArcGIS) gave us a map of the vulnerability to contamination on the whole hydrogeological basin of the springs.

I. Introduction

The richness of agriculture, the population growth and the development of socio-economic activities are mainly due to the quantity and quality of the waters. This increasingly growing development can generate negative impacts on the environment in general and on water resources in a particular way. That is why mapping the vulnerability to pollution subject of this study became obligatory for the protection of this resource.

In this work we present the method EPIK adapted for the first time in Switzerland and considered as the initial method adapted to the study and mapping of the intrinsic vulnerability in karstic environments, case of springs located between El Ksiba and Ouaoumana. It is based on four criteria at stake, namely development of the Epikarst, importance of the protective cover, conditions of infiltration and development of the Karst network, with an indexing system for each of the criteria, then a calculation of Weighting with fixed coefficients is carried out.

This method is used to define a protection factor F using the calculation of the Elemental Criteria index classified as vulnerability slices.

Email* : aliligeomatique@gmail.com

II. Description of the study area

The study area is located in the north of the province of Beni Mellal, it covers the piedmont of the mountain between El Ksiba and Ouaoumana. The Dir is thus the transition zone between the plain and the mountain.

This area is traversed longitudinally by the Oum Er R'Bia river and also by the national road linking Fez to Beni Mellal this has fostered an important demographic (centres of Ksiba, Zaouiat Ech Cheikh and Ouaoumana) and an intensification of Socio-economic activities (livestock, industry, tourism, and quarries). The present study focused on thirteen springs in the Dir that are: Tamda, Ouarnfaa, Bouddilite, Ou Abdellah amont, Ou Abdellah aval, Tamescourte, Tiguemate, N'ougziza, Igly, Boullmatene, Ait atmane, Boutmarsite et Ait laaidi (figure 1).The Lambert coordinates of the study area are carried over to table 1.

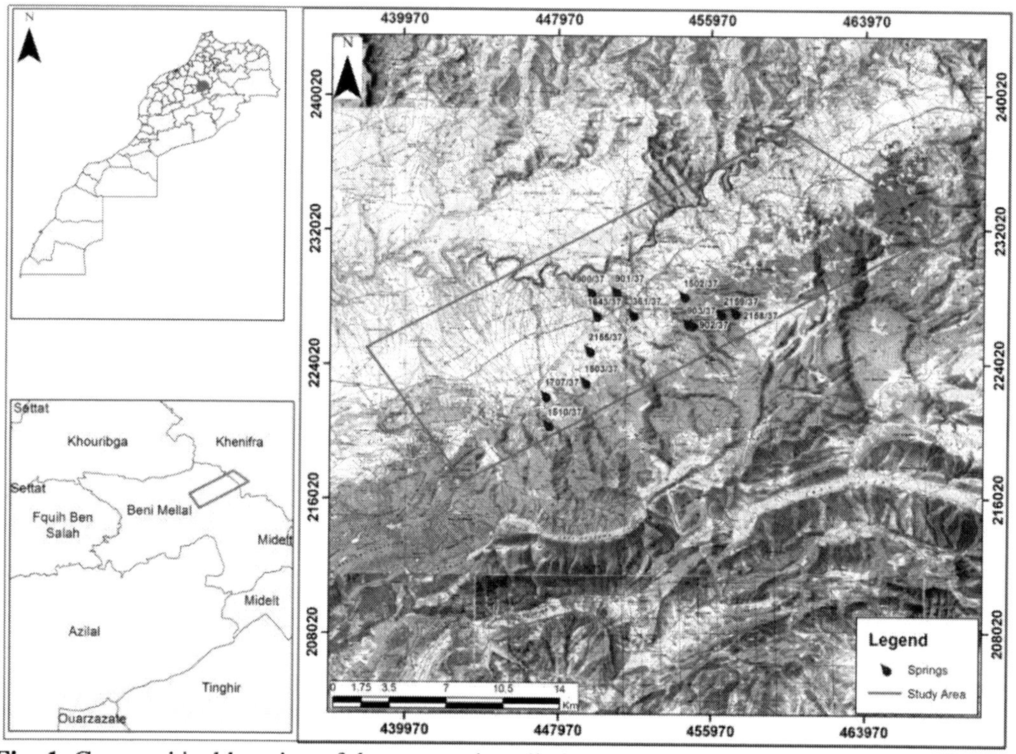

Fig. 1. Geographical location of the area and studied springs

Table 1. Lambert Coordinates limits of the study area.

X (m)	Y (m)
438000	225 000
443000	217000
468000	232000
460000	238000

is characterized by with regard to the hydrographic network, the study area is drained by the Oum Er R'bia river and by some streams, the most important of which is the Ouaoumana river.

From a geological point of view, the Atlas of Beni Mellal has a flattened anticline [1], fractured (principal fracturing directions obtained by the treatment of LandSat 8 satellite images is n 40-70 and n 150-180), overlapping to the northwest and Resting on the Dir. It comprises liassic or tertiary aquifer and travertines veneered on the relief of the high atlas (Figure 2).

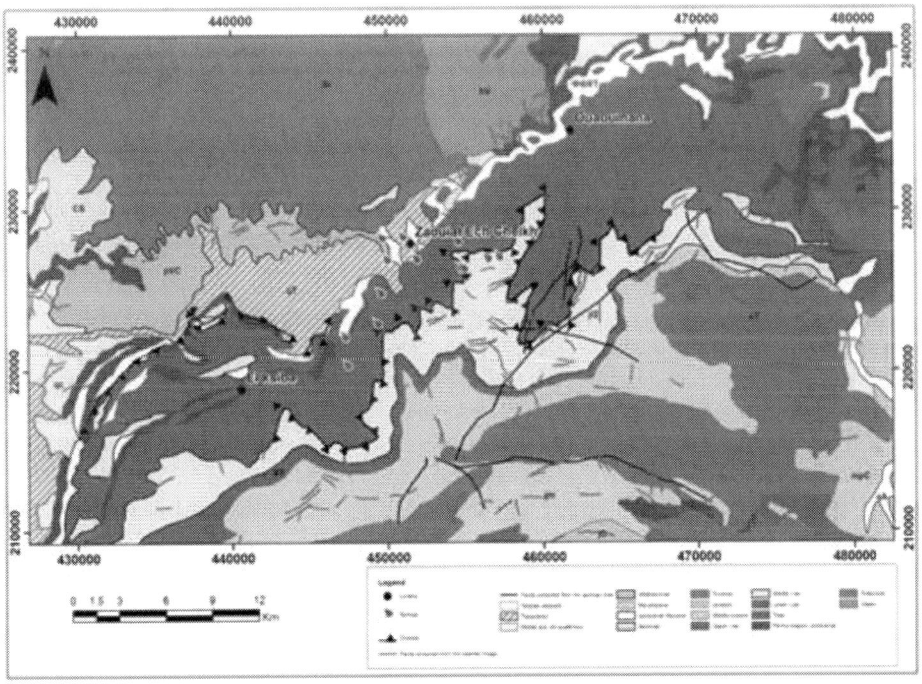

Fig. 2. Geological map of the study area

Triassic and Jurassic form most outcrops in the study area. The first form of the base of scales is represented by marls and red clays in salt. It also contains doleritic basalts in the middle. The Lias or Lower Jurassic includes massive limestones and Dolomites. The Middle Jurassic or Dogger is essentially limestone while the Upper Jurassic is mainly a series of sandstone, continental and red facies. The Triassic and the Jurassic were affected by soft and brittle deformations at the scales origin of El Ksiba [2].

The lower Lias is the most dominant geological formation at the scale of the study area (Figure 2). The Triassic formations are observed locally to the northeast of El Ksiba, flush with the Jbboula river and in contact with the Quaternary formations of the Tadla. The Continental Permo-Triassic is also observed on the right side of the national road from Zaouiat Ech Cheikh to Ouaoumana.

In the mainly Liassic carbonate formations, the phenomena of dissolution resulted in the emergence of a complex system of karstic cavities [3] of varying size and irregular shape.

III. Materials and methods

Karstic aquifers are characterized by geomorphological peculiarities and specific hydraulic phenomena such as the existence of important springs, losses, the absence of a surface drainage system, the existence of networks Karsts resulting from the dissolution of carbonate rocks; The hydrograms of the springs are typical (rapid and violent flooding, rapid flooding, slow recession), and the quality of the water shows changes in chimism as a function of flow [4]. EPIK (Doerfliger and Zwahlen, 1998) is the first method that has been developed specifically for this type of aquifer [6].

This method based on the specific organisation of flows in the karst environment, consists in the evaluation and the combination of four criteria, namely the development of the Epikarst, the thickness of the protective cover, the infiltration conditions and the Karst network development. These criteria are assessed and weighted by a protection index based on the protective role they provide to groundwater.

3.1. Description Criteria E, P, I, K: Dir springs Located between El Ksiba and Ouaoumana Case

3.1.1. Epikarst (E)

E1 (very lowly protective) at E3 (highly protective). Criteria E involves the influence of karstic morphology on the protection of the aquifer. Karstic morphology such as igues, closed depressions and their feeding area will have a very low E1 protection index. The E2 index involves intermediate zones between sinkholes fields and dry valleys. The E3 index will characterize the rest of the catchment (DOERFLIGER N. and ZWAHLEN F., 1998).

For the feeding basin of the springs located between El Ksiba and Ouaoumana, the assessment of the presence and development of the Épikarst was carried out using field observations (karstic modelling and outcrops), geomorphological studies and The examination of high-precision satellites images, In addition to the study of springs recessions.

3.1.2. Protective Cover (P)

P1 to P4: Criteria P involves the protection of soil cover and soil geological formations. The parameter taken into account here is the least costly and most representative of the protection of a soil to be determined, the thickness. The authors distinguish between a soil directly covering the rocky aquifer (Case A) and another, associated with geological soil formations (Case B). These soil geological formations generally constitute a thickness of low hydraulic conductivity (DOERFLIGER N. and ZWAHLEN F., 1998).

Table 2. Determination of protective cover

		Case A	Case B
No protection	P1	0 – 20 cm	
	P2	20 – 100 cm	20 – 100 cm sol
	P3	> 1 m	> 1 m
Very high protection	P4		P4 > 7 m and > 8 m Only soil formations

The evaluation of the Criteria P in the study area is based mainly on the thickness of the soils, determined by the technical data sheets of the boreholes, the outcroppings in the heels of the roads and the éxamination of the satellite images.

3.1.3. Infiltration Conditions (I)

The Criteria I is used to distinguish between diffuse and concentrated infiltration zones. The method provides low-to-moderate protection for areas feeding direct and concentrated infiltration points (igues and losses), and moderate-to-high protections outside these zones. The feeding basin of the concentrated infiltration points is cut into three protection zones I1, I2, and I3. The rest of the system is associated with I 3 and I4. In conjunction with the relationship of each zone to the igues or loss positions, two other parameters are taken into account in the classification of Criteria I. Topography with slope classes (10%/10-25%/25%), and soil type (cropland, grassland and pasture). For the area that is outside the igues feeding basins and losses, the index I 3 corresponds to the slopes > 25% If they are surveyed by grasslands or pastures and at slopes > 10% with cropland. The I4 index therefore corresponds to the lowest slopes that do not have a type of land use used in the classification. This point of definition may provide inconsistencies in interpretations of the final vulnerability for these areas.

This Criteria was evaluated using the numerical Terrain Model (MNT) and topographic maps. The entire feeding basin, hated forests, has been assimilated to grasslands and pastures, largely reflecting the real situation. Therefore, the 25% limit gradient was used for indexing of 1.

The topographic watershed of the losses and streams fuelling these losses was determined on the Geographic Information System (GIS) using the MNT, with a 30 m resolution mesh.

3.1.4. Karst network development (K)

K1 to K3 Criteria K involves the development of the deep karst aquifer. The K1 index defined a well-developed karst, the K2 index a poorly developed karst and the K3 index a cracked or porous aquifer (DOERFLIGER N. and ZWAHLEN F., 1998).

Note that the nature of the geological banking of the aquifer or unsaturated area, as well as the degree of fracturing of the rock, is not taken into account in this method.

This parameter was evaluated from the direct demonstrations of the karst network (caves, igues) and the study of the flows Chronicles. The study of Hydrograms was carried out for the springs of the study area.

3.1.5. Calculation of protection factor F

The protection factor F is calculated with the following basic formula [4-5-6]:

$$F = \alpha Ei + \beta Pj + \gamma Ik + \delta Kl$$

With:

α, β, γ, δ: Weighting factors for each of the criteria;
Ei, Pj, Ik, Kl: Value of each indexed criteria.

In table 3, we have deferred the assigned weights. As for Figure 9, it presents the criteria summary map.

Table 3. Relative weights attributed to the factors α, β, γ and δ

E	P	I	K
α	β	γ	δ
3	1	3	2

IV. Results and Discussion

4.1. Evaluation criteria E, P, I and K

The E map below shows a very lowly protection zone (E1) that extends to areas affected by the morphological and intense tectonic elements (Figure 3), while a moderate protection zone (E2) on Piedmont. The rest correspond to high protection.

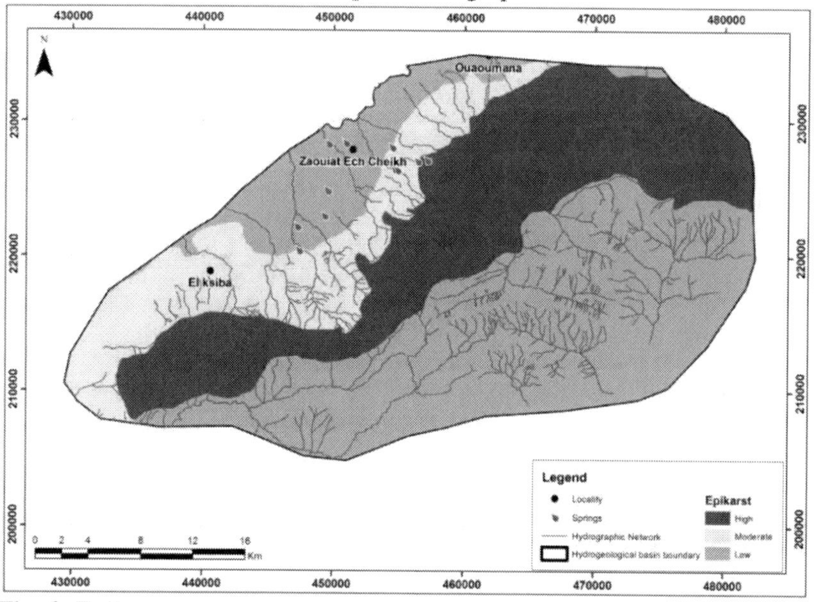

Fig. 3. Epikarst map

The P map (Figure 4) shows a very lowly protective wide area (P1) in the karst outcropping areas of the High Atlas, a Moderate protection zone (P2) in the Dir where a thin layer of clays is placed directly on limestones, areas of West occupied by the Marly and clayey surface layers are considered to be of high protection (P3) and the Quaternary deposit zones are highly protected (P4).

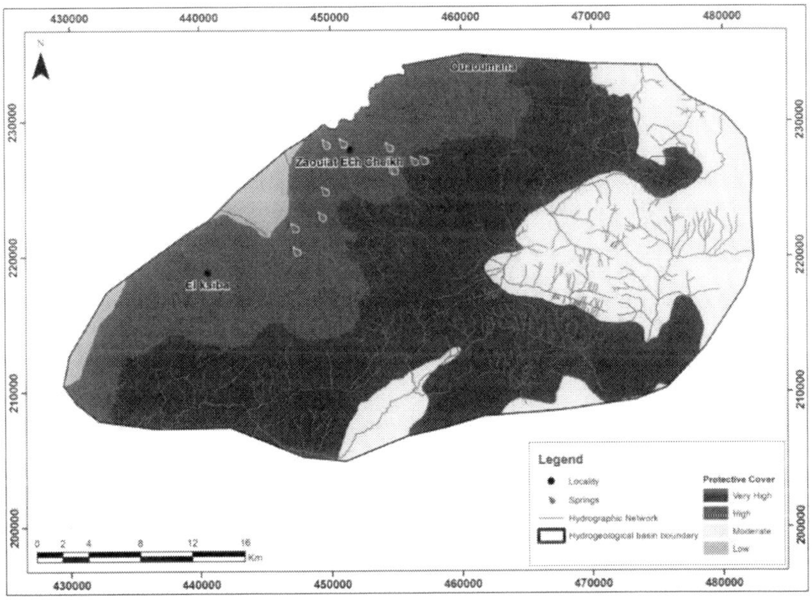

Fig. 4. Protective cover map

The realization of the infiltration map is based on the localization of the morphological elements, the dip and the elements of occupation. As noted above, all losses and sinkholes (Figure 5 and Figure 6) and their feeding pools are mapped as highly vulnerable areas.

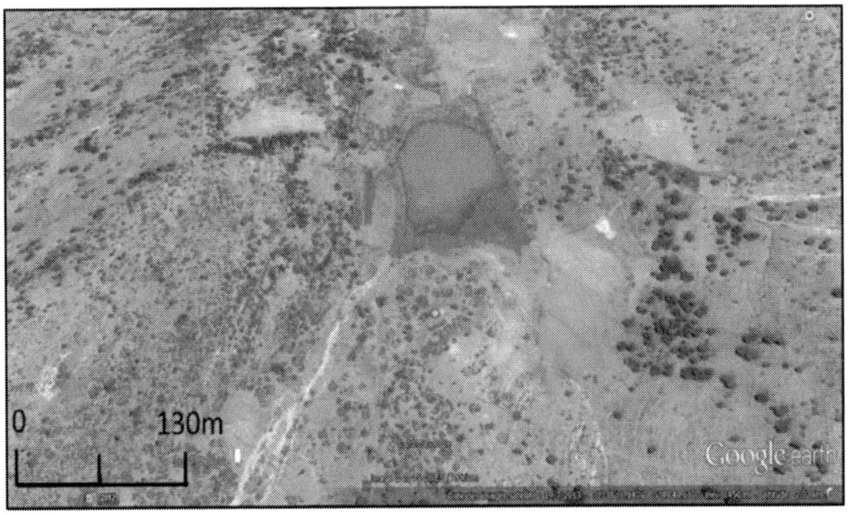

Fig.5. Satellite image of anthropogenic activities around a sinkhole

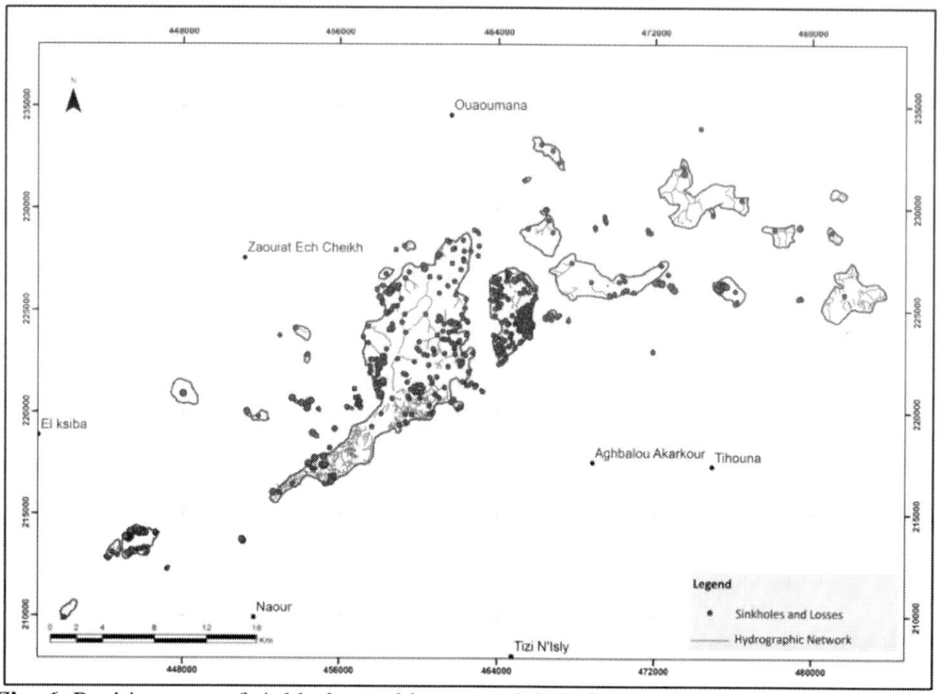

Fig. 6. Position map of sinkholes and losses and their feeding basins

Map I generally presents strong to moderate protection in the study area with the exception of sinkholes and losses and their feeding basins that present low protection (Figure 7).

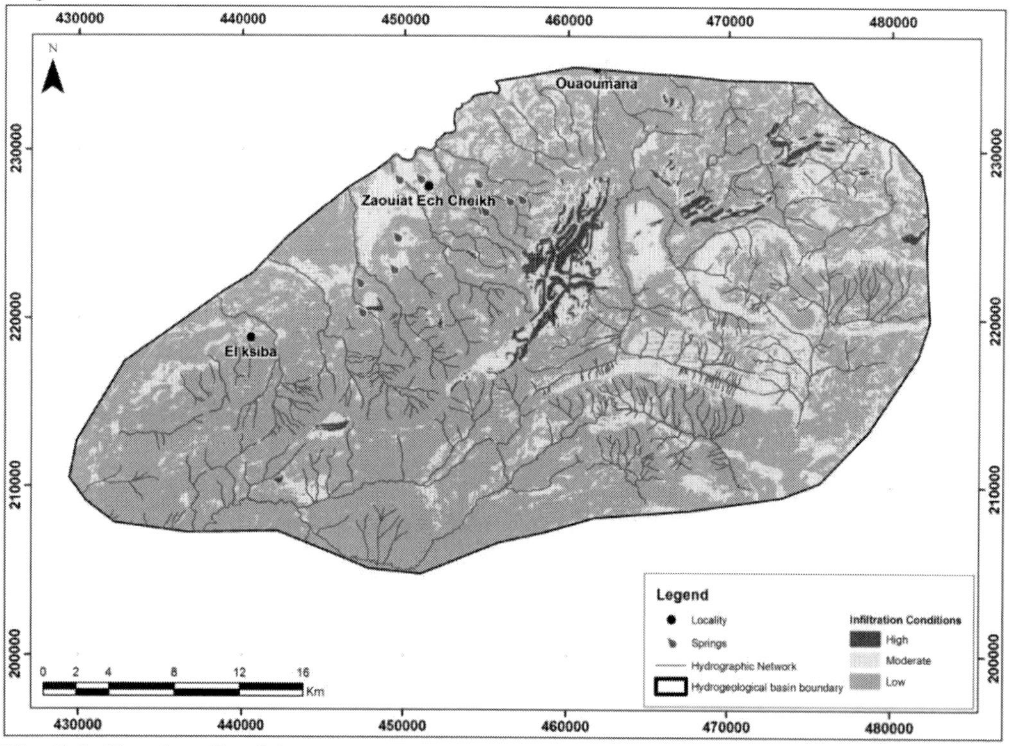

Fig. 7. Infiltration Conditions map

The hydrograms of the Dir springs are similarly reacting. The springs emerging from the entourage of Zaouiat Ech Cheikh (single outlet) demonstrate the presence of a well developed karst network. The flow rates in high water during the tracing tests carried out on the Atlas of Beni Mellal by BOUCHAOU reach some 100 m/h [8].

These speeds indicate the presence of a well-developed karst network.

The K-card has three distinct and homogeneous sets (Figure 8). Areas or surface shapes are classified as K1. The degree of karstification is diminished in passing from the High Atlas to the Tadla Plain, so a K2 indexing has been accepted for Piedmont. The rest of the pond is indexed by K3.

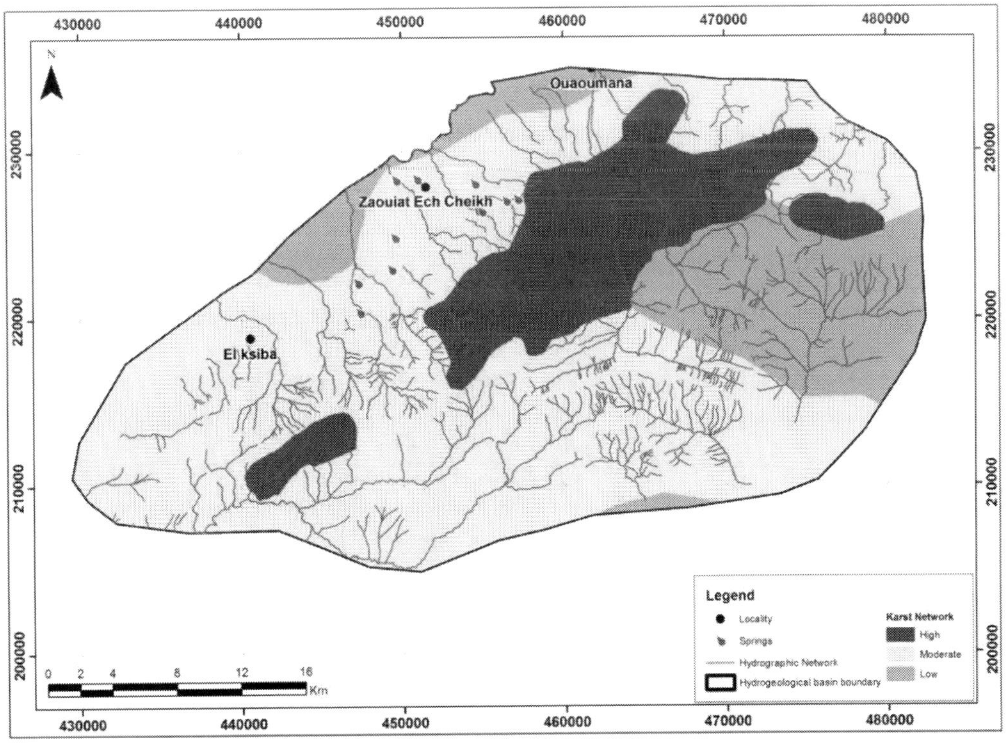

Fig. 8. Karst network development map

4.2. Protection Factor

The vulnerability map in Figure 4 shows that the protection factor values vary between 12 and 31. Areas with very high vulnerability, with a protection factor between 12 and 19, correspond to lapiez fields and areas occupied by sinkholes and losses, affecting the relatively bare limestones of the High Atlas as a result of diffuse and concentrated infiltration.

Large high vulnerability ranges (protection factor of 20 to 25) correspond to the zones of passage between the High Atlas and the quaternary formations of Dir where the karstification is less developed with less infiltration with respect to the slope. The best protected areas are those of the plain, characterized by clay and marly formations. The protection factor associated with these zones is 25 to 31.

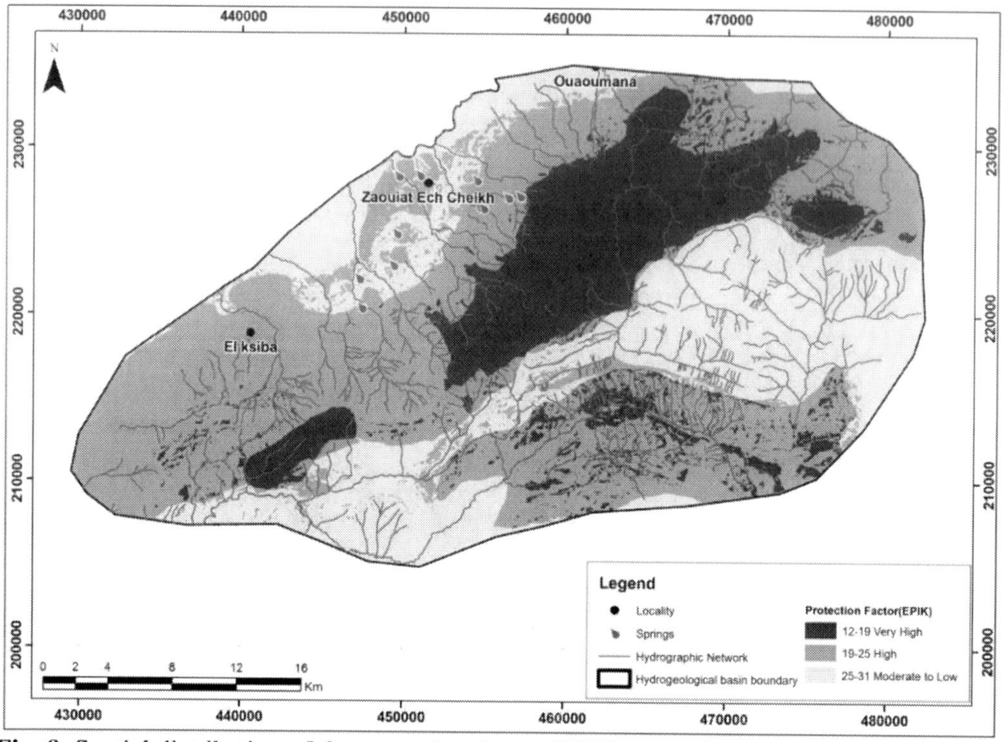

Fig. 9. Spatial distribution of the protection factor (EPIK method)

V. Conclusion

The application of the EPIK method has made it possible to map the vulnerability to pollution of sources located between El Ksiba and Ouaoumana. Three classes of vulnerability emerge from the analysis of this map, namely very high, high and moderate to low. The high class is the most dominant in the source watershed.

Vulnerability mapping is a methodology that has become indispensable for the protection of groundwater resources.

Acknowledgements

We would like to thank all the persons who contributed to the improvement of this work.

References

1. L. BOUCHAOU, P. CHAUVE, J. MUDRY, J. MANIA et Y. HSISSOU. Structure et fonctionnement d'un hydrosystème karstique de montagne sous climat semi-aride: cas de l'Atlas de Beni-Mellal (Maroc). Article, Journal of African Earth Sciences, Vol. **26**, No. **2**, pp. 225-236, (1997).

2. ONEP. Etude d'assainissement de la ville d'El Ksiba. Etude d'assainissement. Mission I, volet d'Etude d'impact sur l'environnement, 158 p, (2003).

3. K. EL KHAMMARI, A. NAJINE, M. JAFFAL, T. AÏFA, M. HIMI, D. VASQUEZ, A. CASAS et P. ANDRIEUX. Imagerie combinée géoélectrique–radar géologique des cavités

souterraines de la ville de Zaouit Ech Cheikh (Maroc). Comptes Rendus Géoscience, Vol. **339**, pp. 460-467, (2007).

4. N. DOERFLIGER et F. ZWAHLEN. Cartographie de la vulnérabilité en régions karstiques (EPIK). Guide pratique. L'environnement pratique. Office fédéral de l'environnement, des forêts et du paysage, Berne, 56 p, (1998).

5. N. Doerfliger , P.-Y. Jeannin , F. Zwahlen. Water vulnerability assessment in karst environments: a new method of defining protection areas using a multi-attribute approach and GIS tools (EPIK method).Article, Environmental Geology **39**(**2**):165-176 (1999).

6. LorienneTHÜLER. Estimation et cartographie de la vulnérabilité des aquifères en milieu forestier. Thèse, Faculté des Sciences de l'Université de Neuchâtel, 183 p, (2014).

7. R. EL bardai, K. Targuisti, A. Maaté, M. Maâtouk et J. Stitou. Evaluation de la vulnérabilité à la contamination des nappes karstiques des environs de Chaouen (Rif septentrional, Maroc) : méthode EPIK. Publications del Instituto Geológico y Minero de España. Hidrogeologia y Aguas Subterraneas N° **18**.239-247. Madrid, (2006).

8. L. BOUCHAOU, P. CHAUVE, J. MANIA, J. MUDRY, J.L. MICHELOT. Apports des traçages artificiel et naturel à la connaissance des circulations et des zones de recharge en région semi-aride cas de l'Atlas de Beni Mellal(Maroc).Article, Revue des Sciences de l'eau **8**(1995) 43-55 pp.

Assessment of groundwater quality in the Al-Burayhi and Hedran sub-basin, Taiz, Yemen – A GIS approach

Ramzy Naser , *Mohammed* El Bakkali , *Nabil* Darwesh, *Khadija* El Kharrim And *Driss* Belghyti

Laboratory of Agro-Physiology, Biotechnology, Environment and Quality. Team Water, wastewater and Health. Faculty of science, university Ibn Tofail, 14000 Kenitra, Morocco

Abstract. In many parts of the world, groundwater sources are the single most important supply for the production of drinking water, particularly in areas with limited or polluted surface water sources. Fresh water has become a scarce commodity due to over exploitation and pollution of water. Many countries and international organizations, including Wolrd Health Organization, are seeking to promote people's access to safe drinking water. The situation in Yemen is no exception. Although we rely on groundwater significantly in our lives and our survival, we do not manage it in a way that ensures its sustainability and maintenance of pollution.

The objective of this study is to determine the suitability of the groundwater in Al Burayhi and Hedran sub-basin (one of the sub-basins of the Upper Valley Rasyan) as a source of drinking water in the shade of the expected deterioration due to natural processes (water interaction with rocks, semi-dry climate) and human activities.

Keywords: assessment of groundwater quality, the spatial distribution of groundwater pollution, Al Burayhi and Hedran sub-basin, Taiz and Yemen.

* Corresponding author:: ramzi_mh1@yahoo.com

1. Introduction

Water is an essential element of life, and without water there will be no life for any living being. Water sources are either surface or groundwater. The source of water supplies in Yemen is mostly from groundwater accumulated during previous and current times. The problem of water in Yemen is that the quantities of water extracted from the ground are much higher than the quantities of recharge water for underground reservoirs. The increase in population and migration from villages to the city leads to pressure on water resources and contribute to their depletion. The situation in Taiz governorate is not different from the rest of Yemen, where groundwater has been depleted without any efforts to feed the aquifers and compensate for what is depleted through the establishment of barriers and dams. Worse still, humans reduce the natural recharge of groundwater as a result of various human activities on the surface of the earth and as a result of pollution that leads to the filling of pores and thus reduce water infiltration downwards and also change the course of water under the ground. Depletion of groundwater resources due to overexploitation and increased evaporation results in enhancement of dissolved constituents(Raju, Ashwani, and Anjali Singh 2017).

The objective of this study is to determine the suitability of the groundwater in Al Burayhi and Hedran sub-basin (one of the sub-basins of the Upper Valley Rasyan) as a source of drinking water in the shade of the expected deterioration due to natural processes (water interaction with rocks, semi-dry climate) and human activities.

2. Material and Methods

2.1 Description of the Study area

The study area is located in the north and north west of Taiz City, Yemen , between (1498000 – 1516000) latitude and (370000 – 395000) longitude (Figure 1). The urbanization of Taiz City has reached some parts of the southern region of the study area (Figure 2). The solid waste landfill of Taiz, and the sedimentation ponds for domestic wastewater (disposed of in the sewerage system) are located in this area, There are also some industrial facilities and poultry farms in this area(Figure 3). Hydrology of study area is an extension of the hydrology of AL-Hawban and AL-Dhabab areas, that means, it is drainage area for these areas. The runoff water collects and flows into the northwestern part of the study area (Figure 3).

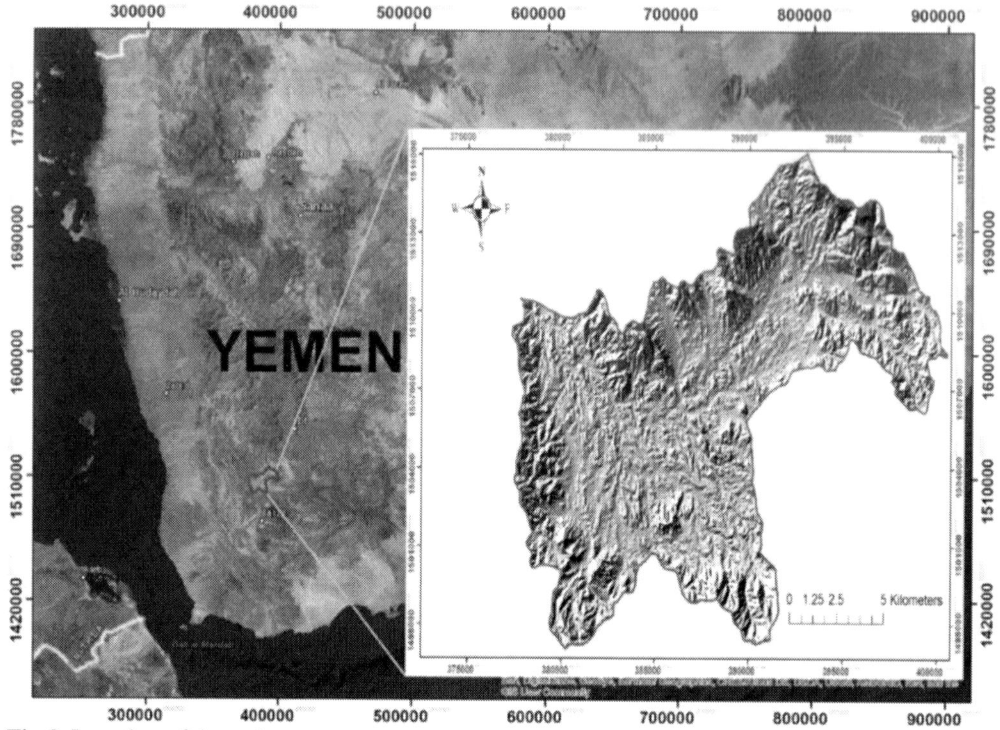

Fig.1. Location of the sudy area

2.2. Inventory of sources of pollution in the study area.

In order to try to understand the relationship between the quality of groundwater and the sources of pollution resulting from human activities, we have made an inventory of the sources of pollution in the study area and then represented them in the map (Figure 3).

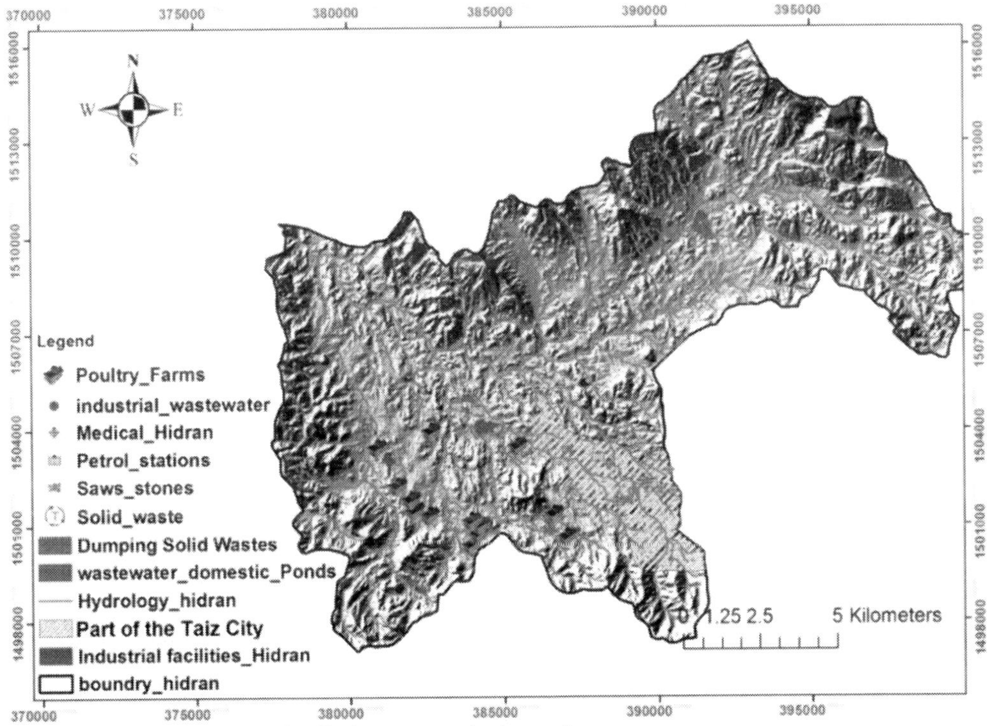

Fig.2. Spatial distribution of pollution sources in the study area

2.3 Collection, transfer and analysis of samples

Twenty-nine sampling points were selected from the different site in Al Burayhi and Hedran sub-basin based on the wells inventory in the study area to analyze their physical and chemical properties. Samples included all types of aquifers (sedimentary and volcanic) and all types of wells (bore well and dug well). Samples were taken only after the water was pumped from the well for a quarter of an hour and the electrical conductivity of the well water stabilized, so that samples could be taken directly from the aquifer. Sampling of wells was accompanied by information on wells. Groundwater samples were collected in new polyethylene bottles of 1000-ml capacity after rinsed with distilled water and the water of the well, and then the bottles were filled with water to the top, practically leaving no air in, through months in August, September and October 2014. All samples were placed in an ice-box container and transferred to the laboratory within a maximum of 72 hours. Laboratory tests were performed for the analysis of samples for : Electrical conductivity (EC), pH, temperature (T C°), Total dissolved solids (TDS), Total Hardness (T.H), Total Alkalinity (T.ALK), Calcium (Ca), Magnesium (Mg) Sodium (Na), Potassium (K) Bicarbonate (HCO 3), chloride (Cl), Sulfate (SO4), Nitrate (NO3), Iron (Fe) and Fluoride (F). The collected samples were analyzed according to the method Assessment of Ground Water Quality described in "Standard methods for the examination of water and wastewater American Public Health Association (APHA).

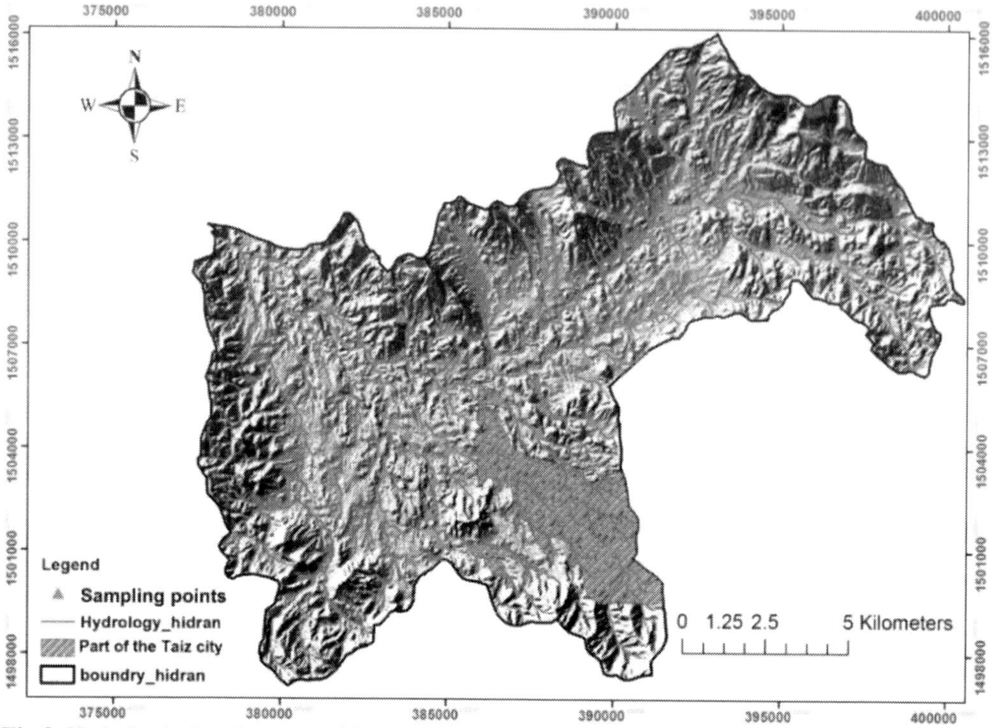

Fig.3. Hydrological study area and Sampling points

3. Results and discussion

Tableau 1 : Groundwater samples of the study area exceeding the permissible limits prescribed by WHO (2004) for drinking purpose.

Water quality parameter	WHO (2004)		Number of sample desirable	Number of samples unfit
	Desirable limit	Maximum permissible limit		
pH	6.5	8.5	29	0
EC (µS/cm)	500	2500	8	21
TDS (mg/l)	500	1500	7	22
TH (mg/l)	100	500	5	24
Ca^{2+} (mg/l)	75	200	18	11
Mg^{2+} (mg/l)	50	150	12	17
Na^{+} (mg/l)	-	200	4	25
K^{+} (mg/l)	-	12	25	4
Cl^{-} (mg/l)	200	600	12	17
SO_4^{2-} (mg/l)	200	400	23	6
HCO_3^{-} (mg/l)	-	500	1	28
$NO3^{-}$ (mg/l)	-	50	26	3
Fe (mg/l)	0.3	1	28	1
F^{-} (mg/l)	1	1.5	5	24

Almost all the samples showed very high concentrations of both Electrical Conductivity; Total dissolved solid; Total hardness; Magnesium; Sodium; Chloride; Bicarbonate and Fluoride. By comparing the results obtained with the standard specifications of Yemen and the World Health Organization standards, it became clear that the Groundwater in most parts of the study area is contaminated and unfit for drinking. The high concentration of contamination at the level of Al Burayhi and Hedran sub-basin could be explained by: The distribution of major ions largely depends on the type of geological formations in contact with the groundwater flowing through, Irrigation return flow, disposal of wastewater domestic and industries, Runoff and evaporation because of climate semi-arid.

The problem of water in the city of Taiz and surrounding areas is a complex problem. There is an extraction of quantities of water in excess of the actual need and there is squandering of these water because of the old and dilapidated network of water supplies (there are a large loss of water), and on the other hand irrigation in various areas by flooding, which consume large quantities of groundwater and beyond the actual need To irrigate the earth. Some of these water evaporate as a result of the semi-arid climatic conditions in the region, especially as irrigation is most often done at times when evaporation is high because of the heat of the sun, some of which go back down and salinity has been concentrated.

The other side of the water problem in and around Taiz is the contamination of groundwater in these areas due to the lack of solid and liquid waste management for various urban and industrial activities. The final outcome of these misdeeds eventually led to the inappropriateness of drinking water from this sub-basin.

References

1. Raju, Ashwani, and Anjali Singh (2017): "Assessment of Groundwater Quality and Mapping Human Health Risk in Central Ganga Alluvial Plain, Northern India." Environmental Processes : p1-23.
2. APHA (1995): Standard methods for the examination of water and waste water, 19th edn. American Public Health Association,Washington, DC
3. WHO (2004): Guidelines for drinking water quality. World Health Organisation, Geneva

Diagnosis of the Ghiss Nekor aquifer in order to elaborate the aquifer contract

Wissal BAITE[1*], *A.* BOUKDIR[1], *A.* ZITOUNI[1], *S. D.* DAHBI[2], *H.* MESMOUDI[2], *A.* ELISSAMI[3], *E.* SABRI[1] et *H.* IKHMERDI[1]

[1] Laboratory of Industrial Engineering, Sultan Moulay Slimane University, Sciences and technologies Faculty, Beni Mellal, Morocco.
[2] Hydraulic Basin Agency of Loukkos, Tetouan, Maroc.
[3] National Office of Electricity and Drinking Water (ONEE), Rabat, Maroc.

Abstract. The Ghiss-Nekor aquifer, located in the north-east of the action area of the ABHL, plays a strategic role in the drinkable water supply of the city of Al Hoceima and of the neighboring urban areas. It also participates in the irrigation of PMH. However, this aquifer has problems such as over-exploitation and pollution. In the face of these problems, the only Solution is the establishment of a new mode of governance, which privileges the participation, the involvement and the responsibility of the actors concerned in a negotiated contractual framework, namely the aquifer contract. The purpose of this study is to diagnose the current state of the Ghiss Nekor aquifer, the hydrogeological characterization of the aquifer, the use of the waters of the aquifer, the Problem identification and the introduction of the aquifer contract, which aims at the participatory and sustainable management of underground water resources in the Ghiss-Nekor plain, to ensure sustainable development.

1 Introduction

Morocco has always made the development of the water sector a priority and a strategic choice. This sector, one of the main levers of economic and social development, faces two key challenges: [1]

- The depletion of water resources in relation to the increase in all-purpose water demand, with a tendency towards absolute scarcity by 2025 [1]

- The degradation of water resources that undergo different forms of pollution. [1]

The Ghiss-Nekor alluvial aquifer is one of the most important aquifers in the northeastern part of the ABHL area. This water table plays a strategic role: it currently provides most of the drinking water supply in the city of Al Hoceima and surrounding urban centers. It also participates in the irrigation of the PMH, especially on the right bank of the Nekor wadi.

However, this aquifer is increasingly exploited because it is the only major aquifer in the area. In addition to overexploitation, groundwater is characterized by high ionic concentrations, due to the geological composition of the lands crossed, the mineralized inputs of the returns of irrigation water and the various polluting anthropogenic activities. This implies the need for a radical change in the management of water resources.

*Corresponding Author: wissalgat@gmail.com

2 Problems of the Ghiss Nekor aquifer

The Ghiss Nekor water table suffers from many problems among which we can mention the pollution. The waters of the aquifer are marked by very high levels of different ions (mainly chlorides and sodium). The high levels come mainly from the accumulation of water supplies of wadis (especially Nekor) rich in dissolved salts from evaporitic and carbonate formations.

The result of the DRASTIC vulnerability index indicates that groundwater resources in the Ghis-Nekor Plain are sensitive to moderate to high levels of pollution [2]. The coastal zone of the study area is characterized by a high vulnerability that could be attributed to the shallow water depth (less than 5 meters below the earth's surface) and high permeability. In addition, the shallow aquifer in this zone consists mainly of interbedded deposits of gravel, sand and silt thin [2]. In addition, aquiferous materials are characterized by a high hydraulic conductivity. Areas with moderate to high vulnerability to groundwater pollution have been found to be correlated with areas of high anthropogenic pollution activity. [2].

The Ghiss Nekor water table has been negatively affected by the succession of years of drought. In recent years, the surface waters of the Ghiss Nekor Plain have been considerably reduced, resulting in high groundwater pressure of the Ghiss Nekor water table, and consequently a decrease in the piezometric level.

Hence the need for immediate implementation of a sustainable management plan based on limiting hazardous activities, restoring groundwater quality and improving environmental good sense [2].

3 Material and methods

3.1 Study zone

The Ghiss-Nekor aquifer covers an area of 100 square kilometers and is located (**fig. 1**) northeast of the ABHL action zone and occupies the downstream part of the two basins of Ghiss and Nekor.

The geographical limits of the study area are: in the North the Mediterranean Sea, in the West the city of Ajdir and the douars of Aït Youssef or Ali and Imrabtene, Douay Bouayach in the Southwest, douars Tessaft and the Dam Mohammed Ben Abdelkarim Al Khattabi in the South, the Southeast Douar of Beni Akki and East the center of Troughout

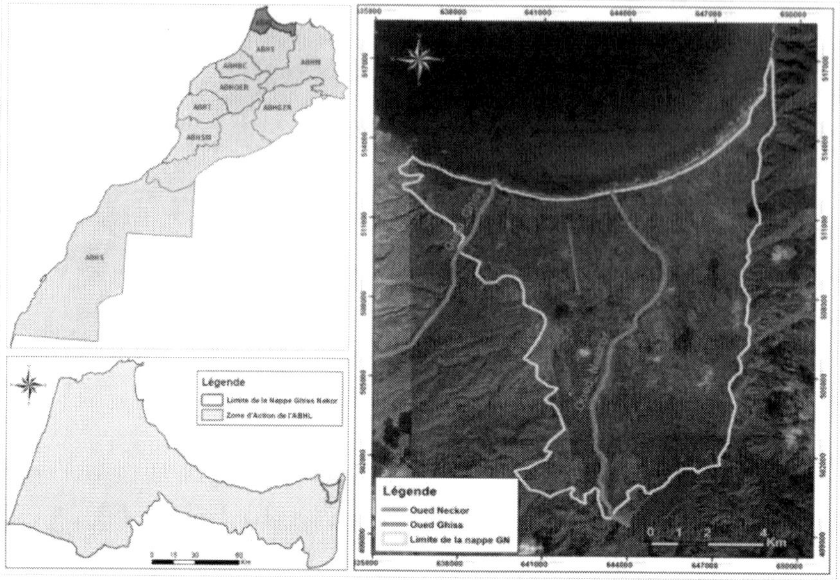

Fig. 1: Ghiss Nekor water table situation map at the national level and at the scale of the ABHL action zone

3.2 Demographic characteristics

The evaluation of the demographic characteristics and the socio-economic conditions of the study area contributes to the valorization of human and natural resources with a view to a better exploitation for a sustainable development adapted to the peculiarities of the region.

According to Morocco's regional breakdown of 2014, the study area falls under two provinces that are part of two different regions Tangier-Tetouan-Al Hoceima region for the province of Al Hoceima and the Oriental region for the province of Driouch.

The population of the Tangier-Tetouan-Al Hoceima region according to the General Population and Housing Census of 2014 is 3556729 inhabitants and the province of Al Hoceima has 399,654 inhabitants or 11.2%.

The population of the Oriental region is 2314346 inhabitants while the province of Driouch has 211059 inhabitants or 9.1%.

The two provinces share similarities despite the fact that Driouch Province is younger, and has a very pronounced rural character, with 68% of the population living in rural areas.

3.3 Geology of the Plio-Quaternary of Ghiss Nekor

Overall, the plio-quaternary Ghiss-Nekor aquifer consists of a complex of alluvial deposits with vertical and lateral facies passages. These deposits store a generalized free web, which in some places is captive. Stratigraphically, we distinguish from bottom to top [3] :

- The Plio-Villafranchian: is represented by a heterogeneous material consisting of conglomerates, sandstones and silts that are outcropping to the west of the plain at the level of the Oued Ghiss and Imzourène plateaus [3];
- The ancient Quaternary: outcrops on the lower Ghiss plateau and is made up of

heterogeneous material with conglomerates at the base [3];

- The Middle Quaternary: is represented by terraces encrusted on gravel Tensiftien-Amirien outcropping to the south of the plain in the region of Beni Bouayach to the west of Oued Nekor, downstream of Oued Tifaouine [3];
- The recent Quaternary: covers almost all of the plain and deposits consist of the gray silts of Rharbien and laterally evolve to the silty deposits of coastal glacis and low terrace (of Soltan age) south of the plain at from the left bank of Oued Nekor and to the east of Plio-Villafranchien d'Imzourène [3].

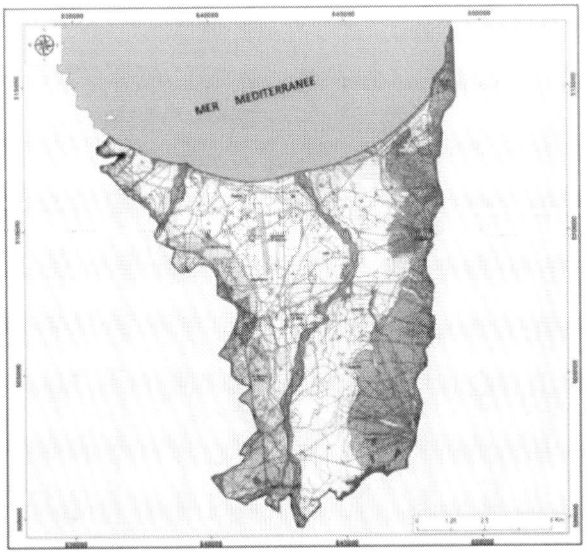

Fig. 2: Geological map of the Ghiss Nekor plain, (Extracted from the geological map 1/50000 of El Hoceima)

The geological, geophysical and hydrogeological study has identified an aquifer consisting of alluvium plio-quaternary age with an average thickness of 240m. Its schistose substratum sinks from south to north under the plio-quaternary cover [9].

3.4 Climatology

The study of the factors governing the climate of the Rhis-Nekor Plain, namely, precipitation, temperature, and evapotranspiration, is necessary because these elements are a strong condition for the expression of the water balance and therefore the volumes exchanged between the aquifer system and its environment [9].

Given its geographical location in the extreme north of Morocco and the variability of the reliefs around its limits, the water table of Ghiss-Nekor is subject to a climate of the Mediterranean type. It is located in the semi-arid bioclimatic stage (dry summers) at relatively temperate winter. The rains usually occur for eight months, from October to May, while the dry season runs from June to September.

Figure n°3 gives the annual rainfall at the SMBAK Dam station. There is a fairly marked year-to-year fluctuation with very critical minimum inputs of 196 mm in the 2013/2014 year and significant rainfall with a maximum of 844 mm recorded during the 2009/2010 rainy year. The annual average is estimated at 396.9 mm.

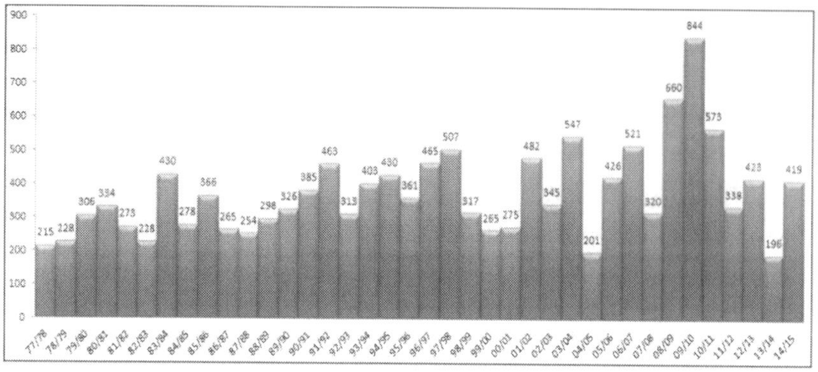

Fig. 3: Annual precipitation variation

3.5 Piezometry

The piezometric map **(fig. 4)** shows feeding areas in the South and East. The feeding follows the superficial flow of the Nekor wadi. This diet is remarkable at the dike dam SMBAK where the diet is due to the released of the latter. Isopipe curves are open downstream, and increasingly spaced (decrease in hydraulic gradient) from upstream to downstream and from east to west.

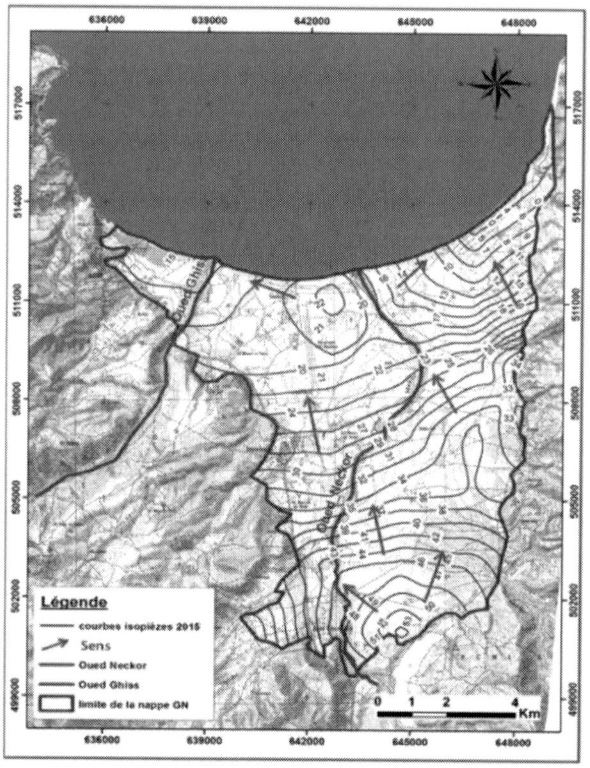

Fig. 4: Piezometric map of the Ghiss Nekor aquifer (2015)

3.6 Hydrodynamic characteristics

The transmissivity values of the aquifer range from $6{,}65*10^{-5}$ to $6{,}4*10^{-2}$ m^2/s around an average of $2*10^{-2}$ m^2/s [3].

The measured permeability values at 25 wells range from $3{,}7*10^{-8}$ m/s to $8{,}5*10^{-4}$m/s around an average of $1{,}8*10^{-4}$m/s [3].

The values of the storage coefficient evaluated at four wells are between 0,65 and $5{,}3\cdot10^{-2}$ [3].

Table 1. Values of the storage coefficient evaluated at four Boreholes

Boreholes	X	Y	S (.10^{-2})
252 /5	638 350	510 140	4.6
256/5	646 650	505 400	5.3
310/5	640 700	510 800	2.2
F4	640 338	500 105	0.65

3.7 Hydric balance of the aquifer

The current balance is deficit of a volume of 2.31 Mm^3 / year [3] following the reduction of wadis' contributions, and the increase of the demand for the drinking water supply and industry and for the irrigation of the irrigated perimeters.

The water balance of the aquifer has a deficit due to natural losses from outflow to the sea and direct evaporation in the littoral zone, and because of water withdrawal from the water table, especially farmers who generally tend to exaggerate pumping time and use water that far exceeds the actual needs of the crop.

Table 2. Water balance of the de Ghiss Nekor aquifer [3]

Inputs (Mm3/ year)		**Outputs (Mm^3/ year)**	
Infiltration of rainwater	**3,16**	**Outings to the sea**	**2,57**
lateral inputs	**5,26**	**Evaporative Outputs**	**5,16**
Inputs from wadis	**2,98**	**Agricultural pumps**	**3,3**
Return of irrigation water	**0,66**	**Water withdrawal by ONEE**	**3,34**
Total	**12,06**	**Total**	**14,37**
Variation of the Stock (m^3/ year)		**- 2,31**	

3.8 Water quality of the water table

Overall, and taking into account chemical and bacteriological parameters, it can be considered that the overall quality of the groundwater is poor to medium.

The dry residue is 1.5 to 5 g / l [3]. The salinity of the water varies from 2100 to 2600 µS/cm. PH values range from 6.7 to 8.1 around an average of 7 to 8 during the period 2004-2014 [3]. The very high sulphate levels in the aquifer are a limiting factor for the exploitation of groundwater for the drinking water of the populations of the city of Al Hoceima and the surrounding centers and douars. The degradation of quality comes mainly from marine intrusion, the presence of salt levels in the aquifer and the return of irrigation water [3].

https://doi.org/10.1051/e3sconf/20183701006

Moreover, it is necessary to protect the collecting fields of drinking water by close protection perimeters to limit pollution caused by human activities.

3.9 Water use

The 2014 inventory identified 299 farms (**Table 3**), almost equally distributed, between the two provinces of Al Hoceima and Driouch. Including 261 sampling points for agriculture to irrigate an area of about 378 ha, which consumes an annual volume of 3.34 Mm^3, 9 boreholes for the National Office of Water and Electricity that use an annual volume of 721 172 m^3 and 69 sampling points for rural drinking water with a volume of approximately 183 887 m^3. With a total of 0.91 Mm^3 for drinking water [4].

Table 3. Sampling points from the Ghiss Nekor water table according to the inventory 2014 [4]

Total number of points	Designation				
	Agriculture			Drinking Water Supply (DWS)	
	Number of points	Area (ha)	Volume (Mm^3/ year)	Number of points	Volume (m^3/ year)
299	261	378	3.34	ONEE-BE 9	721 172
				rural DWS 69	183 887

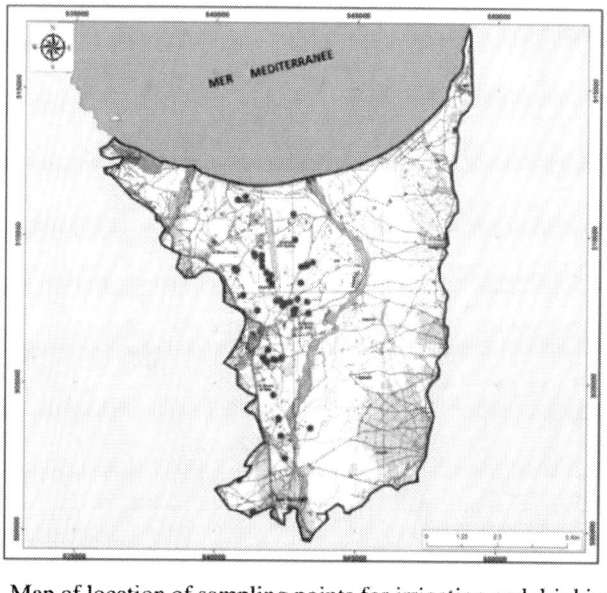

Fig. 5. Map of location of sampling points for irrigation and drinking water

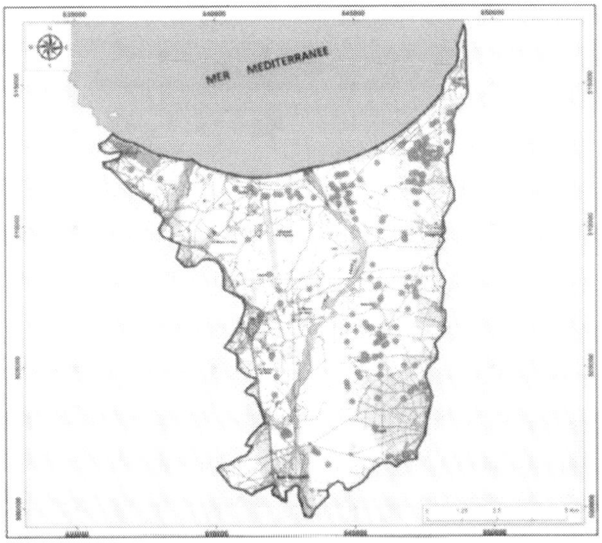

Fig. 6. Map of location of sampling points for Irrigation

3.9.1 Drinking water supply (DWS)

The groundwater resources from the fields of Ghiss wadi and Nekor wadi consist of 9 boreholes and two wells with a global flow rate of 185 l/s [4].

The adduction, coming from the catchment field of Ghiss wadi, includes the following works:

- A discharge pipe, DN 400 mm diameter, connecting the Ghiss wadi station to the connection with the pipe from the Ajdir booster station;
- A discharge pipe, connecting borehole N°IRE 1083/5 to the Imzouren reservoir.

The adduction from the Nekor catchment field consists of discharging the water from the boreholes to the existing loading tank of 3500 m^3, with two pumping stations and a discharge pipe, diameters DN 315 and DN 400 mm. , and total length of about 12 Km.

The following graph (**fig. 7**) shows the evolution of the volume produced by the wells and boreholes (385/5, 1768/5, 1085/5, 573/5, 510/5, 576/5, 514/8, 514/9, 514/10, 515/2, 1086/5, 1083/5, 578/5, 1691/5, 797/10, 673/10, 1692/8) intended for the supply of drinking water during the period April 2016 to January 2017, which tends towards the increase of pumping.

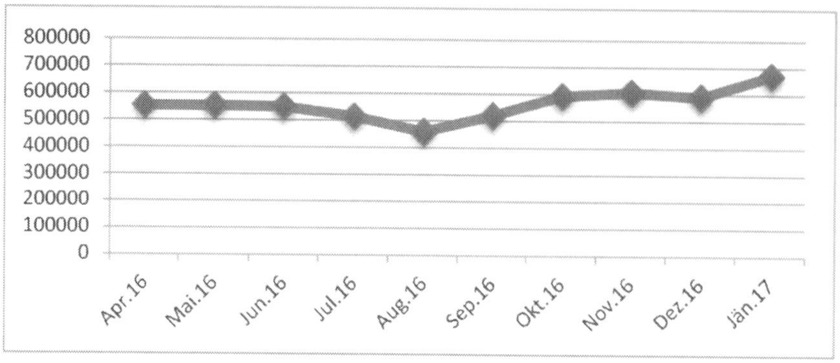

Fig. 7. Evolution of the volume of water produced by wells and boreholes during the period April 2016 - January 2017

3.9.2 Irrigation

Agriculture is the sector that consumes a large share of water resources, hence the need to study irrigation water consumption and possible measures to better manage this use.

Table 4 represents the areas occupied by the irrigated perimeters located in our study area.

Table 4. Areas of irrigated perimeters of the Ghiss Nekor Plain (Provincial Directorate of Agriculture of El Hoceima)

N°	Perimeter	Area (ha)
1	Boukhlifa	70
2	Tazaghine	12
3	Boujbour	20
4	Boumassous	25
5	Handoune	40
6	Takhsaste	10
7	Boumakek	30
8	Touna	15
9	KhmisAghzar	15
10	Arnas	22
11	Bougzour	35
12	Tighza N'romane	10
13	Ait Hicham Tourthoute	10
14	Aarbi	75
15	AzgharTafournoute	16
16	Ghis	900
Total		1305

According to the inventory made by the Loukkos Watershed Agency (2014), the number of water fetching points for irrigation is 214 points for purely agricultural use, in addition to 63 water fetching points used for drinking water supply and irrigation at the same time. The irrigated area is 1305 ha (El Hoceima PDA) and the annual volume consumed is of the order of 3.34 Mm^3.

3.9.3 Industry

According to the 2014 inventory there is only one point of water withdrawal from the aquifer that is intended for industry, located in the municipality of Imzouren, and which uses a flow of 15 l/s.

Otherwise, the waters of the Ghiss-Nekor aquifer are used for cleaning the gravels extracted from the Nekor wadi bed, which is carried out by excessive pumping. The residues of the sands are thus a source of pollution of the aquifer but also lead to a saturation of the soil by the fines and reduces the infiltration of rainwater into the aquifer [6].

The quarry and the pumping area are 400m from the sea; this creates a high risk of invasion of the aquifer by marine waters [6].

3.10 Aquifer contract

Independently of any climate change, water management is one of the major problems that shape the future of Morocco [1]. The Nekor Ghiss aquifer is among the most exposed aquifers to the problem of pollution and overexploitation. Theoretically, the exploitation of groundwater results in the modification of the state of the aquifer and its dynamics in a more or less extended space according to the impulses exerted and the nature and parameters of the aquifer [1].

To face these problems the only way out is to set up a new mode of governance of groundwater which privileges the participation, the implication, the responsibility and the commitment of the actors concerned and the regional and local partners in the management of water resources, using a participatory approach and in a negotiated contractual framework.

A water table contract can be defined as a technical and financial agreement between the different users of the water table for the purpose of exploitation and especially of sustainable management. And must be signed between the different partners: Local Authorities, Water Basin Agency, water users and also scientific institutions.

The aquifer contracts constitute, at the scale of the aquifer, the operational tools best adapted in terms of programming. They set targets for a given water table in terms of the quality and quantity of water, the development of the aquatic environment and the balanced management of water resources. It provides in an operational way (5-year action program, designation of project owners, financing method, deadlines for works, etc.) the arrangements for carrying out the work required to achieve these objectives [7].

3.10.1 The objectives of the Ghiss-Nekor aquifer contract

The long-term objectives of the Ghiss-Nekor aquifer contract are:

- The diagnosis of the current state of the water table;
- The fight against water scarcity in the region of Al Hoceima and the surrounding area
- Preserving groundwater in the long term and rebalancing the aquifer balance;
- Prevention of water resources pollution;

- Guarantee the security of water supply for the local population and the agricultural, industrial and tourist sector;
- Reducing vulnerability to risks related to extreme events and climate change;
- Guarantee sustainable socio-economic development especially in the agricultural sector;
- Engaging a reflection to all stakeholders for setting up a convention aquifer contract;
- Improving the monitoring of water resources and controlling their use.

3.10.2 The strategic axes of the aquifer contract

Table 5. Strategic axes of the aquifer contract

Axes	Specific objectives
Development of the supply of water resources	- Mobilization of additional water resources: Dams. - Gradual reduction of the use of groundwater and its replacement by surface water: drinking water supply and irrigation.
Demand management and protection of groundwater resources.	- Strengthening control: water police. - Rationalization of water uses. - Valorization of water resources. - Preservation of the quality of the water. - Reuse of treated wastewater in irrigation. - Artificial recharge of the water table.
Improvement of the knowledge framework.	- Monitoring the qualitative and quantitative status of water resources. - Establishment of an information exchange system - Development of scientific research.
Strengthen communication and training.	- Development of cooperation and partnership. - Awareness, communication and coaching. - Capacity Building.

4. Conclusion

The Rhiss Nekor aquifer is considered to be the most important alluvial aquifer in the Mediterranean zone. Nevertheless, it is confronted with major natural and anthropogenic pressures, namely the aridity of the region, the development of agricultural activity, the vulnerability to marine intrusion as well as its chloride-sodium and / or potassium sulphate facies. What constitutes a major challenge for its water resources [5].

In the face of this critical situation, the only solution is to implement a sustainable, integrated and participative management of water resources, which involves all stakeholders, who generally have conflicting interests, and which takes into account the particularities of the region.

Finally, insofar as integrated management is always an attempt to reconcile different or even contradictory interests or rights, conflicts of use, territories with multiple legitimacies, it seems interesting to look more closely at the way in which groundwater has been progressively integrated into public water management policies [8]. And subsequently make the changes that make this management more suitable to the particularities of each region, and based on a partnership and contractual approach.

References

1. M. BAHIR et A. MENNANI, EstudiosGeol., **58**, 103, 2002
2. A. SALHI et All, EuroMediterranean Scientific Congress on Engineering, Algeciras, 186, (2011)
3. ABHL, étude d'élaboration du schéma d'aménagement des ressources en eau du bassin de Ghiss-Nekkor mission 1 : évaluation des ressources en eau sous-mission 1.2 : évaluation quantitative des ressources en eau. (2015)
4. ABHL, Etude d'actualisation de l'inventaire des redevables d'utilisation de l'eau du domaine public hydraulique dans les bassins du Ghiss, Neckor, Lou et à partir de l'oued Loukkos, 2013
5. N. IOUZZI et All, Conférence Méditerranéenne Côtière et Maritime, édition **2**, 345 (2011)
6. A. SALHI,Géophysique, hydrogéologie et cartographie de la vulnérabilité et du risque de pollution de l'aquifère de Ghis-Nekor (Al Hoceima, Maroc) thèse de doctorat, 2008.
7. Schéma d'Aménagement et de Gestion des Eaux (SAGE)
8. O. PETIT Economie Rurale **309**, 50, 2009
9. H. JOUNAID, Hydrogéologie et modélisation de la nappe phréatique de Rhis-Nekor (Nord du Maroc). Thèse de doctorat en sciences appliquées, Ecole Mohammadia d'Ingénieurs. (2002).

Theoretical study of the dissolution kinetics of galena and cerussite in an abandoned mining area (Zaida mine, Morocco)

*Lamiae*El Alaoui, *Abdelilah*Dekayir*

UR Geotech, Département de Géologie, Université Moulay Ismail, Meknès, Maroc

Abstract. In the abandoned mine in Zaida, the pit lakes filled with water constitute significant water reserves. In these lakes, the waters are permanently in contact with ore deposit (cerussite and galena). The modelling of the interaction of waters with this mineralization shows that cerussite dissolves more rapidly than galena. This dissolution is controlled by the pH and dissolved oxygen concentration in solution. The lead concentrations recorded in these lakes come largely from the dissolution of cerussite.

Introduction

In the worldwide, abandoned mine sites are known by significant degradation of water quality and contamination of aquatic systems by heavy metals, which are known to be very harmful to human health due to the acid mine drainage phenomena causes by the oxidation of the sulphides, in particular pyrite and arsenopyrite. The oxidation of the sulphides causes a significant decrease in the pH, which causes the dissolution of the mineral phases and the release of metals in solution.

In the upper Moulouya, water is a rare commodity. In fact, the aridity of the climate and the low rate of precipitation require the preservation of lakes waters which represent important reservoirs for the irrigation of the crops and the watering of the livestock.

In the Zaida mine, primary mineralization consists mainly of cerussite (70%) and galena [7]. In the absence of sulphides such as pyrite and the presence of a granitic geological environment, the pH of surface-flowing water is often alkaline.

In the pit lakes, water is in permanent contact with the remains of galena and cerussite. In fact, the concentrations of lead and arsenic found in these waters are controlled by chemical reactions between the mineralization and the aqueous solution.This work deals with the theoretical study of the dissolution kinetics of galena and cerussite considered as the main mineral phases source of heavy metals. This study was conducted using the PHREEQC modelling program.

*Corresponding author: dekayir@yahoo.frTel: (+212668105079)

1 Studied site

The abandoned Zaida mine is located in the centre of the upper Moulouya watershed, at 26 km north of Midelt. The mining activity was done from 1972 to 1985. The main ore extracted is lead, mainly carried by cerussite (70%) and galena [7]. The climate in this locality is arid to semi-arid with rainfall of 300mm / year and mean annual temperatures ranging from 18 ° C to 29 ° C, while minimum winter temperatures are in the range of 14 to 21 ° C [2]

The studied lake ZA located near the tailing I (mining waste), is filled with surface water with an area of about 50Ha accumulating a volume of water of about 1Mm^3; its chemical composition sometimes shows concentrations of metals in lead and Arsenic [1, 3]

Table1. Characteristic of the lake

Perimeter	1,36 Km
Area	50 Ha
Depth	5 à 12 m
Volume of water	1Mm3

Fig.1. mining area of Zaida and location of the lake ZA.

2 Material and methods

2.1 Collect and analysis of ZA lake waters

The main physical-chemical parameters (temperature, pH, redox potential, dissolve dioxygen concentration, conductivity) of the water were studied in situ using a portable multiparameter of the HQ40d type. The samples taken were filtered at 0.2 µm and stored in Teflon bottles. Water samples for cations chemical analysis were acidified by the addition of nitricacid HNO3. The other samples were prepared for anion analysis. These samples were analyzed by ICS-MS in the Hydrosciences Montpellier laboratory (France).

2.2 Mineralogical analyzes

In Zeida, a sample of arkose ore deposit was crushed with an agate mortar and was analyzed by X-ray diffraction (XRD) in the Research Center at Meknes university. The used device is a Shimadzu type diffractometer. The mineralogical analysis allowed calculating the proportions of the mineral phases contained in the arkose.

3 Results

3.1 Mineralogical analysis of mineralized arkose

Mineralogical analysis by X-ray diffraction of mineralized arkose shows the presence of typical minerals of a granitic heritage such as quartz, orthoclase, muscovite with galena and cerussite (Table 2).

Table 2. Mineralogical composition of the analyzed arkose in mole and in percentage

Elements	Galena	Muscovite	Kaolinite	Cerussite	Quartz	Orthoclase	Baryte
Composition	0,19	0,07	0,08	0,16	0,17	0,17	0,17
%	18,81	6,93	7,92	15,84	16,83	16,83	16,83

3.2 Physical-chemical analysis of the lake ZA waters and calculation of chemical speciation

To examine the degree of contamination of Lake ZA by heavy metals and arsenic, a field survey was conducted in this area in October 2010 (dry season), to sample the surface pit lake waters of the ZA to determine the concentrations of trace elements and arsenic. An example of these analyzes is given in Table 3.

Table3. Chemical analysis of ZA lake waters

Physicochemical parameters							Total alkalinity(ppm)		
Conductivity (µS/cm)	Temperature (°C)		pH	ORP (mV)	pe	Dissolve doxygen (ppm)	$[CO_3^{2-}]$	[TAC]	$[HCO^{3-}]$
5630	15.6		9.02	21	0.35	8.32/104%	82.82	616.30	450.7
Concentration in traced elements (ppb)									
Li	B	Al	Ca	V	Cs	Mn	Fe	Co	Cu
1851	55.95	11.09	29000	5.92	0.219	25.2	3.621	0.584	1.844
Zn	As	Rb	Sr	Mo	Sn	Sb	Ba	Pb	U
29.61	43.89	25.78	2823	18.88	0.018	1.867	64.37	0.022	0 .022

These hydrogeochemical analyses of the samples were simulated using the PHREEQC numerical code (version 3.3.12) [5], to evaluate the chemical speciation of the dissolved constituents and calculate the saturation of different mineral phases. The thermodynamic database wateq4f.dat has been used for chemical equilibrium calculations. The results of calculation of the speciation of lead and arsenic showed that in the case of lead, the bulk of the metal in solution is in $PbOH^+$ form, whereas arsenic exists in its form $HAsO_4^{-2}$ at 57, 37%, Zinc in its Zn $(OH)_2$ form (39.76%) and copper in Cu $(OH)_2$ form (2.85%) (Fig. 2)

Table 4. Chemical speciation of lead, zinc and arsenic

		Pb	Zn	As	Cu
Lake ZA	Molality	$7,32*10^{-11}$	$4*10^{-7}$	$5,77*10^{-7}$	$2,87*10^{-8}$
	%	0,0073 %	39,76%	57,37%	2,85%
	Elements	$PbOH^{+}$	$Zn(OH)_2$	$HAsO_4^{-2}$	$Cu(OH)_2$

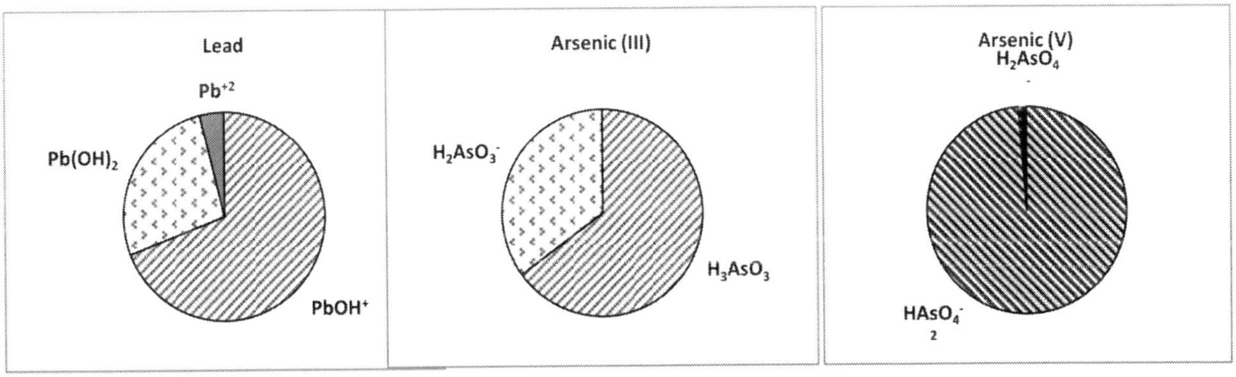

Fig. 2. Chemical speciation of lead and arsenic in the lake ZA

The calculation of the saturation indices shows that the waters of Lake ZA are under-saturated for most compounds of Lead, Zinc and Arsenic, but supersaturated in Iron with a possible precipitation of ironoxides such as hematite. In the case of lead, the compound Pb $(OH)_2$ is in equilibrium with the aqueous solution.

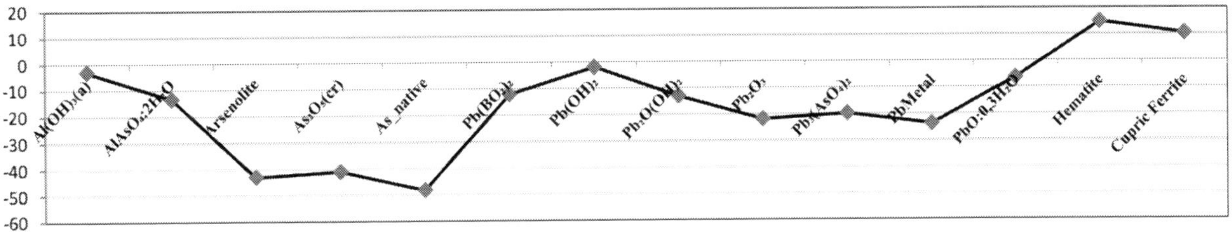

Fig. 3. Water saturation indices of Lake ZA

3.3 Simulation of the dissolution reactions of galena and cerussite

The dissolution of galena has been the subject of several works [6], Indeed, its dissolution rate can be quantified by the following equation:

$$PbS + 2H^{+} = Pb^{2+} + H_2Saq \text{ [8]}$$

$$R = \frac{V}{S} \cdot \frac{dC_{Pb}}{dt} = k.[H+]^{n} et \log R = \log k - n\, pH \text{ [8]}$$

R : dissolution rate (mol/l/s) ; t : reaction time elapsed (s) ; S : surface area of galena (m^2) ; C_{Pb} : concentration of lead ; V : volume of solution (l) ; k : dissolution rate constant ; $[H^+]$: activity of H^+ ; n: Reaction order

While the dissolution of cerussite has been little discussed.

$$PbCO_3 + 2H^+ = Pb^{2+} + CO_2 + H_2O \quad [5]$$

To understand the interaction of Lake ZA waters with mineralized arkoses, the dissolution rates of galena and cerussite were simulated at pH = 9 over a 31 years period corresponding to the mine abandonment period. The alteration solutions used in the simulation correspond to rainwater and water from the lake ZA.

The obtained results in both cases show that the dissolution rate of galena remains much lower than that of cerussite (Fig. 4). Consequently, very low concentrations of Pb^{2+} in solution obtained in both cases are similar and of the order of 10^{-13}mol/l, which corresponds exactly to the concentration of dissolved lead measured in the same lake ($1{,}06*10^{-13}$mol / l). In summary, the lead concentration is essentially related to the dissolution of cerussite (Fig. 4). It is important to note that the waters of the lake ZA have very low electron potential values (pe), resulting in stagnant and less oxygenated water, which slows down the dissolution rate of metal phases.

Conclusion

In the abandoned mine of Zaida, the pit lakes store significant amounts of water used by local residents for crop irrigation and livestock watering. The waters of these lakes are permanently in contact with the mineralized arkose in cerussite and galena which, by dissolution, release metals in solution, especially lead and arsenic.
The dissolution models of galena and cerussite elaborated by PHREEQC show that the dissolution of these minerals is essentially controlled by the pH and oxygen in solution. Calculations show that cerussite dissolves more rapidly than galena inducing an increase in the pH of these waters. The lead concentrations recorded in these waters are largely controlled by the dissolution of cerussite.

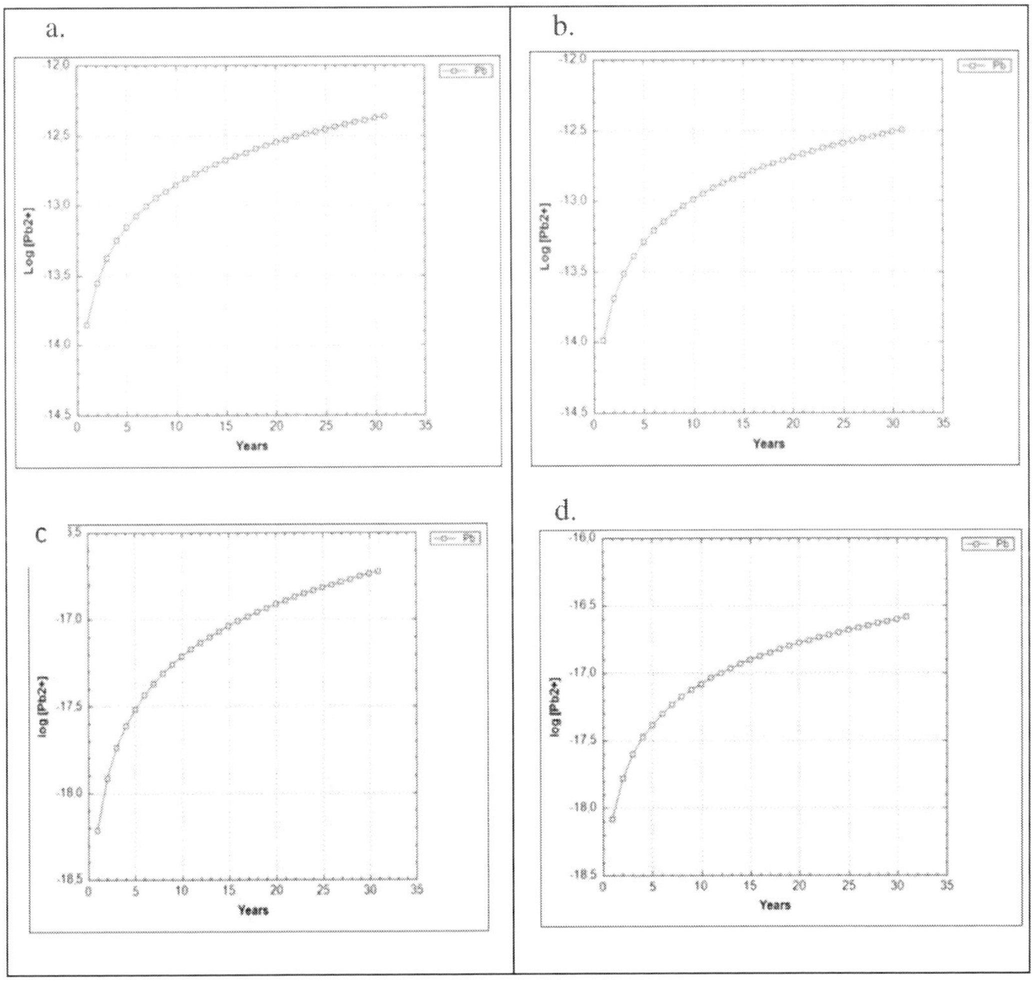

Fig. 4. Concentrations of Pb^{2+} in aqueous solutions calculated at pH = 9 for 31 years, in the case of the dissolution of cerussite with (a: rainwater, b: lake water) and Galena with (c: rainwater, d: lake water)

This work have been granted by CNRST under the project priorities No 34/2015

References

1. E.M. Benyassine, A. Dekayir, C. Casiot,. I.J.E.R.T. 11 (2013)
2. M.Combe ,M. Simonot Notes & M. Serv.Geol. Maroc, N° 231(1971)
3. M.L. ElHachimi, L. ElFounti, A. Bouabdli, N. Saïdi, M. Fekhoui, N. Tassé, R.S.E. 20, (2007)
4. R. Nasanen, P. Merilainen, K. Leppanen, ActaChim. Sinica.15, (1961)
5. D.Parkhurst, C.Appelo, USGS, (1999).
6. A. Salih, Murat, Erdemoglu , U.Gökhan , A.Ali, Hydrometallurgy. 88,(2007)
7. J. M. Schmitt Thèse de Doctorat. ENSM, (1976), Paris
8. Z. Sheng, L. Jianping, W. Yurong, H. Guangqian, Ap. Geochem, 19 (2004)

Impact of wastewater from the rural commune of Jmaa Moulblad on the physico-chemical quality of the waters of the Grou River (Rabat region, Morocco)

Karim ARIFI [1,2;*], *Latifa* TAHRI [2], *Abdallah* EL ABID [4], *Fatima Zahra* HEFIANE [2], *Souad* ELBLIDI [2], *Ahmed* YAHYAOUI [3] and *Mohammed* FEKHAOUI [2]

1, 3 University Mohammed V – Faculty of Science, Ibn Battuta Av, PB : 1040 Agdal, Rabat.
2, University Mohammed V – Scientific Institute, Ibn Battuta Av, PB : 703 Agdal, Rabat.
4 National Institute of Hygiene, Hydrology and toxicology Laboratory of the Environmental and Industrial Hygiene, Ibn Battuta Av, Agdal, Rabat.

Abstract. The Grou River is one of the main rivers that feed the reservoir of the Sidi Mohammed Ben Abdellah (SMBA) dam in Morocco. However, this river is particularly threatened by the pollution problem. It is in this context that this study aims to study the effect of wastewater from the rural commune of Jmaa Moulblad on the physicochemical quality of the waters of this river. We analyzed 11 physicochemical parameters (T° of area, T° of water, pH, EC, mV, Sal, TDS, BOD, COD, SS and TOC), with a monthly frequency since the month of December 2014 until November 2015. The results obtained show a worrying situation of the state of this watercourse. To remedy this problem, we recommend pre-treatment of wastewater before discharge and the implementation of a management plan and sustainable management of the Grou River watershed.
Keywords : Pollution, wastewater, physicochemical, development, SMBA dam, Jmaa Moulblad, Morocco.

*Karim ARIFI : E-mail: arifi_karim@hotmail.com

1. Introduction

Water is an indispensable resource for life. It deserves special attention in view of the dangers to which it is subjected by human activities (urban discharges, industrial activities, agriculture, etc.) [1]. These activities tend to diminish the potential of good quality water resources [2]. But also, have a negative impact on the health of the population, the uses of water (water abstraction, bathing, etc.) and subsequently compromises socio-economic development [3,4]

In Morocco, a country with a semi-arid climate, the water potential is relatively limited and uncertain [5]. This potential is affected by the pollution problem in all regions of the kingdom, especially in the Rabat region, following a growing population increase, which has almost doubled in recent years and the continued development of the sector industrial and agricultural.

At the level of the rural commune of Jmaa Moulblad, located a few kilometers from the reservoir of the Sidi Mohammed Ben Abdellah dam (SMBA), the Grou River plays a very important role in the region; its waters are used in livestock watering, the production of drinking water, bathing and other activities of the local population. In recent years, the region has experienced rapid urbanization and a very significant development of agriculture and industry. This situation is accompanied both by a very high demand for water and by the increase in discharges of wastewater discharged without any prior treatment and also the degree of pollution of the surface waters, in particular those of the Grou River.

It is in this context that this study aims to evaluate the impact of wastewater from the rural commune of Jmaa Moulblad on the physicochemical quality of the waters of the Grou River to remedy this problem. Several water quality parameters were measured and analyzed (T° of area, T° of water, pH, EC, mV, Sal, TDS, BOD, COD, SS and TOC), since the month from December 2014 until November 2015 with a monthly frequency.

2. Materials and Methods

2.1. Study site and sampling stations

This study was conducted at the rural commune of Jmaa Moulblad (Rabat region in Morocco). Two sampling stations were selected on the bed of the Grou River (Figure 1).

- Station S_1 : Located upstream, has been chosen as the reference point.
- Station S_2 : Located immediately downstream after the confluence zone with wastewater from the rural commune of Jmaa Moulblad without any prior treatment.

2.2. Sampling and analysis

Sampling is done once a month (December 2014- November 2015). The water sampling is carried out in 1.5l plastic bottles, previously rinsed with water from each of the selected stations. At each sampling, the temperature of the area, the temperature of the water, the electrical conductivity (CE), the potential Hydrogen (pH), the redox potential (mV), the Salinity (Sal) and the total hardness (TDS) were measured in situ in the field using portable waves. The other physicochemical parameters are measured later, immediately upon our return to the laboratory. The transport is done in a cooler at low temperature (+ 4° C). The analytical techniques followed are those described by AFNOR [7] and Rodier [8].

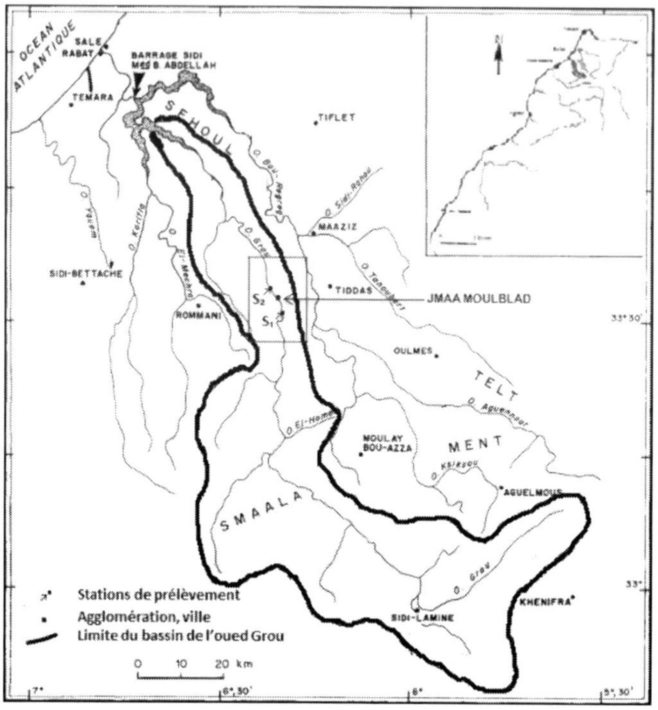

Fig. 1. Location of sampling stations (Source: QNINBA 1988 Slightly modified).

3. Results and Discussion

3.1 Air temperature (°C)

The air temperatures measured in this study show markedly irregular spatial and temporal variations (Figure 2a). The lowest were measured during the winter period, with a minimum of 9°C recorded in both stations in December 2014 and January 2015, while the highest were measured during the summer, with a maximum of 38°C recorded in July 2015 at the station S_2.

3.2 Water Temperature (°C)

The results of the water temperature obtained in this study show very large spatio-temporal variations (Fig 2b), generally hot in summer (maximum 28.6 ° C) and cold in winter (minimum 10, 4 ° C). It is related to local conditions (climate, duration of sunshine, flow) [9].

In conclusion, the temperature of the water closely follows that of the atmospheric temperature in relation to the climate of the region. According to the Moroccan standard of surface water [10], the temperatures measured in the two stations belong to the middle class.

3.3 Hydrogen potential (pH)

The results obtained in this study show seasonal variations in pH between 7.69 and 9.03 (Figure 2c). This pH is relatively alkaline due to the effect of the discharges that this stream receives throughout its course and in particular in the rural commune of Jmaa Moulblad. The low values are recorded in winter because of the acidity of the precipitations while the highest values are recorded in the summer period because of the bacterial activity which degrades the abundant organic matter at the bottom and because of the activity photosynthetic algae, our results align with those of [11] in the waters of the Boufkrane River.

In conclusion, the spatio-temporal variations in the pH of the waters of the Grou River are characterized by an alkaline pH of over 8.0 during all months except for (December, January and February), thus exceeding the value threshold recommended by the World Health Organization (WHO) which should be less than 8. According to [12] water that has a pH of more than 8.0 is difficult to treat.

3.4 Electrical conductivity (µS/cm)

The electrical conductivity reflects the degree of global mineralization, and provides information on the salinity rate [13].

The results obtained in this study fluctuate between 454 µS/cm in March and 1080 µS/cm in May, thus showing very large spatio-temporal variations (Figure 2 d). These values remain below the Moroccan standard for surface water (2700 µs/cm) [10]. This made it possible to classify the water quality of this zone as good to medium. This mineralization is much lower than that shown in some Moroccan rivers such as, the Khuan River [14], the Tizguit River [15] and the Boufekrane River [16].

3.5 The redox potential (mV)

In our study, the values of the redox potential vary between a minimum of -148 mV in S_2 and a maximum of -67 mV in S_1, thus show very significant spatio-temporal variations. The lowest levels were recorded during the summer, while the highest levels were recorded during the winter (Figure 2 e). While going from the station S_1 to the station S_2, a reduction in the redox values is observed. This could explain the fact that this river receives directly the wastewater generated by the rural commune of Jmaa Moulblad without any prior treatment, thus causing the decrease of the oxygen contents and consequently the deterioration of the quality of the water. This finding is consistent with [17]; the high redoxes coincide with the high oxygen levels while the low redoxes are in agreement with the lack of oxygen.

In conclusion, the values of the redox potentials obtained clearly show the polluting effect of the wastewater at the station S_2.

3.6 Salinity (Sal)

In this study, the salinity values vary between 0.1 g/l in December (S_1) and 0.5 g/l in May and October (S_1 and S_2), thus show different spatio-temporal variations (fig. 2 F). This could be explained by the effect of the wastewater coming from the rural commune of Jmaa Moulblad without any prior treatment and by the effect of the weak slope which makes the speed of the current almost nil by inhibiting the phenomenon of self-purification. On the other hand by the effect of strong evaporation in the absence of any precipitation promotes the elevation of salinity in summer and by the nature of the soil crossed and the origin of water. According to [18]; Salinity is mainly related to the origin of the waters.

In addition, there is a high salinity in the waters of both stations with a gradient increasing while going downstream.

3.7 Total Dissolved Solids (TDS)

The total dissolved solids of the water vary at the level of the study area according to the same gradient as that of salinity, thus showing very large spatio-temporal fluctuations. Values range from 199 mg/l in December to 570 mg/l in May (Figure 2 g). Indeed, the total dissolved solids is all the higher as we move from S_1 to S_2. This translates, once again, the effect of the lands crossed and the wastewater, provoking the precipitation of Ca^{++} and Mg^{++} cations. In addition, there is a high total dissolved solids (TDS) which reflects a state of pollution of the waters of the Grou River.

https://doi.org/10.1051/e3sconf/20183701008

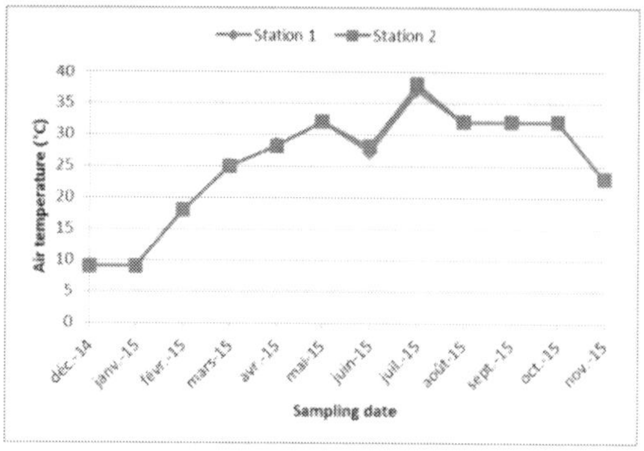

(a)

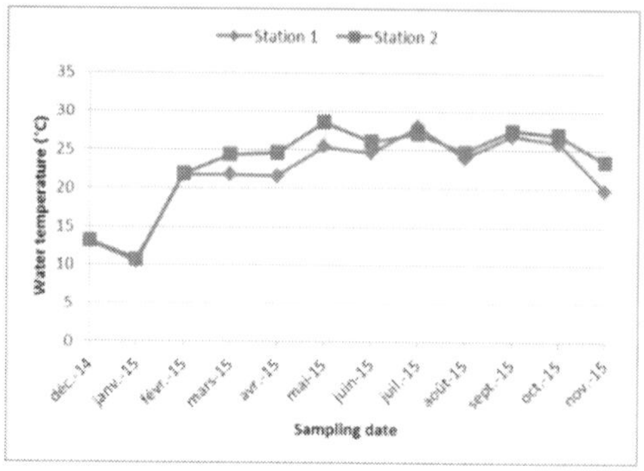

(b)

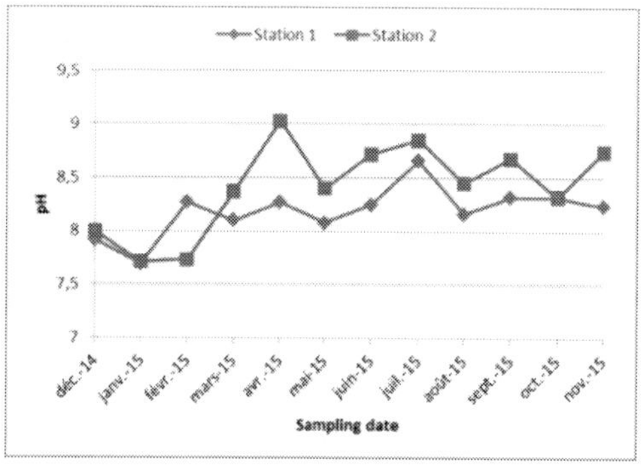

(c)

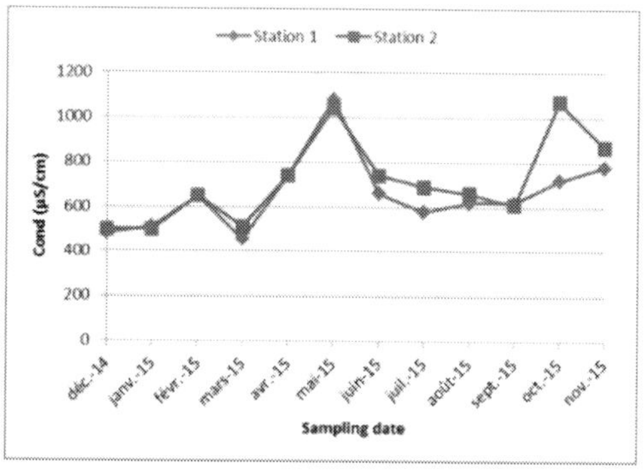

(d)

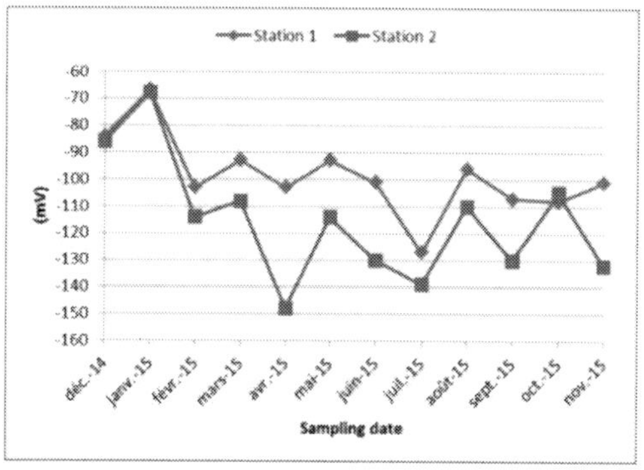

(e)

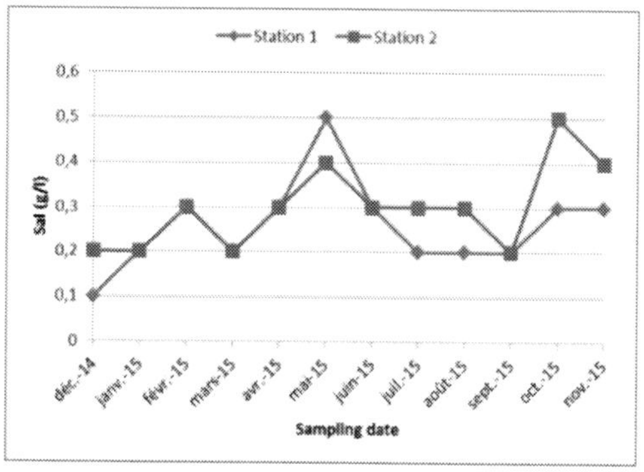

(f)

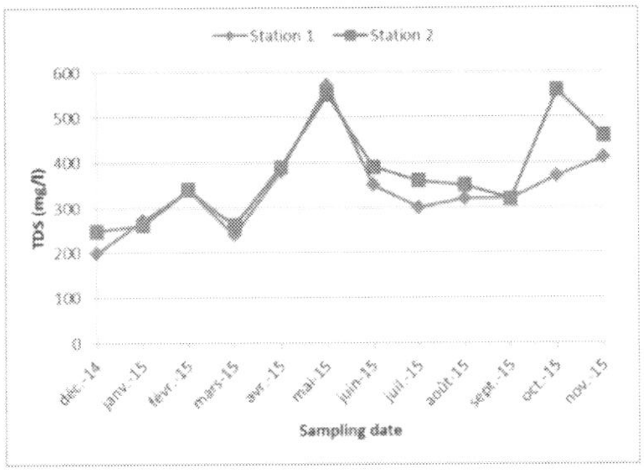

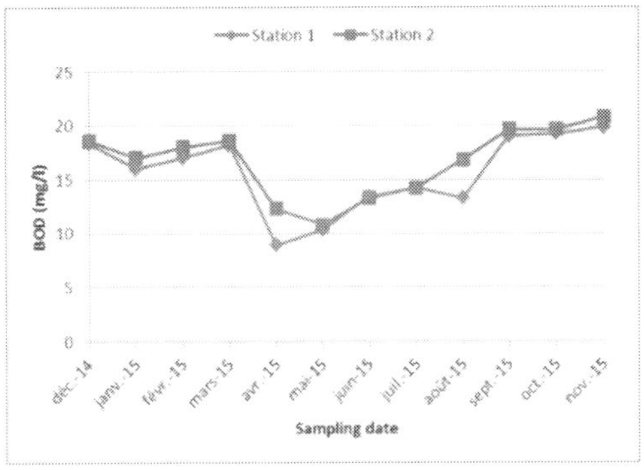

(g) (h)

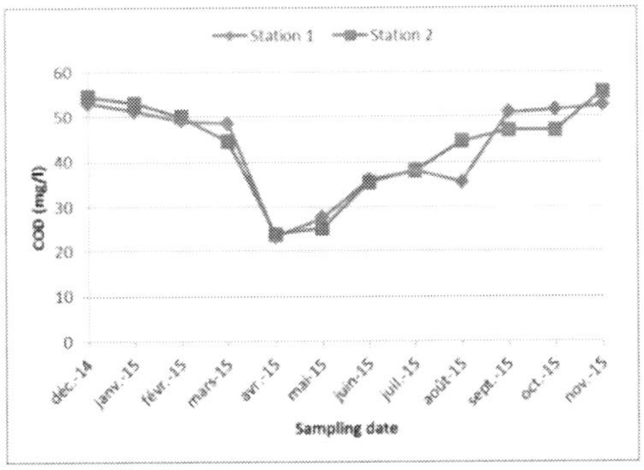

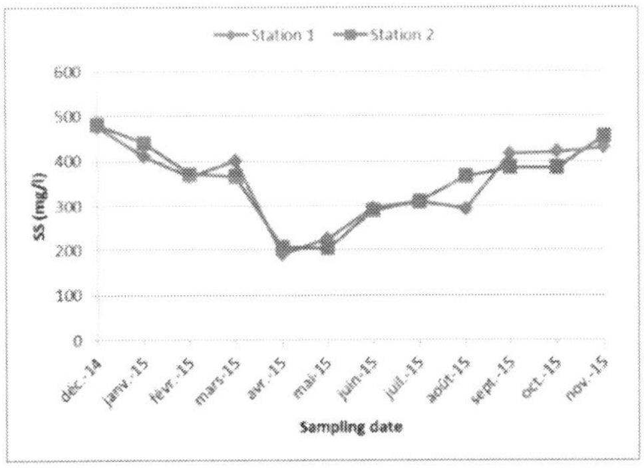

(i) (j)

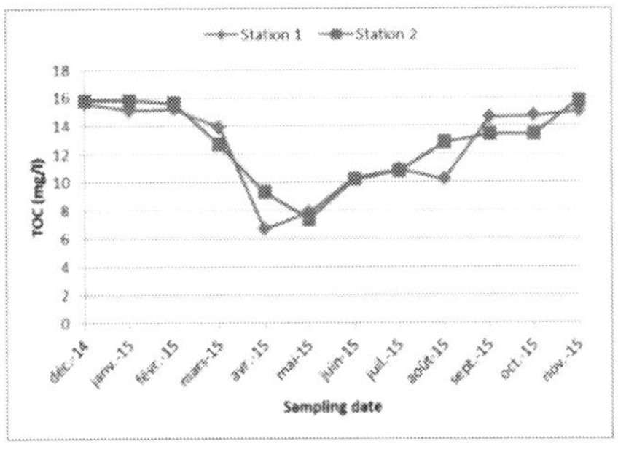

(k)

Fig. 2. Spatio-temporal variation of various physicochemical parameters of the Grou River under the effect of wastewater from the rural commune of Jmaa Moulblad

3.8 Biochemical Oxygen Demand (BOD)

In our study, the BOD contents ranged between 8.90 mg/l (S_1) in April and 20.8 mg/l (S_2) in November, thus showing very significant spatio-temporal variations (Fig 2 h). In fact, at the level of the S_1 station, the BOD contents are between 8.90 mg/l and 19.8 mg/l with an average of 15.63 mg/l which exceeds the guide value. This can be explained by the fact that as the river flows downstream, it receives more and more water from anthropogenic activities in charge of organic matter. Then, at the S_2 station the levels are slightly increased, due to the effect of wastewater from the rural commune of Jmaa Moulblad without any prior treatment.

In conclusion, the average values of BOD, obtained both in summer and in winter, remain above the guideline value of European and Moroccan standards (NMES) (from 6 to

10 mg/l). As a result, the waters of both stations are classified in the category of poor quality water.

3.9 The Chemical Oxygen Demand (COD)

In this study, the recorded COD content is between 23.2 mg/l (S_1) in April and 55.5 mg/l (S_2) in November, thus showing very large spatio-temporal variations (figure 2 i). In winter, the values obtained in the two stations are higher than those measured in the summer period, this could be explained by the leaching and the inflow of runoff water. Moreover, in summer, the levels recorded at station S_1 are slightly lower than those recorded at station S_2. This could be explained by the effect of wastewater from the rural commune of Jmaa Moulblad.

In conclusion, the averages of the COD content recorded at the two stations in the downstream part of the Grou River slightly exceed the Moroccan standard of 40 mg/l [10]. As a result, the waters of both stations are classified as medium quality waters.

3.10 Suspended matter (SS)

In this study, suspended matter (SS) levels, ranging from 190 mg/l (S_1) in April to 480 mg/l (S_2) in December, show very large spatio-temporal variations (Figure 2 j). The values obtained in winter are higher than those measured in summer. This observation is in agreement with other studies made on other rivers such as, the Moulouya River in Morocco [19], the Mellah River in Algeria [20], and the Senegal River [21]. Indeed, in winter, high levels are the result of a sudden hydrological event (flood), whose load in SS is attributed to an intense erosion of the watershed. In summer, the concentrations of SS are lower in station S_1 than in station S_2, due to the effect of wastewater discharges from the rural commune of Jmaa Moulblad.

In conclusion, the SS values recorded at the two stations far exceed the Moroccan standard for surface water quality [10] set at 50 mg/l. As a result, the waters of this area of the Grou River are classified in the category of poor quality water.

3.11 Total Organic Carbon (TOC)

In our study, TOC levels ranging from 6.7 mg/l (S_1) in April to 15.8 mg/l (S_2) in January, December and November, show significant spatial and temporal variations (Fig. 12 k).

In winter, the values obtained are higher than those measured in summer, this can be explained by the inputs of runoff and leaching. In addition, there is a high concentration of TOC that reflects a state of pollution of the waters of the Grou River. This finding is consistent with [22]; which explained the high TOC load of the lower Loukkos estuary (8.33 and 19.96 mg/l), in the upstream-downstream direction through intense erosion of the watershed, following brutal storm rains and contributions of erosion and pollution.

In conclusion, the average TOC values recorded at the two stations of the Grou River, 12.5 mg/l in S_1 and 12.7 mg/l in S_2, are higher than those recorded at Bou Regreg River 7,8 mg/l; [23] and those recorded at the Port Jackson Estuary 4.54 mg/l; [24]. This increase can also be attributed to the effect of the discharges that this stream receives along its course and to the effect of wastewater from the rural commune of Jmaa Moulblad at the station S_2.

4. Conclusion

In the light of the results obtained in this work, there is a deterioration of the water quality of the Grou River at the level of the study area, particularly at the station S_2 located downstream of the rural commune of Jmaa Moulbad generates very large quantities of wastewater and discharges them without any prior treatment.

The results obtained vary widely according to physicochemical parameters, seasons and stations. The analyzes revealed levels that exceed Moroccan standards for surface water. The causes result from the non-responsible behavior of the population who live along this watercourse, the exploitation of sands, the discharges and the influx of runoff during periods of floods following the increase of the leaching in coincidence with the agricultural season where soils are rich in nitrogen-based products.

To remedy these problems, we recommend that wastewater discharges be treated before arriving at the Grou River and that a watershed management and sustainable management plan be put in place. And it must be borne in mind that any pollution of its surface waters, no matter how small, will have serious consequences, and will be a real crime for the region.

Thanks

This study is part of the scientific research, is the result of a collaborative work between the University Mohammed V, Rabat (Faculty of Science & Scientific institute) and National Institute of Hygiene, Rabat, Morocco.
My heartfelt thanks to the entire research team and to those who have improved this manuscript.

References

1. M.L. Mc Kinney, Urbaization, biodiversity and conservation. *Biosci.*, 52 (2002) 883-890.

2. R.M. Mulliss, D.M. Revitt, R.B.E. Shutes, *Water Sci. Technol.*, 36 (1997) 195-199.

3. G.A.JR. Burton, R.E. Pitt, Stormwater effects handbook, a toolbox for watershed managers, scientists, and engineers. *CRC/ Lewis Publishers.* (2001) 875.

4. A. Chahlaoui, P. Cugny, S. Lek, A. Zaid, M. Ramdani, *Bulletin de la Société d'Histoire Naturelle.* Toulouse, 133 (1997) 71-76.

5. J. El Addouli, A. Chahlaoui, A. Berrahou, A. Chafi, A. Ennabili, Approche de la qualité biologique de l'oued Ouislane, au voisinage des effluents bruts de la région de Meknes. *Larhyss Journal* 09 (2011) 21-33.

6. A. Qninba, M.A. El Agbani, M. Dakki, A. Ben Houssa, *Bull. Inst. Sci.* 12 (1988) 149-156.

7. AFNOR., Qualité de l'eau. Recueil des Normes Françaises Environnement. Tomes 1, 2, 3 et 4. (1997) 1372.

8. J. Rodier, L'analyse de l'eau naturelle, eaux résiduaires, eau de mer, 8ème Edition, *Dunod*, Paris, (1996) 1383.

9. A. Brahimi, A. Chafi, *J. Mater. Environ. Sci.* 5 (2014) 1671-1682.

10. N.M. Norme Marocaine de qualité des eaux. *Bull. Off.* n°5062 (2002).

11. L. Karrouch, A. Chahlaoui. Bio-évaluation de la qualité des eaux de l'Oued Boufekrane (Meknes, Maroc). *Biomatec Echo* 3 (2009) 6-17.

12. JM. Anderson, Influences of pH on release of phosphorus from lake sediments. *Arch. Hydrobiol.* 76 (1975) 411-419.

13. J. Rodier, Analyse de l'eau : eaux naturelles, eaux résiduaires, eau de mer. 7éme édition, *Dunot*, Paris (1984).

14. A. Ben Moussa, A. Chahlaoui, El.H. Rour, *J. Biol Chem. Sc.* 6 (2012) 7096-7111.

15. M. Touabay, N. Aouad, J. Mathieu, *Ann Limnol* 38 (2002) 65-80.

16. H. Lamrani, A. Chahlaoui, J. El Addouli, A. Ennabili, *Science Lib.* 3 (2011) 111-112.

17. R. Delhi, F. Benzha, A. Hilali, M. Tahiri, A. Kaoukaya, L. Baidder, H. Rhinane, J.P. Hangouet, *ScienceLib.* 4 (2012) 2111-4706.

18. S. El Blidi, Rizières du Gharb : Qualité physico-chimique, composition faunistique et évaluation de la contamination métallique des différentes composantes de l'agrosystème. Thèse de doctorat Fac. Sci. Rabat, (2005) 194.

19. M. Makhoukh, M. Sbaa, A. Berrahou, Van M. Clooster, *Larhyss Journal*, 09 (2011) 149-169.

20. A. Maoui, M. Kherouf, F. Derradji, *Afrique science*, 07(3) (2011) 49-54.

21. A.D. N'diaye, M.S. Khadijettou Mint, M. Ould sid'Ahmed, M. Ould Kankou, *Larhyss Journal*, 12 (2013) 71-83.

22. M. El Morhit, M. Fekhaoui, A. Serghini, S. El Blidi, A. El Abidi, A. Yahyaoui, M. Hachimi, *Bull. Inst. Sci.* 34 (2) (2012) 151-162.

23. F. Zidane, K. Cheggari, J-F. Blais, N. Khlil, P. Drogui, J. Bensaida, *J. Mater. Environ. Sci.* 3 (2012) 99-108.

24. V. Hatje, *Environ. Geol.* 44 (2003) 231-239.

Study of the technical performance of localized irrigation and its environmental and agro-economic impact in the first areas of collective reconversion at the irrigated perimeter of the Tadla - Beni Moussa perimeter of the west – Morocco

ABDELLAH MOURADI [1*], ZEHOR AIT YACINE [2], ABDERRAZAK EL HARTI [3]

1 PhD student FST/ Sultan Moulay Slimane University, Beni Mellal, Morocco
2 Professor at FST/ Sultan Moulay Slimane University, Beni Mellal, Morocco
3 Professor at FST/ Sultan Moulay Slimane University, Beni Mellal, Morocco

Abstract- The evaluation of the performance of the localized irrigation system involved a selected sample of farmers to reflect the diversity of the study area. The hydraulic diagnosis revealed the absence of apparent malfunctioning anomalies of the installations studied (Coefficient of Distribution Uniformity ≥ 90% with average application efficiencies and overall of 90.54 and 86.83% respectively). In terms of the combined use of surface and underground irrigation water this new technique has saved about 30% compared to conventional irrigation. The agro-economic evaluation revealed that the crops practiced have high value-added and optimize the value of irrigation water. The environmental impact has resulted in an average drawdown of the static level of groundwater of 2.59 m due mainly to the new irrigation method introduced, which limited the percolation of water to the aquifer. The drip-to-drip transition resulted in an increase in salinity relative compared to the reference situation (+ 0.59 %, or 0.01 mS / cm) but to different degrees depending on the prospecting soil horizon. The practice of fertilization remains the major and probable cause of soil salinization of aquifers. The effect of soluble salts on the soil was investigated through the risks associated with sodium, which showed that the soil permeability problem does not arise at this time (SAR ≤ 15). The residual sodium carbonate remains less than 1.25 meq / l thus not causing soil dispersion.

Keywords: Transition, localized irrigation, Collective conversion, water saving, valorization of Irrigation water, environmental impact, efficiency, Uniformity distribution, Tadla perimeter.

[*] Corresponding author: mouradi_abdellah@yahoo.fr

1 - INTRODUCTION

In the context of the scarcity of water, the agricultural sector mobilizing more than 80% of water resources (Belghiti 2012) is called to use irrigation water wisely through a better technical, economic and social valuation. of this resource, and especially to preserve it for future generations (Moghli, E., Benjelloun, T., 2000). The organization of agricultural users around collective irrigation projects aimed at reconversion to localized irrigation represents a promising solution to deal with this new situation (Dionnet M. et al 2006). In the wake of these reconversion processes, Vidal et al. 2001 cited by Dionnet M. et al 2006 indicated that if the conditions for implementing such techniques are not met, particularly in terms of support for new users, reconversions in localized irrigation can lead to disappointing agronomic performance. For this purpose, the post-project evaluation of the technical performance of installations is a tool for detecting whether the practices of agricultural users are in line with good practice, namely ecological intensification, the use of scarce resources and the increase in productivity and added value.

2 - MATERIAL AND METHODS

To achieve the objectives assigned to this work, a methodological progression has been adopted. To do this, a representative sample of 32 farms was chosen to undertake this investigation. Data collection and information was based on interviews with irrigators on various aspects of exploitation, resource use and productivity. Field visits were made to measure certain technical performance parameters in situ, in this case the technical efficiencies and uniformity of watering of the facilities. The environmental impact was addressed through the analysis of environmental parameters from the ORMVAT analysis laboratory. The material used consists of volumetric water meters of the block plug, manometers, graduated test tubes, chronometers, auger, precision balance and an oven. The analysis of the survey sheets, the production of graphs and the processing of the data were carried out on a Microsoft Office Excel spreadsheet (version 2016) with the use of all its statistical processing functions (calculation of mean, standard deviation standard and coefficient of variation). Some schematics have been done using the Auto CAD version 2009 software.

3 - RESULTS AND DISCUSSIONS

3-1 HYDRAULIC DIAGNOSTICS

3-1-1 QUALITY OF SURFACE IRRIGATION WATER

By analyzing the April 2016 data from the ORMVAT analysis laboratory related to the ionic balance of surface irrigation water and taking into account the guidelines for assessing the physicochemical quality of water from irrigation, it has been shown that this water is classified in the category requiring no restriction for its use in crop irrigation (low salinity: 0.51 mmhos / cm, ie 326.4 mg / l, remaining below the threshold of 1500 mg / l advanced by Phocaides A., 2008, moderately basic pH found in the normal range of water suitable for irrigation "6.5 to 8.4, Phocaides A., 2008", the risk of soil sealing is zero due to low salinity and the Sodium Adsorption Rate (SAR: 1.1) below the threshold of 15 (Phocaides A., 2008) The ions that may present risks of toxicity namely Na^+ and Cl^+ are in normal ranges (less than 10 meq / l, Phocaides A., 2008. The Nitrate content is very low (0.1 meq / l i.e 6.1 mg / l, remaining below the 10 mg / l threshold, Labomag, 2012).

3-1-2 TEST OF THE UNIFORMITY OF THE DISTRIBUTION OF IRRIGATION WATER

According to Benouniche et al, 2014, the Distribution Uniformity Coefficient (CU) represents the ability of the facility to provide the same irrigation dose to an entire plot. At more than 90%, uniformity is considered excellent; the lower it is, the more there are problems of distribution (clogging of the drippers for example). The study of the homogeneity of the water distribution at the irrigation station was based on the uniformity test and the uniformity coefficient proposed by Keller J. and Karmeli D. 1974. This Coefficient is defined as follows:

$$CU(\%) = (q_{25} / q\text{-}) * 100$$

With: $q_{25} = \Sigma qi / 4$ (i = 1 to 4): Average flow rate of the lower quartile (4 lower values in l / h)
$q\text{-} = \Sigma qi / 16$ (i = 1 to 16): average flow (l / h) of the 16 measured drippers.

The uniformity test consists of the measurement of the flow rate for a sample of 16 drippers uniformly distributed over the sector and on the irrigation ramp (R_o, $R_{1/4}$, $R_{1/2}$, R_1) as illustrated on the diagram. experimental set-up below (Cemagref, 2003 cited by A .El Amri A. et al., 2012) following the protocol described in FAO bulletin 36 (Vermeiren L. 1983).

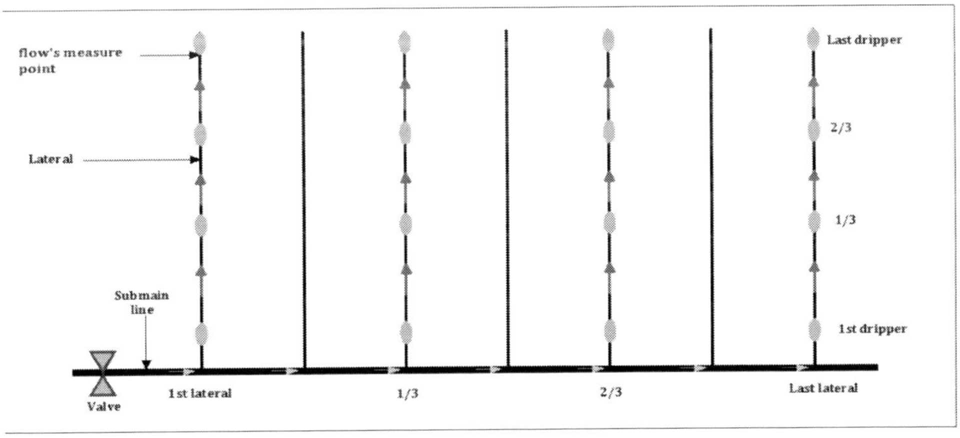

Fig 1: Experimental device of measurement points

The results obtained with regard to the uniformity of the distribution of irrigation water are very promising oscillating between 90% and 98% with an average of 96% and a small inter-farm variation (CV of 2.12%). Referring to the interpretation standards adopted by CEMAGREF, it can be concluded that the facilities operate normally and do not show any apparent anomalies at the time of the evaluation (CU> 90%). In addition, Ammari, O., and Bayali, N., 2016 reported that 10% of the cases studied at Tadla show poor irrigation uniformity (old installations).

3-1-3 EFFICIENCY OF THE IRRIGATION SYSTEM

- EFFICIENCY OF APPLICATION

According to Kambou D. et al 2014, This indicator is defined as the ratio between the amount of water retained in the root zone (Ws: net dose) and that brought by irrigation (Wd: gross dose) :

$$Ea(\%) = (Ws / Wd) * 100.$$

According to Hanafi S., 2011, WS is the variation of the water stock in the root profile before and after irrigation, determined by gravimetric method, Ws ≤ SWD.

$$\textbf{SWD (mm)} = (\textbf{Hcc} - \textbf{Hpf}) * \textbf{da} * \textbf{Z} * \textbf{10}$$

With: Hcc (Humidity at field capacity in%), Hpf (Moisture at the point of wilting in%), da (apparent density in g / cm^3, z (depth of soil in m). SWD: useful reserve storable in the root zone at the time of irrigation (SWD being 50.72 and 127.62 mm for z = 30 and 60 cm respectively). Efficiency was calculated assuming percolation losses are negligible (low soil permeability: 0.32 cm / h, Labomag 2012) and have a high water retention capacity (Ammari O. and Bayali N., 2016) and also losses due to leakage in the valves are absent. The quantities of water brought in during the irrigations were mastered by reading the meter readings at the beginning and at the end of the irrigation. Thus, the average efficiency obtained is 90 and 91% respectively for the horizon of the soil z = 30 cm and z = 60 cm and revolves around an average of 90.5%, which corroborates with the results reported by Elattir, H., 2005 and Ammari, O., and Bayali, N., 2016 (Ea = 90% and 87% respectively).

- GLOBAL EFFICIENCY

The overall efficiency (Eg) of localized irrigation, also referred to as hydraulic efficiency, is defined as the product between the Irrigation Efficiency (Ea) and the Uniformity of Watering Coefficient (CU):

Eg = Ea * CU; (Ammari, O., and Bayali, N., 2016).

The average overall efficiency obtained is 87% (Min: 82, max: 89% and CV: 2.12%). In this regard, Ammari O. and Bayali N., 2016 reported that only 60% of cases had a satisfactory Eg (80 to 86%).

4- AGRO-ECONOMIC PERFORMANCE

Data from techno-economic surveys are the basis of the production of various classical agro-economic indicators.

4-1 ANALYSIS AT THE SCALE OF EXPLOITATION

Harbouze R. et al. 2009, reported the existence of a coherence between the ratio r: product / charges and the efficiency levels of farms in the Gharb region of Morocco. Thus, in our study three classes have been identified: C_1 ($r \leq 2$), C_2 ($2 < r < 2.5$), and C_3 ($r \geq 2.5$). The analysis of the results of the typology of the farms shows that the highest ratios (class C_3) are generally recorded by those which consume fewer inputs per hectare (average ratio of 2.8) but generate outputs per hectare the highest (48.000 Dh / ha). These farms, which are considered efficient, consume less water (7.300 m^3 / ha) and value it at best at 5 DH / m^3 and 6.4 kg / m^3. The farm class (C_1) had the lowest values of the Outputs / inputs ratio of 1.7, which means that the farmer made a profit equal to 17/10 of the production costs spent against 28/10 for the C_3 class. These inefficient farms consume more irrigation water (12.700 m^3 / ha) and make little use of it at 2.6 Dh / m^3 and 4 Kg / m^3.

4-2 ANALYSIS AT THE SCALE OF CULTURE

By analyzing the scores obtained by speculation, it appears that citrus, sugar beet, fodder and vegetable systems produce the highest added values, with a better valuation of irrigation water (2.3 at 9 DH / m^3 and 5 to 11 Kg / m^3). This type of farming is characterized by intensive use of factors of production because of their ability to diversify widely and intensify their production methods. On the other hand, the production systems based on field peas, fodder barley, cereals and beans have little choice of diversification and play more on management oriented towards the economy of inputs: management by the charges. These results confirm those obtained by Harbouze R. et al., In 2009 in the Gharb region of Morocco.

5- CHANGES CAUSED BY TRANSITION TO LOCALIZED IRRIGATION

5-1 EFFECT ON THE CONSUMPTION OF IRRIGATION WATER

With respect to the combined use of surface and groundwater, the transition to localized irrigation, as reported in Figure 2 below, has resulted in an irrigation water saving of order of 29.52% by moving from the reference situation to that of after - project. This gain is mainly attributed to a significant decline in the use of groundwater which before approximates 52% of the quantities of water brought (Hammani A. 2015, reported to Tadla-Morocco that private pumped water accounts for 66% of the perimeter of Beni Moussa groundwater exports), although the use of surface water has increased due to an intensification of cultivation from 107 to 125% currently. The farmers' declarations go in this direction and confirm the desire to favor the use of currently available surface water at the lowest cost and of the best quality. Fusillier and Saque 2001 quoted by Petitguyot T. 2004 have thus reported situations in which the water consumption per hectare has increased with localized irrigation techniques, due to a poor technical mastery.

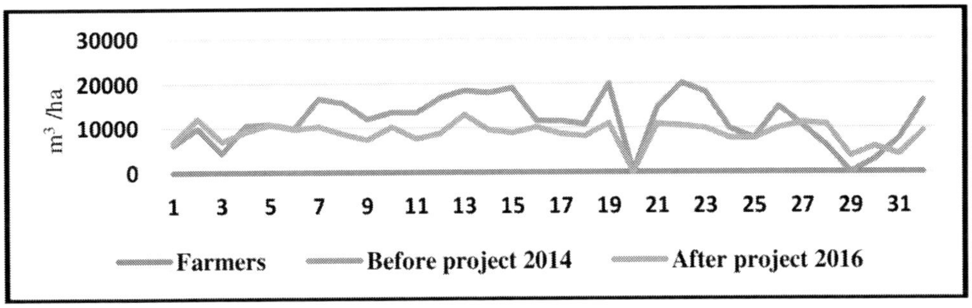

Fig 2: Variation in the combined use of surface water and groundwater Between the reference situation and the post-project situations.

5-2 ENVIRONMENTAL IMPACT

The environmental parameters data analyzed were retrieved from the water, soil and plant analysis laboratory of the Regional Agricultural Development Office of Tadla.

A- SALINITY OF THE SOIL

The post-project data for soil salinity measured by horizon were compared to the pre-project data as shown in Figure 3 below.

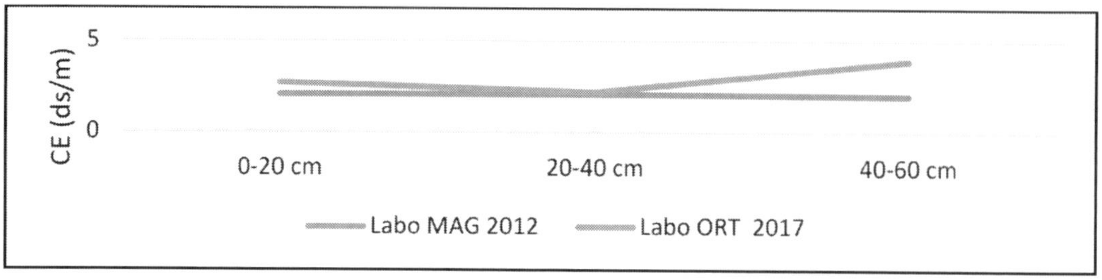

Fig 3: Variation of the electrical conductivity of the soil between the Reference and post-project situation

The analysis of the graph showed an increase in salinity compared to the reference situation 2012, but to different degrees depending on the horizon of the soil prospected (+32, +6 and + 95% respectively for the horizon 0- 20, 20-40 and 40-60 cm. Irrigation and evaporation of water from the soil surface are responsible for the accumulation of salts and the gradual increase of salts in the root zone. intensive agriculture, as in the case of the Tadla perimeter, where the irrigation water is of good quality (0.51 mS / cm for surface water and 2 mS / cm for groundwater), the practice of fertilization remains the major and probable cause of soil salinization. The current values recorded remain below the salinization thresholds advanced by different authors (4.7 dS / m, Noomene H. 2011, up to 8 dS / m, Phocaides A. 2008).

B - QUALITY OF GROUNDWATER

- The salinity of water: The electrical conductivity of the groundwater recorded an increase between 2014 and 2016 from 2.02 to 2.03 mS / cm, ie + 0.01 mS / cm. This level of salinity remains low to medium and at present does not constitute any constraint in localized irrigation.

- Nitrates: Nitrate ions decreased by 20% between 2014 and 2016, from 0.46 to 0.37 meq / l: 22.57 mg / l thus constituting no constraints at the moment (average quality range of 25 at 50 mg / l, Labomag, 2012)

The effect of soluble salts on soil was approached through:

. Risks related to sodium **(SAR = [Na $^{+}$ * ((Ca^{2+} + Mg^{2+}) / 2) $^{-1/2}$]).** The content of Ca^{2+} and Mg^{2+} ions remains high compared to that of sodium, for the time being eliminating all the risks of permeability by soil sodisation. The current SAR is 2.33 and remains below the threshold of 15 (Phocaides A., 2008).

. Residual sodium carbonate **[HCO3^{-} - (Ca^{2+} + Mg^{2+})]** remains below the standard 1.25 meq / l (Phocaides A., 2008) from which water becomes dangerous (calcium and magnesium may react with bicarbonate and precipitate in the form of carbonate, the relative concentration of sodium in the exchangeable complex increases and causes the dispersion of the soil).

C- STATIC GROUNDWATER LEVEL: The comparison of the static level of pumping points of underground irrigation water of the studied sample of the preliminary project year 2014 (avg: 12.56 m, Variation Type: 2.94 and CV: 19.44%) to that of 2016 (avg: 15.15 m, ET: 2.85 and CV: 22.67%) showed a drawdown of 2.59 m due mainly to localized irrigation having limited the infiltration of water towards the aquifer (Hammani

A. 2015 reported that surface irrigation contributes 71% to the recharge of the Beni Moussa aquifer, thus constituting the main source of aquifer recharge).

5-3 EFFECT ON AGRO-ECONOMIC PARAMETERS

The effect of the transition to localized irrigation on agro economic parameters generally led to an improvement in these indices in 2016 for most crops [the case of sugar beet: +142 and + 125% respectively for Value Added and Valuation of m^3 of water] in comparison with 2014. These results are consistent with those obtained by Ammari O. and Bayali N. in 2016 at Tadla- Morocco, having reported a clear superiority of localized irrigation compared to old gravitational technique.

6- CONCLUSIONS AND PRACTICAL RECOMMENDATIONS

At the end of this investigation, it appears that an irrigation modernization project cannot achieve all the objectives assigned to it without good management being tied up because generally the problems of reconversion are more related to management that at the installation of localized irrigation equipment. The initial objectives of promoting greater sustainability of the use of irrigation water defined in terms of reduction of the use of groundwater and exclusive use of surface water is achieved. However, it is necessary to match development support with localized irrigation safeguards (annual quotas, dissuasive tariffs and optimal irrigation management) for a sober use of factors of production, able to better deal to the risks environmental impacts, shortages, and loss of human capital and invested equipment. At the present time, there are no saline soil surface markings in the area or a problem of destruction of the soil structure by alkalization.

7- ACKNOWLEDGMENTS

I cannot finish this work without thanking my supervisors Zehor Ait Yacine and Abderrazak El Harti, professors at the FST Beni Mellal for their foresight and the valuable advice that he gave me during this work. My thanks also go to the ORMVAT for providing us with data relating to the quality of the soil and water of the conversion zone. Also my thanks are addressed to the scientific committee of the EDE6 / 2017 BM congress for their framing and their guidelines for the elaboration of the final draft of this article.

BIBLIOGRAPHIC REFERENCES

1 - Ammari Ouafa et Bayali Najoua, 2016, *Etude des pratiques d'irrigation et évaluation des performances de l'irrigation localisée dans le cadre de reconversion collective : Cas du secteur pilote du Tadla.* MFE- diplôme d'Ingénieur GR, IAV Hassan II, Rabat, 119p.

2 - Belghiti, M'Hamed, 2012, *L'économie et la valorisation de l'eau en irrigation : un impératif pour un développement agricole durable au Maroc,* Conférence internationale pour le dessalement et le développement durable, Casablanca.

3 - Benouniche Maya, Marcel Kuper, Ali Hammani. 2014, *Mener le goutte-à-goutte à l'économie d'eau : Ambition réaliste ou poursuite d'une chimère ?* Alternatives rurales (2) Novembre 2014.

4 - Dionnet, M., Kuper, M., Garin, P., Hammani, A., Eliamani, A., Saaf, M., 2006, *Accompagner les acteurs dans le changement de leur système. Un jeu de rôles pour des projets collectifs d'irrigation au Tadla, Maroc, Actes du séminaire Wademed, Cahors, France.*

5 - El Amri A., Majdoub R., M'Sadak Y, Aouichaoui, 2012. *Appréciation expérimentale de l'uniformité de distribution de l'eau dans le périmètre irrigué ZAAFRA II (Tunisie centrale), Larhyss / journal n°11, 11/2012.*

6 - El Attir H., 2005. *Conduite et pilotage de l'irrigation goutte à goutte en maraichage.* Bulletin mensuel n°124 d'information et de liaison du PNTTA, transfert de technologie en agriculture.

7 - Hammani A. 2015. *Techniques de gestion des nappes d'eau souterraines en zones interne et externe du réseau hydro-agricole relevant d'un périmètre irrigué.* Module de formation ORT/2017.

8 - Kambou D., Xanthoulis D., Donkora K., Ouattara & Aurore Degré, 2014, *Concepts d'efficience et de productivité de l'eau,* (synthèse bibliographique), Base En ligne, Volume18, Numéro1, 108-120.

9 - Keller J., Karmeli, D.,1974, *Trickle irrigation design parameters,* ASAE Transactions, 17(4): 678-68.

10 - Labomag 2012, *Caractérisation de référence des paramètres environnementaux au niveau de la zone de reconversion collective à l'irrigation localisée.* Rapport fin hiver *2012.*

11 - Moghli, E., Benjelloun, T. M., 2000. *Valorisation de l'eau d'irrigation par les productions végétales dans les grands périmètres irrigués au Maroc.* Hommes Terres et Eaux, Issue 116, pp. 30-38.

12 - Petitguyot T., T. Rieu, A. Chohin-Kuper, R. Doukkali. *Modernisation de l'agriculture irriguée et durabilité des ressources en eau dans le périmètre du Tadla au Maroc.* Ali Hammani, Marcel Kuper, Abdelhak Debbarh. Séminaire sur la modernisation de l'agriculture irriguée, 2004, Rabat, Morocco. IAV Hassan II, 13 p., 2005. <cirad-00189662>

13- Phocaides A., 2008. *Manuel des techniques d'irrigation sous pression. Chapitre 7.14, FAO, Rome, seconde Edition 2008.*

14 - Vermeiren L., 1983- Bulletin FAO *d'irrigation et de drainage n°36- l'irrigation localisée : calcul, mise en place, exploitation, contrôle du fonctionnement- FAO, Rome, seconde Edition 1983, 220p*

Public services for distribution of drinking water and liquid sanitation in urban zones in Morocco Relevance of introduction the performance indicators for preservation water resources.

AKKA *Habib*[1] *, BOUZIDI *Abdelhamid*[2], HOUSNI *Said*[3]
[1]* PhD student in Environmental Science and Development laboratory- BSE- University Hassan I. FST -Settat.
Email: akka.habib@gmail.com
[2] Laboratory of the Sciences of the Environment and the Development – Training: Biology, health and Environment- University Hassan I. Faculty of Sciences and Techniques of Settat. BP 577 Settat - Morocco
[3] Directorate of Communal Operators and Conceded Services- Ministry of Interior- Rabat- Morocco

Summary. Because of the absence of regulations and specific national norms, the unilaterally applied indicators for performance evaluation of water distribution management services are insufficient. This does not pave the way for a clear visibility of water resources. The indicators are also so heterogeneous that they are not in equilibrium with the applied management patterns. In fact:

1- The performance (yield and Linear loss index) of drinking water networks presents a discrepancy between operators and lack of homogeneity in terms of parameters put in its equation. Hence , It these indicators lose efficiency and reliability;
2- Liquid sanitation service has to go beyond the quantitative evaluation target in order to consider the qualitative aspects of water.

To reach this aim, a reasonable enlargement of performance indicators is of paramount importance in order to better manage water resource which is becoming scarce and insufficient.
Keywords: Service - Distribution –Drinkable Water – Sanitation – Indicators – Performance – Water Resources

Email Correspondant : akka.habib@gmail.com

I- INTRODUCTION

Drinking water and liquid sanitation public services constitute a communal prerogative (Loi organique 113-14, 2014)). Communes have the freedom to create and manage or delegate their management to public or private operators. In addition to their responsibility in relation to other townships, the operators are obliged to respect three fundamental principles while exercising their management. These principles are inscribed in the constitution of the Moroccan Kingdom namely, equal citizen access to services, continuous service supply and fair territory (Constitution du Royaume du Maroc, 2011).

These principles can only be truly guaranteed through two main axis for durable water resource management e.g., resource saving and its protection against pollution. Indicators are set up to evaluate the performance of these two management aspects which are still regarded as insufficient to figure out the reality.

This article comes within the scope of a national research entitled ‹‹ Public service management of water and liquid sanitation: performance measures between communal operators and private operators››

This work is supervised by Professor Abdelhamid Bouzidi, within the environmental science and development laboratory at the scientific and technical university of Settat, Morocco. This work is also supervised by Mr. Said Housni, head of the technical department at the Directorate of communal operators and Concession Services of the Ministry of the Interior, Morocco.

The objective is to analyze the suitability of performance indicators for the management of water resource, which is intended for urban distribution. This can be realized through a national assessment and review of some applied models worldwide. An indicator panel of water resource management will be suggested at the end of this report.

II- STATEMENT OF FACTS

The indicator assessment, ‹‹Performance of drinkable water distribution system›› which is applied in drinkable water distribution service without a legal basis, demonstrates an average value of 74.5% consolidated by the communal operators and private operators. (Ministère de l'Intérieur, 2016)). A deposit of 58 Mm^3[(1)]of drinking water is lost every year with regard to an optimal feasible distribution performance of 82%. The amendments made to this indicator between 2005 and 2015 are 4.3 points for private operators and 5.9 points for commune operators. Nevertheless, these values remain biased due to the absence of standardization of calculation methodology and performance indicators. In this respect, we notice that in the 2015 database, a 1 point improvement performance would help to gain 8 million m^3 of drinking water every year.

II.1. Multiplicity and interference of contributors: factors against good water resource mastery.

The institutional and functional schema of water in Morocco is characterized by a vast number of participants. The attributions are redundant and stand as an obstacle against a positive management of public water service. In fact, as outlined by the economic, social and environmental council (CESE, 2014): ‹‹ In a macroscopic view of the participants' role in the water sector in general, the multiplicity of the contributors, and the global scheme of interoperability become extremely complex, risk generator and inefficient». These interferences are manifested in two main levels:

1- Perimeter of urban planning [(2)] does not always coincide with perimeter of management. This leads incoherence in the establishment of directing schemes and causes, for the communes, a lack

[(1)]The total volume manipulated by the communal operators and private operators in 2015 is 774 Mm3 while that actually recorded and consumed by the users is of the order of 577 Mm3

[(2)] The Perimeter of urban planning is fixed by statutory way fixing the spatial extent and the rules of its urbanization. This perimeter covers inevitably the territory of the territorial community.

of visibility on the resource;

2- In the case of private operators' management, small townships are likely to be demographically and economically stifled by other big ones due to the absence of a strong institutional authority. The managers invest their money in projects which are highly profitable and where there is strong social pressure. Any interference is constraining especially when two operators intervene in the same perimeter.

II.2. Terminology ambiguity and performance indicator construction.

II.2.1. Drinkable water service

The glossary used and calculation parameters differ from one operator to another. This lack of conformity is manifested at the level of variable indicators calculation. Concerning independent public companies in particular (communal operators), the product specifications do not demonstrate any indicators that can measure performance. However, and as a consequence of the competition in the distribution sector together with the introduction of private operators in management, annual reports of public companies indicate that the technical and financial ratios are similar to performance indicators. Three prominent dissimilarity examples:

- ✓ The designation of "access to service" is discordant. The denomination access rates, the rate of connection coverage are used to give information about special network coverage. Still, this can mean for certain operators the availability of water in taps.
- ✓ Network performance is calculated according to different approaches. It is defined as being the volume of water sold and added to the volume of water mobilized (produced volumes and imported volumes). Because sold water is not defined beforehand, it creates a confusion concerning the volumes that are not changed but used with full knowledge of the facts (consumed volumes for free, used volumes for service requirements, etc). The majority of operators don't give enough precision to this subject.
- ✓ Linear loss index follows the same approach as performance without the presence of standardization reference rate. Although the technical performance systems (yield of network), which are run by communal operator and private operators, have improved by 80%, linear loss index is so high that it makes this result superfluous. (16.9 m^3/day/km in 2015) (Figure n°4).
- ✓ Users' water consumption tendency is declining. It was the range of 7.5% from 2005 to 2015 (all operators and all customer segments are taken together). The tendency has gone from 115.8 to 107.1 l/inhabitant/day.(AKKA H and Bouzzidi A, 2017) (Figure n°1). Temporary evaluation indicators of consumption in the users' tap, which are absent in the current indicator set, are relevant in measuring when a drop or rise persists, and inevitably impacts the whole production-distribution process.

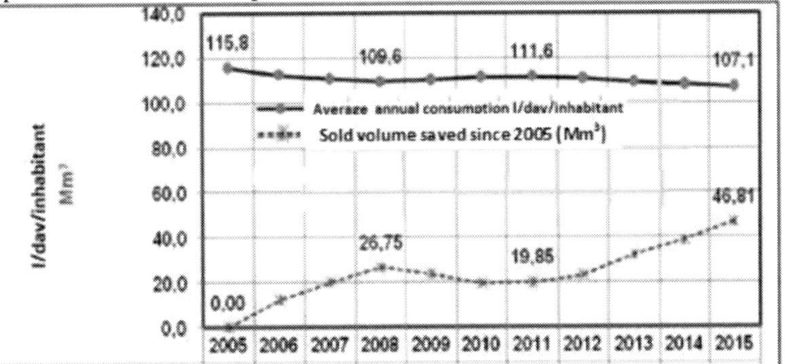

	2005	2006	2007	2008	2009	2010	2011	2012	2013	2014	2015
Average annual consumption l/dav/inhabitant	115,8	112,8	111,0	109,6	110,5	111,5	111,6	111,0	109,4	108,4	107,1
Sold volume saved since 2005 (Mm³)	0	12,15	19,94	26,75	23,88	19,53	19,85	23,38	32,18	38,73	46,81

Figure n°1: Consumption dynamics per unit (Communal operators and private operators)

 https://doi.org/10.1051/e3sconf/20183701010

II.2.2. Liquid sanitation.

Wastewater, which is manipulated by communal operators and private operators, accounts for 510 Mm^3 volume in 2015. This number represents approximately 70% of wastewater discharges at the national level. Treatment rate is at 80% for both communal operators and private operators. All this water is rejected either in the sea or in hydraulic public land (figure n°2). Only 37 Mm^3, among 63 Mm^3 is reused (Ministère de l'Intérieur, 2016)). The most known project is that of Marrakech city which recycles part of the used water for golf irrigation, the irrigation of palm grove and green municipal .places. It has to be underlines that the reuse of raw sewage has always been a common practice. In 2009, it was estimated that about 70 Mm^3 of raw sewage was reused to irrigate an area of at least 7200 ha in the outskirts of some big cities like Marrakech, Meknes, Oujda, Fes etc)(CORDOM NICOLAS and AL, 2015). This shows that the reuse of used water is relatively accepted.

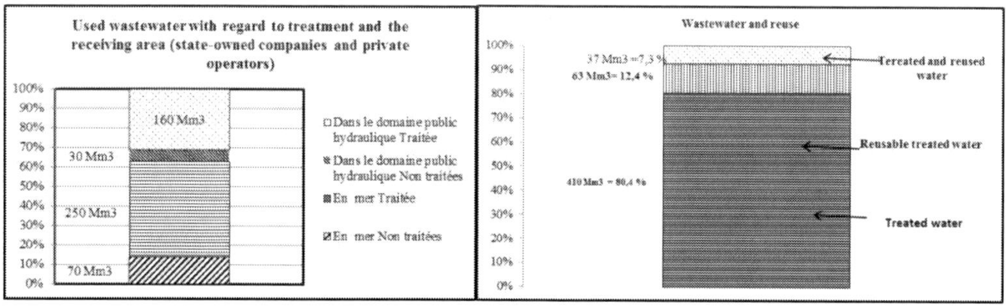

Figure n°2: Waste water rejected according to the treatment and the receiving environment (Communal operators and private operators)

II.3. Absence of national norms:

II.3.1. Resorting to benchmarking for indicator qualification.

In 2006, water and sanitation management laboratory of (ENGREF) established a frame of reference concerning linear loss index depending on the covered area (rural or urban) (table n°1). The correlation is done by taking into account subscribers' size per kilometer network (DREAL-CERA, 2013). In this frame of reference, the consolidation of the last five year results qualifies the performance of communal operators and private operators as poor. Obviously, the values exceed the reference in the urban area of GEA. The national average is 20.8m^3/day/km. By applying this ratio on the total network length of communal operators and private operators' area, we are surprised by the big amount of water lost; either 556 500m^3 per day[(3)] (with regard to an optimal LIL average of 3,3 m^3/day/km). In order to line up with the diagram norms in figure n°3, these loses have to be brought back to 104 940 m^3/ year.

N.B: The mathematical correspondence between production and LLI is not obvious. When it's about estimated or measured variables, the bias are numerous. The marked difference in the values of volume loss using the two indicators shows the suitability of the consistency of the two equation indicators.

[(3)]The length of the total drinking water network (communal operators and private operators) is of the order of 31 800 km

 https://doi.org/10.1051/e3sconf/20183701010

Table n° I: GEA frame of reference for loss index (RNAUD, 2009)

Type of area	Rural	Intermédiate	Urban
Criterion	D≤20	20<D≤40	40<D
Excellent	LIL<0.7	LIL<1.5	LIL<3.3
Average	0.7≤LIL≤2.5	1.5≤LIL≤5.2	3.3≤LIL≤12.8
Weak	2.5<LIL	5.2<LIL	12.8<LIL

D = density of subscribers / linear of the network in km excluding connections), LIL = Linear index of loss (loss index) in m^3 / day / km.

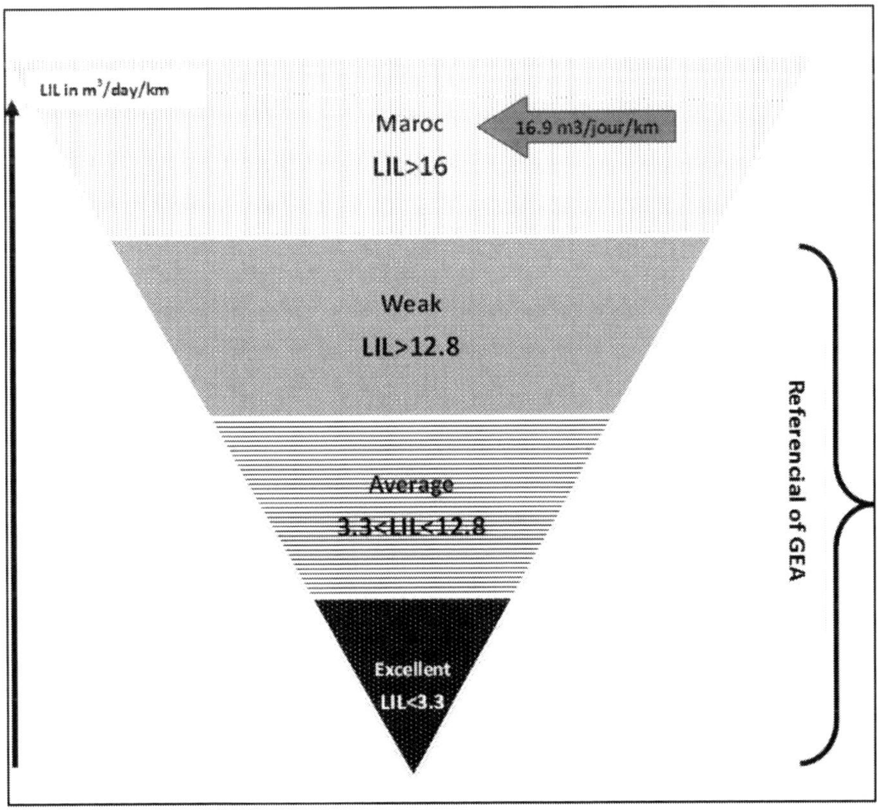

Figuren° 3: LIL in Morocco and norms of GEA

This signifies that drinkable water is wasted five times more that it is authorized for an efficient network in terms of losses.

II.3.2. The case of Maghreb countries and Jordan

Drinkable water system in Tunisia (SONEDE, 2016) is characterized by a linear index of loss that is 50% less than that wasted in Morocco (8.7 versus 16.9m³/day/km in 2015). The average yield of network is also higher (figure n°4). The indicators of water resource management in Tunisia particularly cover all the economic aspects of water. There are twelve established indicators. The consumption ratios, which give information on users' behavior, are also established (SONEDE, 2016)

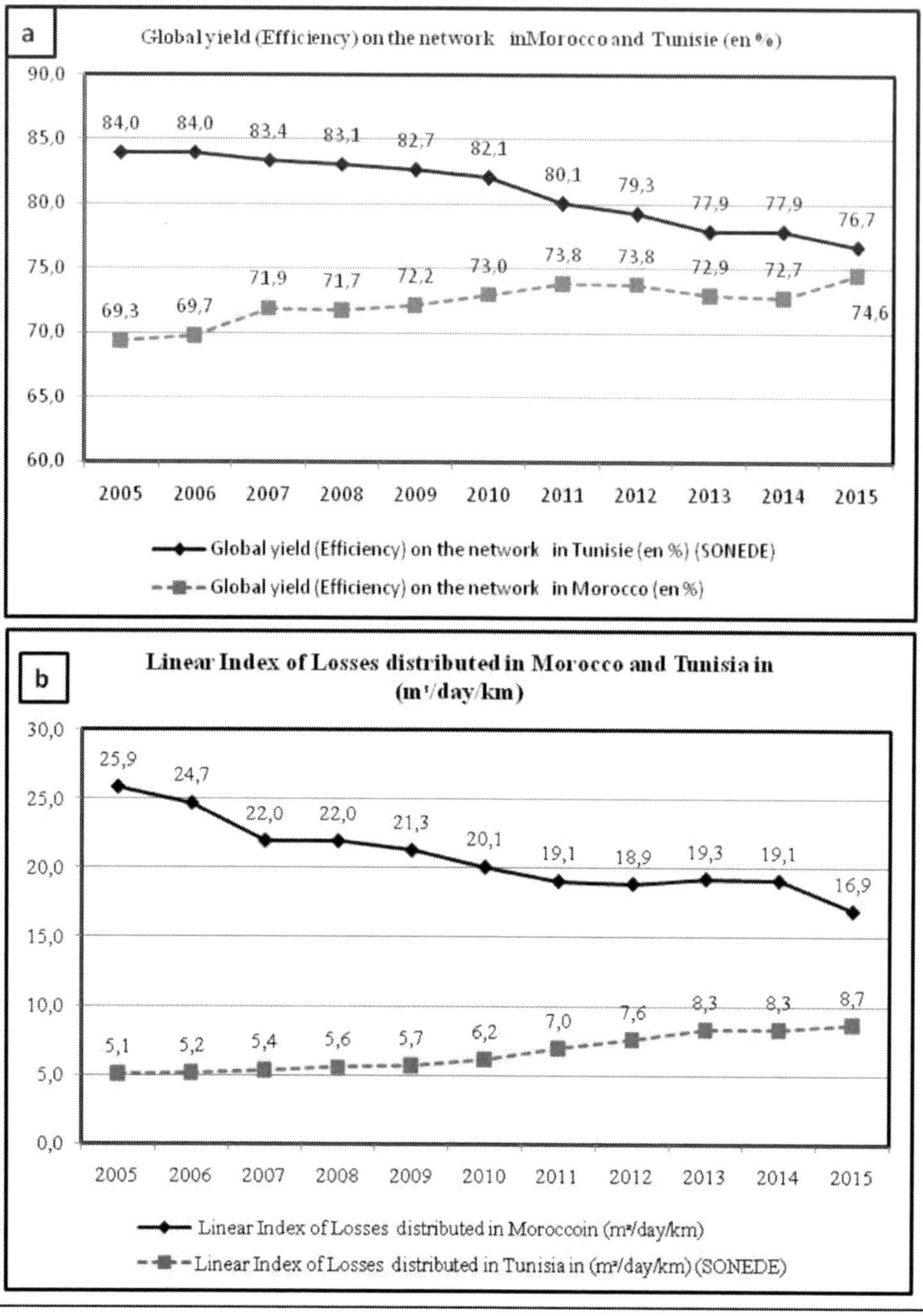

Figure n° 4: Yield (a) and LIL (b) of water distribution network in Morocco and Tunisia

In Algeria, individual demand for drinking water was at 120 l/inhabitant in 2011 and the estimated performance was at 55% in 2010 (BENBLIDIA, 2011).

Jordan, another country in the MENA region, uncalculated drinkable water volumes or non-revenue

water constituted in 2015, 51.3 % of the total distributed volume. (JORDAN WATER SECTOR, 2015).
The reuse in Tunisia, however, is more developed that in Morocco. If the latter reuses 7% of wastewater manipulated by communal operators and private operators, Tunisia will succeed in reusing 26%, which is equivalent to 60 mm³ (figure n°5). As for liquid sanitation, the current indicators in Morocco cover quantitative aspects. Used, collected and treated volumes of water as well as the number of STEP are the only indicators that appear in management reports. We also note the total absence of information concerning sludge management.

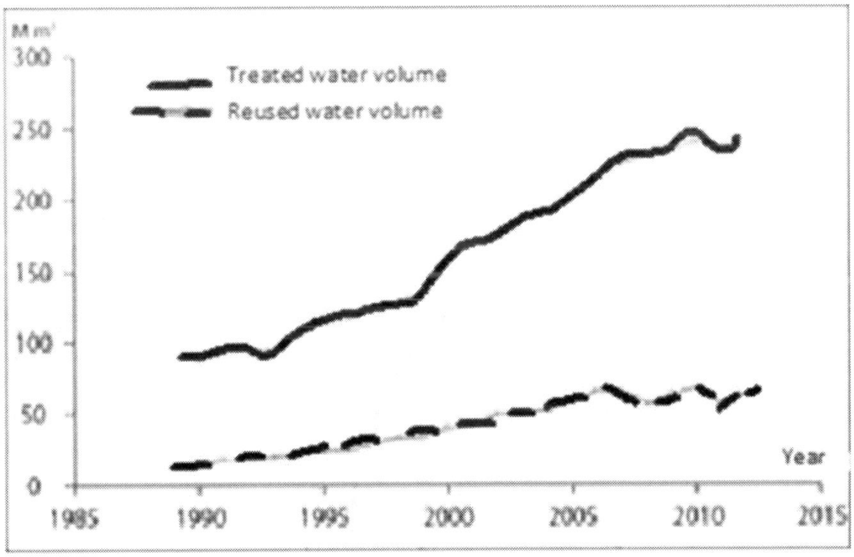

Figure N° 5: used water volume treated and reused in Tunisia (2013) (ONAS, 2013)

II.4. Urban rain water.

The problem of urban rain water continues to create differences between territorial community and operators. In fact, there is an important rain water flow; thus, it necessitates investment to be collected. In addition, the last is not recovered by applied tariff model. Their reuse is still low. The lack of regulatory and financial incentives seems to be a determining factor in the negligence noted in this regard.

III- The relevance of introducing performance indicatorstomanagingdrinking water and sanitation services in Morocco.

III.1. Particularities of Morocco

The following specifications have to be taken into account while adopting the performance indicator approach in public service management of drinking water distribution and sanitation in Morocco.

1- Direct management by the communes: a lot of townships in Morocco themselves manage water services and/or sanitation. When there is lack of means, communes cannot assume the maintenance of performance indicators in a detailed and precise manner.

2- Urbanization degrees of management perimeter: in a rural area, the density of individuals connections is very small. Diameters of canalizations are reduced and the network is wide in space. Concentrating on connection runaways will have no sense. The dysfunction frequency indicator of public fountains is more significant than dysfunction detection frequency in the pipe system. The latter is generally installed in places where road traffic is infrequent; hence, it does not provoke breakage (even if installation conditions aren't always optimal).
3- Traditional systems of drinking water: Including indicators for traditional systems of drinkable water is so required. A lot of villages are provided with water through schemes which are composed of well, a small tank and some public fountain. These systems have always existed since the PAGER program started and still function in areas where there are active users associations.
4- Information and communication: The transparency need and participation in public life are emerging dynamics in Moroccan society. This dynamic has to be taken into consideration while monitoring the dashboard. It is about designing indicators that translate users' expectations with regard to these two aspects. Information campaign or joint bulletin attached to the invoice including details about the services as well as comparisons with neighboring communities etc. the recent law 31 -13 about access to information supports this suggestion.
5- Utilities of information and communication between actors: the implementation of indicators is not solely about setting variables. The discussions about setting up performance indicators are always arduous. This is mainly due to the different possibilities of perceiving an indicator and also of the conflicting actors' objectives. Consensus, assessment and clamping are of paramount importance in this process. The phase of normalization is there only the crowning.

III.2. Admonition and suggested indicators panel

III.2.1. Drinkable water distribution service:

1- The necessity to review mathematical construction efficiency (yield).

a- Productivity is an environmental and economic indicator before being an indicator of the network state. For some authors, it is primarily an indicator of the waste of the resources (GUERIN SHNEIDER, 2001). For others, only the net return (performance indicator of the distribution network) reflects the use of a resource with or without waste (IGD. Fondation d'entreprises, 2004.)
b- A number of specialists touch upon the necessity of supplementing the production with other indicators ((IGD. Fondation d'entreprises, 2004.)) ((GUERIN SHNEIDER, 2001)). In addition to that, the interpretation is always difficult. The difficulty resides in the extent to which the management perimeter is qualified; urban, semi urban or rural. In the case of Morocco, frequent construction of villages and suburban districts in cities makes this idea reasonable. The indicators, the density of subscribers and Linear Consumption Index have to be included in the matrix in order to be able to judge whether it is a rural, urban or intermediate perimeter ((Renaud, 2009)). The employment of output as an exclusive indicator of the system condition in France was discussed and questioned (long ago) ((GUERIN SHNEIDER, 2001)). Furthermore, we notice that IWA didn't include the network efficiency (yield) as an indicator in its report about performance of water distribution services. (IWA, 2003).
c- Water public service operators demonstrate the performance on their management dashboard. In fact, the network efficiency intervenes in the economic calculation: the yield indeed intervenes in the economic calculation: the more the yield is low, the more the volume produced for a given consumption will have to be important, and the more the proportional loads in the produced volume will be raised. This approach has been particularly theorized in England together with the "economic level of the leaks" (GUERIN SHNEIDER, 2001). The comparison helps to guide the choices between different technical solutions (realization or non realization of leakage research, system rehabilitation or construction of new production infrastructure...). In Morocco however, it would be very difficult to ignore the indicator of yield of the network, given its simplicity and popularity to all actors. It is however necessary to see to it to separate between yield in the production (adduction) and yield in the distribution, and not to neglect the index of the losses in network.
d- The examination of the common Yield on the network calculation formula draws the following

observations:

$$R = \frac{Consumed\ Volume}{At\ the\ head\ of\ network\ volume} x100$$

- An additional distortion of the yield that manifests itself in the processing of the volumes produced by the operator: a surface resource requires more treatment compared to an underground resource. The quantities rejected in the first case are obviously larger. The overall yield would be affected without any change in consumption.
- The supplementary distortion of production is manifested at the level of the processed volumes which are produced by the operator: a surface quantity necessitate further treatment as opposed to ground water resources.
- Distinguishing between technical yield and commercial yield.
- Some authors prefer to use net yield which eventually intervenes with uncounted consumed quantities, diverted volumes and service volumes. However, this raises the difficulty of uncounted and diverted volumes. This feature renders the principle of net output in contradiction with the principles of measurability and verifiability. For this reason, some authors prefer and encourage the appellation "primary yield instead of net yield". In order to preserve water resources, the authors suggest adopting production performance (yield) (COUSQUER et Al., 2005)

2- *Linear index of losses*

$$LIL = \frac{volume\ put\ in\ distribution - consumed\ volume\ by\ ordinary\ users}{Network\ length\ (individual\ connection\ off)\ _{x}365}$$

This indicator is better that the yield so as to translate the network state. It strongly depends on the linear density of connections (number of connections per km of network) and of uncounted authorized volumes that are to be consumed: numerous connections multiply the losses in the downstream of these and the authorized consumed volumes sidestep real loses in the network.

a) A lot of authors recommend analyzing the following indicators simultaneously: a- primary yield- b- number of leaks-c- replacement rate-d- leakage finding rate.
b) When the network comprises a part in a rural area, the indicator will be divided into two sub-indicators (rural and urban). A rural network is generally dispersed and extended. LIL is necessarily low in comparison with urban areas.

3- *Resource protection index.*

This indicator gives information on the development of actions taken in order to protect water resources against major factors of pollution. This has a percentage explanation. It is calculated through a scale-variant from 0 to 100%

No action	study launch	defined perimeter	signed regulatory order	implementation and completed jobs	existence of periodic follow-up procedures
0%	**20%**	**40%**	**60%**	**80%**	**100%.**

The indicator "water resource protection index" is rarely or never used by the operators in Morocco. The reason seems to be linked with the fact that water resource management, belongs the water department prerogative. The indicator's utility is recommended for operators who produce at least part of the distributed water

https://doi.org/10.1051/e3sconf/20183701010

4- Charged consumption unit (sold) (l/day/inhabitant) (consumer segment)

$$\text{CCUc} = \frac{\text{annual charged volume (sold)}}{\text{Served population x 365}}$$ Per liter/day/inhabitant (The index "c" refers to the considered category

This indicator serves to know the effective consumption tendency of users. It is preferable to divide the population in terms of consumer segments because the trend can be low for one category and high for another. In order not to have biased results, it is not recommended to consider service consumption and that of agents because it is not charged. Everything that is cost-free encourages excess.
The aim is to appreciate water consumption variation of the consumer and to study her tendency with regard to awareness campaigns and pricing system. Eventually, the allocations used for the design of the networks equipments' can be revised.

III.2.2. Liquid sanitation service: Widening indicator matrix of resource management performance

In Morocco, the regulations as far as the standardization of used water rejection is concerned relate certain specific activities to limited values of rejection. (Domestic used water and industrial activities of sugar, cement, surface treatment, paper pulp, paper and cardboard). In the same respect, a decree related to individual sanitation system was promulgated in 2006 listing the sanitation systems that belonged to this group. Joint application texts should specify the technical arrangements of these systems construction.
In 2015, a new law about water was promulgated (law 36-15) amending the law 10-95 which has arranged a whole section to liquid sanitation, treatment and reject (article 106 to 110). This legal framework is waiting to be completed by executive acts for an appropriate start-up of performance measurement mechanisms.
Nevertheless, at the present stage, it would be possible, through contractual or conventional means, to insert specific indicators for the preservation of the water resource against water pollution. The two aspects to be evaluated are the liquid discharges and the fate of the treatment products. The proposed indicators are:

- *Ratio of the volume of used and treated water (in %): RUT*

$$RUT = \frac{Volume\ of\ used\ and\ teater}{Voulme\ of\ used\ and\ collected\ water}\ x\ 100$$

- *Performance of Treatment (in %)*

$$PT =. = \frac{Polluant\ load\ at\ station\ exit}{Poluant\ load\ at\ station\ inlet}\ x\ 100$$

This indicator is the denomination of being in compliance with sewage discharges which refers to regulatory limits of rejected, used and treated water.

- *Ratio of Water volume that is treated and Reused (in %)*:it portrays the percentage of used water volumes valued with respect to collected volumes by the system.

$$RWR =. = \frac{Treated\ and\ reused\ water\ volume}{Used\ and\ collected\ water\ volume}\ x\ 100$$

- *Ratio of treated sludge:*

$$RTS =. = \frac{Treated\ sludge\ quantity}{collected\ sludge\ quantity}\ x\ 100$$

➢ *Ratio of Reused Sludge (in %)*

$$RRS =.= \frac{Reused\ sludge\ quantity}{collected\ sludge\ quantity}\ x\ 100$$

IV- Conclusion

The performance indicators, which are generated by drinkable water and sanitation public service operators, are generally of internal type. They pave the way for exploitation management objectives. Moreover, their discrepancy and heterogeneity complicate their efficiency. Because of the lack of standardization, at the national level, these indicators do not allow a macroscopic visibility of the public service. At the regulatory level, the governing law 13-14 which is related to communes stated the necessity for local public services to establish performance indicators (article n°12). This text should constitute the basis for engagement in this process.
In the meantime, the suggested indicators in this research paper can be introduced as evaluative and analytical tools for water and liquid sanitation services. This can be done through contracts and agreement established with the operators. For the latter, the three-year revision constitutes an appropriate occasion for them to be introduced in the system. As for the communal operators, board of directors is in a position to adapt performance indicators in the management of public water and sanitation services. However, a cushion system has to be realized in the presence of operators. This approach will provide scope for testing the indicators and appropriating them by service providers.

SPECIAL THANKS

The authors of this work warmly thank Mr. EL HASSAN BOUKOUTA, Governor Director of the Communal operators and Conceded services of the Ministry of the Interior in Morocco. His repeated interest in the research theme was an additional source of motivation for the entire team.

BIBLIOGRAPHIE

1. AKKA H et BOUZIDI A. (2017). Dynamique de la consommation d'eau potable en milieu urabin au Maroc. Maroc: Editions Universitaires europpéennes. ISBN 978-3-330-87606-4.
2. BENBLIDIA, M. (2011). L'efficience d'utilisation de l'eau potbale et aproche économique. Etude nationale. Algérie: Plan Bleu- Centre Régional PNUE/PAM.
3. CESE. (2014). La gouvernance par la gestion intégrée des ressources en eau au Maroc : Levier fondamental de développement durable.p 11. Rabat: Auto-saisine n° 15/2014.
4. Constitution du Royaume du Maroc. (2011). BO n° 5964 bis. p. 1926 . article 154. Royaume du Maroc.
5. CORDOM et AL, N. (2015). Réutilisation des eaux usées pour l'irrigation agricole. Rapporten zone péri-urbaine de pays en développement. Ecofilae AFEID COSTEA AFD.52015.Rapport EcofilaeRéutilisation des eaux usées pour l'irrigation agricole. Rapporten zone péri-urbaine France: Ecofilae AFEID COSTEA AFD.
6. CORDOM Nicolas et AL. (2015). Réutilisation des eaux usées pour l'irrigation agricole. Rapporten zone péri-urbaine de pays en développement. France: Ecofilae-AFEID-COSTEA-AFD.
7. COUSQUER et Al., Y. (2005). Les indicateurs de performance appliqués aux services publics de l'eau et appliqués aux services publics de l'eau et de l'assainissement. Constats et propositions. France: Conseil Général du Géni Rural des Eaux et des Forêts, Conseil Général des ponts et Chaussées. Inspection Générale de l'Environnement.
8. DREAL-CERA, R.-A. (2013). Etude des pertes d'eau potable dans les réseaux. Analyse des performances du réseau d'eau potable en Rhône-Alpes. France.
9. GUERIN SHNEIDER, L. (2001, Avril 2004). Laëtitia Guerin Schneider. Introduire la mesure de performance dans la régulation des services d'eau et d'assainissement en France. Instrumentation et organisation. Gestion et management. ENGREF (AgroParisTech), 2001.

Français. 〈tel-00005754〉. Consulté le Juin 30, 2017, sur https://tel.archives-ouvertes.fr/tel-00005754: https://tel.archives-ouvertes.fr/tel-00005754
10. IGD. Fondation d'entreprises. (2004.). Charte des service ublic locaux Indicateurs de performance Eau otable et Assainissement. France: IGD.
11. IWA. (2003). Indicateurs de performance des services de distribution. Version traduite en français par GESA-ENGREF et FNCCR. IWA.Version traduite en français par GESA-ENGREF et FNCCR.
12. IWA. (2003). Indicateurs de performance des services de distribution. Version traduite en français par GESA-ENGREF et FNCCR. GESA-ENGREF et FNCCR.
13. JORDAN WATER SECTOR. (2015). Facts & Figures. Jordan: Kingdom of Jordanie. Ministry of water and irrigation.(2015 J. ordan Water SectorFacts & Figures.
14. Journal Officiel de lLa République Française du 19 décembre 2013 texte 26 sur 163. (2013). Arrêté du 2 décembre 2013 modifiant l'arrêté du 2 mai 2007 relatif aux rapports annuels sur le prix et la qualité des services publics d'eau potable et d'assainissement.
15. Loi organique 113-14. (2014). Maroc: Bulletin Officiel N° 6440 du 09 Joumada I 1437 (18 Février 2016).
16. citoyens et de l'Environnement. Maroc: Ministère de l'Intérieur du Royaume du Maroc. DRSC.
17. ONAS. (2013). Rapport annuel. Tunisie: ONAS.
18. Renaud, E. (2009). Valeurs de références del'indice linéaire de pertesdes réseaux d'alimentationen eaupotable. Application dans le contexte du SAGE. Nappes Profondes de Gironde. France.
19. SONEDE. (2016). Rapport des statistiques. Tunisie: SONEDE.

LIST OF ABBREVIATIONS

AFD	: Agence Française de développement
AFEID	: Association Française pour l'Eau l'Irrigation et le Drainage
CEMAGREF	: Centre du Machinisme Agricole du Génie Rural, des Eaux et Forêt France (IRSTEA)
CERA	: Cellule Economique Rhône-Alpes
COSTEA	: Comité Scientifique et Technique Eau Agricole
DREAL	: Direction régionale de l'environnement, de l'Aménagement et du Logement. Rhône-Alpes
DRSC	: Direction des Régies et des Services Concédés
ENGREF	: Ecole Nationale du Génie Rural, des Eaux et des Forets
FNCCR	: Fédération nationale des collectivités concédantes et régies, France
IGD	: Institut de la Gestion Déléguée, France
IRSTEA	: Institut national de Recherche en Sciences et Technologies pour l'Environnement et l'agriculture. (Old naming: CEMAGREF)
IWA	: International Water association
MENA	: Meadle East and North Africa
ONAS	: Office national d'Assainissement, Tunisie
PAGER	: Programme d'Alimentation Globale en eau Rurale
SAGE	: Schéma d'Aménagement et de Gestion des Eaux, France
SISPEA	: système d'information des services publics d'eau et d'assainissement
SONEDE	: Société nationale d'Exploitation et de Distribution des eaux, Tunisie
STEP	: Station d'Epuration

New Tubular Ceramic Membranes from Natural Moroccan Clay for Microfiltration Application

*A.*Ait Taleb[1]*, *N.* El Baraka[1]*, *N.* Saffaj[2], *A.* Laknifli[2], *R.* Mamouni[2], *A.* Fatni[1], *A.*El Hammadi[1] and *N.* El Qacimi[2]

[1] Faculty polydisciplinaire,University Ibn Zohr, BP 271, Taroudant, Morocco, .
[2] Faculty polydisciplinaire,University Ibn Zohr, BP 638, Ouarzazate, Morocco,
[3] Faculty of Sciences,University Ibn Zohr, BP 8106, Cite Dakhla, Agadir, Morocco,

Abstract. This paper is devoted the preparation of low cost microfiltration membranes using Moroccan clay powder. The preparation of membrane was composed with two steps: First a macroporous tubular support with a pore diameter 10 µm and porosity 43%. Secondly a microfiltration layer was performed by the slip casting method. A deflocculated slip was obtained by mixing mineral powder of ZrO2, PVA (polyvinyl alcohol) and water, after drying at room temperature for 24 h, the microfiltration layer was heated to 800°C for consolidation. Scanning electron microscopy observation showed homogeneous layers without cracks with an average pore diameter of 0.19 µm for the active layer. Water permeability obtained is about 841 L/h.m2.bar. The membranes have been tested to cleaning of colored wastewater.
Keywords: Wastewater treatment; Ceramic membrane; Tubular support; Microfiltration; Clay.

* Corresponding author: a.aittaleb@uiz.ac.ma / n.elbaraka@uiz.ac.ma

1. Introduction

In the last 10 years the separation by the use of membranes ceramic has rapid increasing. The membrane separation technologies have found a use in various fields: chemistry, food, urbanization; industrial, agricultural and domestic wastewater treatment [1]. For environmental and economic consideration, numerous researches have been focalized to development of new type of low inorganic membranes, which present good chemical stability, favorable mechanical strength, good antimicrobial ability and high separation efficiency.

Generally, commercial ceramic membranes are prepared from amphoteric oxides such as alumina, titania, zirconia [2-6], and sometime also with carbone which is in expensive precursor to fabricate the membrane, Yachino 2005 [7]. To avoid the issue of membrane cost, resent research in the fabrication of inorganic membrane is focused towards of utilization of cheaper raw materiels.

A lot of local natural materials were used for membrane preparation, in our laboratory, such as natural and artificial clays like kaolinite, cordierite [8-10], perlite [11] and animal bon [12] to prepare support which were coated with $ZnAl_2O_4$–TiO_2 ultrafiltration layer, Rakib et al. [13] who developed the microfiltration membranes with natural clays, Elmodden et al. [14] and Weir et al. [15]. In our previous studies, we developed a new ceramic support for microfiltration membrane, with flat disk configuration, based on natural Moroccan clay [16]. The support was elaborated by extrusion of the plastic paste and sintering at 800 °C. The structural properties of clay support are satisfying in terms of porosity and pore diameter. It presents a well chemical resistance in acidic and basic medium.

In order to reduce energy costs of membranes we have developed in this present work a new microfiltration monotubular membranes prepared from natural clay.

The choice of natural clay is mostly due on its abundance in Morocco, its low cost than other commercial powders, to its mechanical resistance, their chemical and thermal stability [17-18].

In the current work reports the fabrication of tubular macro-porous clay support by extrusion followed by sintering, the goal of this work is now focused on the use of ZrO_2 as possible intermediate layer material for microfiltration membrane.

In particular, the preparation of microfiltration membrane made with natural Moroccan clay, will be presented successively and application of these membranes for the clarification of synthetic solutions the methylene bleu.

2. Expérimental

2.1. Clay powder

Natural clay was collected from Ouled Taima at 40 Km of Agadir, Morocco. The powder was crushed at 125 μm with disc mills. The chemical composition of this powder is given in weight percentage (wt.%) as follows : 43% Al_2O_3, 35% SiO_2, 15% Fe_2O_3, 3.5% Na_2O, 1.2% K_2O and 0.03%MnO, it reveals that the clay is essentially formed of alumina with silica [16].

2.2. Elaboration of porous tubular supports

For this study, the process to elaborate tubular support is similar like that used in our previous work to prepare flat disk support [16]. The plastic pastes are prepared from the clay powder at 125 µm mixed with organic additive and water. Plasticizer and binder are required to prepare a paste with rheological properties allowing the chapping by extrusion. The optimal formulation of the paste after several test is: 81.7% w/w of clay, amidon 10% w/w (Amidon de maïs RG03408, Cerestar) as porosity agent, Methocel 4% w/w (The Dow Chemical Company), cellulose derivative, as a plasticizer, Amijel 4% w/w (Cplus 12072, Cerestar) as a binder and PEG 1500 0.3% w/w (Prolabo) as a binder. The mixture was aged (250 tr/min) during 30 min in order to obtain a good homogeneity. After adding the water (32% w/w of powders) and Zusoplast 0.24 % w/w (Zschimmer and Schwartz), we pugging the mixture for 30 min and the paste is kept in a closed box for 2 days under high humidity to avoid premature drying and to ensure complete diffusion of the water and organic additives.

The tubular support is then obtained by extrusion of the prepared ceramic paste. After extrusion, the wet support is kept on rollers of filter paper to obtain a homogenous drying at room temperature. The extruded pieces were sintered at 800 °C in furnace (Figure 1). Process of the ceramic support preparation was discrebed in Figure 2.

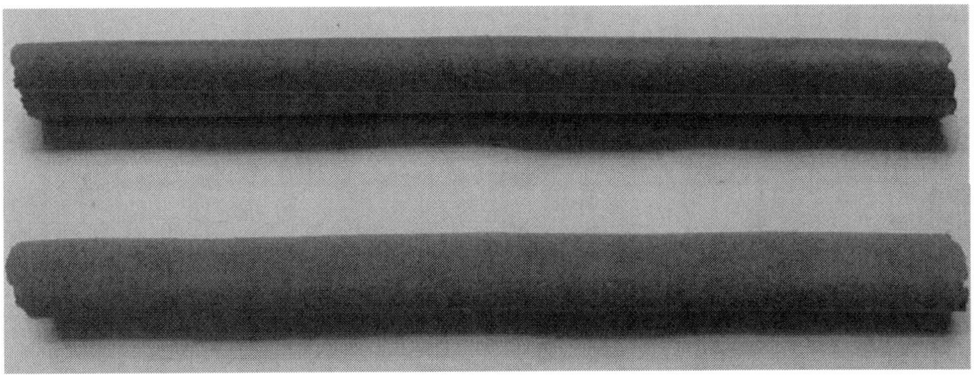

Fig. 1. A photograph of porous tubular ceramic supports.

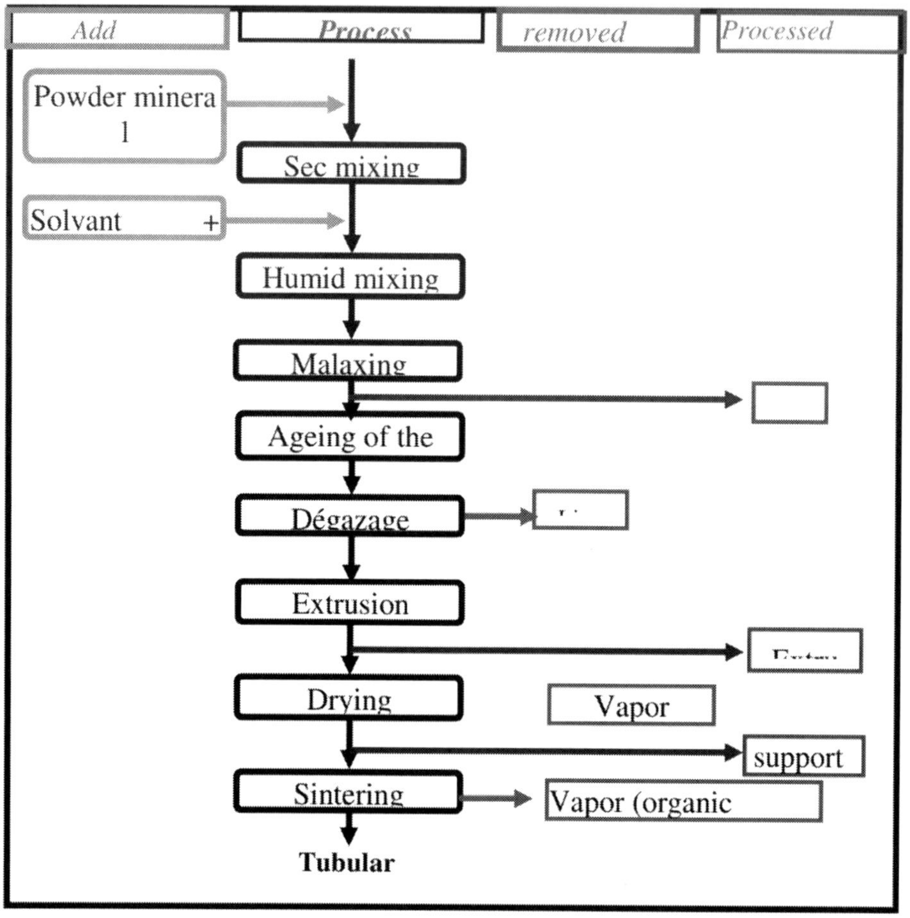

Fig. 2. Process of elaboration porous tubular suuport

2.3. Development of microfiltration layer

The microfiltration layer was obtain by slip casting (Figure 3) by means of a deflocculated suspension which is prepared by mixing 10 wt % of mineral powder ($ZrO2$), 30 wt % of PVA (12 wt % aqueous solution) and water with dispersant DOLAPIX PC 67 at 60% (0.2 wt % aqueous solution). Dispersion of the apatite particles was achieved by applying ultrasonic agitation for 10 min. The viscosity is an important parameter to determine the slip optimal composition.

The tubular support was coated with the $ZrO2$ dispersion, dried at room temperature and fired at a temperature of 800 °C for 2h. Total porous volume and pore size distribution are measured by mercury porosimetry.

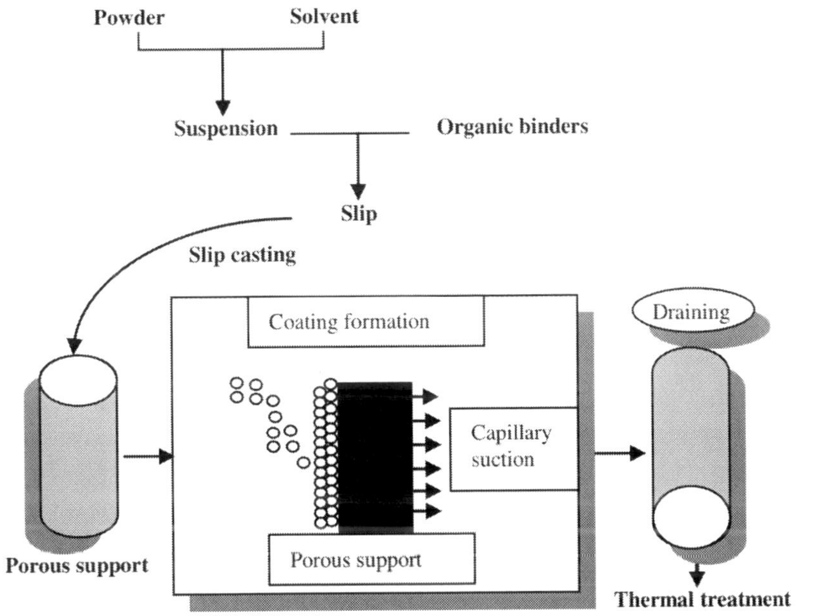

Fig. 3. Slip-casting process [19].

2.4. Tests of filtrations

Tangential filtration tests were performed on a laboratory scale filtration pilot using a recycling configuration (Figure 4). It was equipped with an adjustable out-flow pump, a thermostatic feed tank and a membrane module of 15 cm length. The filtration area developed by the membrane is 1.9*10-3 m2. A valve was used to control the pressure in the system, it varies between 0 and 2 bar. Membranes were conditioned by immersion in pure deionised water for a minimum of 24 h before the filtration tests.

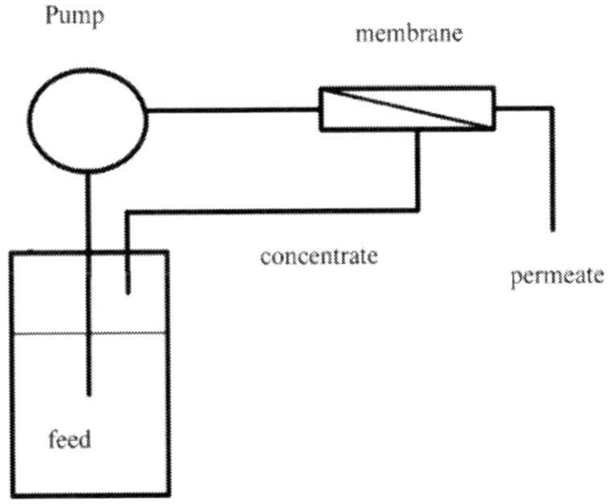

Fig. 4. Scheme of the pilot.

3. Results and discussion

3.1. Characterization of the support

3.1-1 Scanning electron micrograph

The SEM micrograph reported on Figure 5 show the surface and the cross section view of the tubular support. SEM image show there is a good consolidation obtained at sintered temperature; the appearance of the sample surface is homogeneous and does not show defects.

Fig. 5. Scanning electron micrographs of the support :(a) surface view.,(b) cross section view.

3.2-1 Water permeability

Water permeability is always a key factor on membrane applications, experiments show that the water flux depends on the applied pressure. The average support permeability determined using pure distilled water is 1926L/h m2 bar (Figure 6).

The other characterizations of the support like the pore volume Vp, the pore diameter Dp, chemical resistance were detailed in our previous paper [Me].

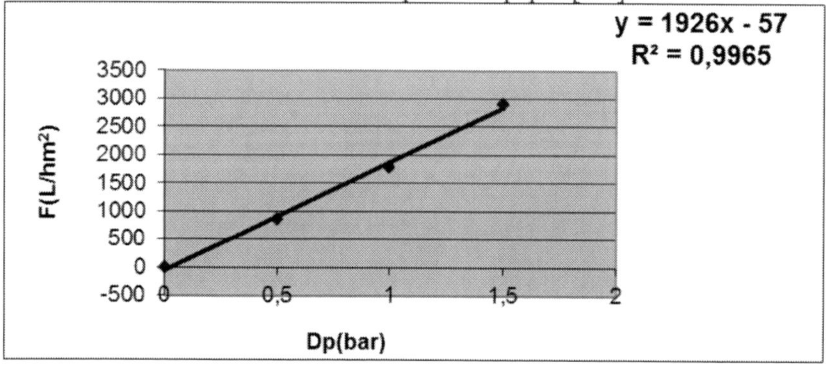

Fig. 6. Water flux versus working pressure.

3.2. Characterization of the microfiltration layer

3.2-1 Scanning electron micrograph and mercury porosimetry

The morphology, surface quality and thickness of top layer membrane were examined by scanning electron microscopy. Figure 7 shows the cross-section and the surface views of the prepared microfiltration layer; the surface of the top layer is homogenous without defects and microcracks. Moreover, it has a good adherence with the support. Pore size distribution of the top layer measured by mercury porosimetry is centered on 0.19 μm that confirms we achieved to prepare a microfiltration layer.

Fig. 7. SEM views of surface (a) and cross-section (b) of zirconia layer.

3.2-2 Microfiltration tests

The water flux through the membrane depends on the applied pressure but the stabilization of the flux needs about 1 hour due to formation of a polarization layer. The permeability was determined from the values of flux measured after stabilisation for each working pressure. The curve given the flux vs. pressure is a straight line with a slop equal to about 841 L/h.m2 bar which corresponds to the permeability of the membrane.

3.2-3 Filtration tests

Filtrations are performed in the tangential mode to have no deposits on the membranes can cause clogging; the working temperature is kept constant by a cooling by circulating water around the feed tank.

The filtration is carried out without re-circulation of the permeate which results ultimately in accordance with the flow of the membrane used, an increase in the concentration in the feed tray. The concentration of the filtered solution is regarded as constant throughout the filtration. Between each filtration experiment, we systematically perform washing with water under pressure until the flux value at the initial membrane water.

Efficiency of microfiltration process was also tested for the treatment of methylene blue. The results presented in (Figure 8) confirm the possibility to reduce the concentration of the methylene blue by microfiltration treatment.

After a filtration test on a microfiltration membrane, we found that the rejection rate (Figure 8) reaches a maximum value of 3.8% after 40 min.

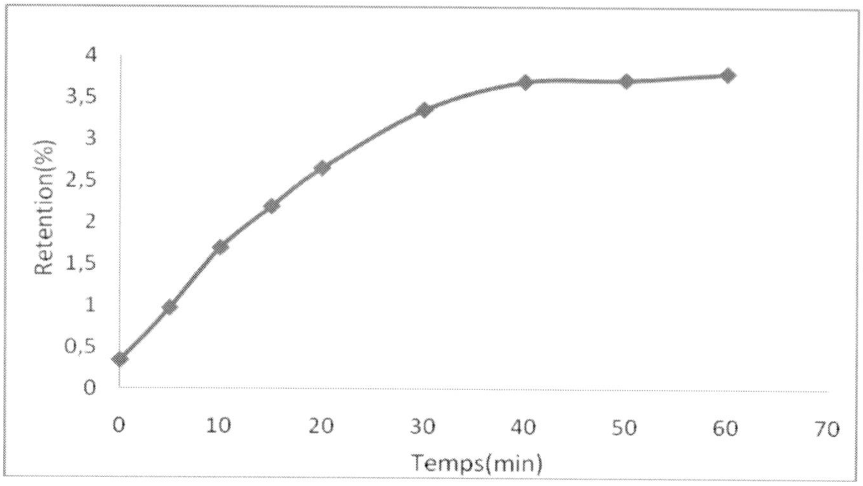

Fig. 8. The rejection rate of methylene blue solution.

4. Conclusion

The preparation and characterization of a new microfiltration membrane based on Moroccan clay has been described in this work. The membrane support was prepared by the extrusion of ceramic paste made with Perlite powder. The development of a microfiltration membrane requests a fine grinding of the Perlite powder. The microfiltration layer was deposited on the support by slip casting. The resulting filters present reasonable characteristics of permeability for water.

Notably interesting are the results on retention of methylene blue which allow envisaging these membranes to design a first clarification step of wastewater treatment.

Références

1. Kosinov N., Gascon J., Kapteijn F., Hensen E.J.M., J. Membr. Sci. 499 (2016) 65.
2. Yeung K.L., Sebastian J.M., Varma A., J. Memb. Sci. 131(1997) 9.
3. Lee S.Y., Lee S.J., Kwon S.J., Yang S.M., Park S.B., J. Memb. Sci. 108 (1995) 97.
4. Nagasawa H., Shigemoto H., Kanezashi M., Yoshioka T., Tsuru T., J. Memb. Sci. 441 (2013) 45.
5. Ahmad A.L., Abdullah Sani N.A., Sharif Zein S.H., Ceram. Int. 37 (2011) 2981.
6. Bouzerara F., Harabi A., Ghouil B., Medjemem N., Boudaira B., Condom S., Procedia Eng. 33 (2012) 278.
7. Elomari H., Achiou B., Ouammou M., Albizane A., Bennazha J., Younsi S.A., El Amrani I., Desal. Water Treat. (2015) 1.

8. Saffaj N., Persin M., Younsi S.A., Albizane A., Cretin M., Larbot A., Appl. Clay Sci. 31 (2006) 110.

9. Bouazizi A., Saja S., Achiou B., Ouammou M., Calvo J.I., Younsi S.A., Appl. Clay Sci. (2016).

10. Majouli A., Younssi S.A., Tahiri S., Albizane A., Loukili H., Belhaj M., Desal. 277 (2011) 61.

11. N. Saffaj, N. El Baraka, R. Mamouni, H. Zgou, A. Laknifli, S. Alami Younssi, Y. Darman, M. Aboulkacem, O. Mokhtari. New BioCeramic Support Membrane from Animal Bone. j. Micribiol. Biotech. Res, 3 (2013) 1-6.

12. S.Rakib, M Sghyar, M .Rafiq, A Larbot, L. Cot, New porous ceramics for tangential filtration. Sep. Purif. Technol. 25, 385-390, (2001).

13. N.Elmodden, A.Elghazouali, S.Rakib, M.Sghyar, M.Rafiq, A Larbot, L.Cot, Nouveaux supports membranaires à base de chamotte d'argile. Ann.Chim. Sci. Mat. 26, 5-11, (2001).

14. M.R.Weir, E.Rutinduka, C. Dettellier, C.Y.Feng, Q.Wang, T.Matsura, R.LeVan Mao, Fabrication, characterization and preliminary testing of all inorganic ultrafiltration membrane made of a naturally occurring sepiolite clay mineral. J.Membr, (2001).

15. N. El Baraka, N. Saffaj, R. Mamouni, A. Laknifli, S. Alami Younssi, A. Albizane, M. El Haddad, .Elaboration of a new flat membrane support from Moroccan clay. Desalin. Water Treat. 52, (2014) 1357–1361.

16. M.Slansky, Géologie des Phosphates Sédimentaires, Mémoire du BRGM N° 114, (1980).

17. R.A.Young, J.C.Elliot, Atomic-scale bases for several properties of apatites J.Arch.Oral.Biol 11, 699, (1966).

18. 27. Barrouk I., Younssi S.A., Kabbabi A., Persin M., Albizane A., Tahiri S., Desal. Water Treat. 55 (2015) 53.

Adsorption of Cr (III) from aqueous solution by two forms of a superabsorbant polymer : parametric study and effect of activation mode

A. OUASS[1,2*], *Y.*ESSAADAOUI[1], *L.* KADIRI[1], *I.* LEBKIRI[1], *C.* LAFREME[1], *M.* CHERKAOUI[2], *A.* LEBKIRI[1], *E. H.* RIFI[1]

1. Laboratory of Organic Synthesis and Extraction Processes
2. Laboratory of materials, Electrochemistry, Environment
Department of Chemistry, University Ibn Tofail, Kenitra - Faculty of Sciences, PO Box. 133 – 14000- Morocco

Abstract. The adsorption of chromium ions Cr3+ using two forms of a superabsorbant polymer PANa from aqueous solution was investigated. On one hand the equilibrium pH with the distilled water and the pH of point of zero charge have been studied in order to characterize the acid-base behavior of both of PANa forms. On the other hand, the effect of contact time between PANa and the metallic solution and stirring speed of aqueous solution on the adsorption rate were established to highlight the importance of PANa as an efficient adsorbent of chromium ions Cr3+.
Keywords: adsorption, Chromium, ICP, superabsorbant polymer.

1. Introduction

The presence of heavy metals in the environment caused by discharges of industrial wastewater is behind considerable damage to the aquatic ecosystem, when they are above certain concentrations. Literature shows that it is widely used in the surface treatment industry and in the tannery[1, 2]. To remedy this issue, various removal techniques have been applied[3], for example: chemical precipitation, ion exchange, membrane filtration, electrocoagulation [4]... Last decades, attention was focused on adsorption method[4–7] and the use of new adsorbents based on abundant natural materials, natural [8, 9]and synthetic zeolites, clays [11, 12] and especially hydrogel [10, 11].

Sodium polyacrylate PANa superabsorbent polymer is a part of the compounds of polyelectrolytes ionizable groups capable of dissociating in a polar solvent (usually water). The presence of charged groups increases solubility chains and has an important role in industrial applications and in biological processes, agriculture, hygiene products, medicine, pharmaceuticals, the cosmetics industry and the paper industry or also the oil industry, recovery of elements harmful to the environment, as heavy metals is an important application fields [15].

The objective of the present work is to investigate the possible use of PANa in two forms (beads and powder) as a removal of chromium (III) ions from contaminated solutions. Some adsorption parameters such as contact time and stirring speed werestudied using Inductively Coupled Plasma (ICP) in order to evaluate the adsorbent-metal ion interactions.

* Corresponding author: karim.ouass.smp@gmail.com

2. Material and methods

2.1. Adsorbent material

The adsorbent used in this study is a superabsorbent polymer based on sodium polyacrylate. PANa is able to retain up to 200 times their weight in water, which are in the form of transparent spherical beads, non-porous surface appearance and in the form of a fine white powder, sold under the reference SJQ-007, is provided by Xinchang Chengtan Magic Bean & Grass Artware Factory of Origin Doll: Zhejiang, China [16].

2.2. Aqueous solution

Aqueous solutions were prepared from hydrated chromium nitrate salt (Cr (NO3)3, 9H2O) purchased from Sigma Aldrich. The stock solutions were prepared and then used to obtain the working solutions through dilution with distilled water. Fresh dilutions were prepared and used for each experiment. 10-1 mol /L HNO3 or NaOH solutions were used to adjust the initial pH of working solutions[7].

2.3. Adsorption experiments

The adsorption process was followed by adding 7.5 g the gel (PANa) corresponding to 0.035 g of the dry polymer placed in a thermo-regulated cell at room temperature (25°C) containing 100 ml of the 10 ppm chromium metal solution with magnetic stirring. All the adsorption tests were carried out at an initial pH of 5 and the evolution of the pH during the reaction was followed by a pH meter. The reaction time was set at 3 hours except in the case of kinetic studies. At the end of the reaction, the suspension is filtered and the filtrate is assayed by inductively coupled plasma (ICP) emission spectroscopy.

2.4. Data evaluation

The adsorption process of chromium ions by PANa was evaluated using the following parameters, which were calculated from experimental results:
Rate of copper ions removed (R, %):

$$R = \frac{C0 - Ce}{C0} * 100$$

C0 is the initial concentration of Cr3+ in the solution (ppm).
Ce is the equilibrium concentration of heavy metal ions in the solution (ppm).
All the data are the mean values of three replicate measurements, and the standard deviation calculated in each case was lower than ± 1.0%.

3. Results and discussion

3.1. Behavior of PANa in contact with distilled water (equilibrium pH)

The equilibrium pH was determined after putting 1 g of each form of the material (0.315 mm) in contact with 100 ml of distilled water (pH = 5.2 ± 0.2). The sample was stirred at 300 rpm for 6 hours. The pH measurements were made at regular time intervals minutes using a pH meter pHM210 (Standard pH meter, MeterLab).

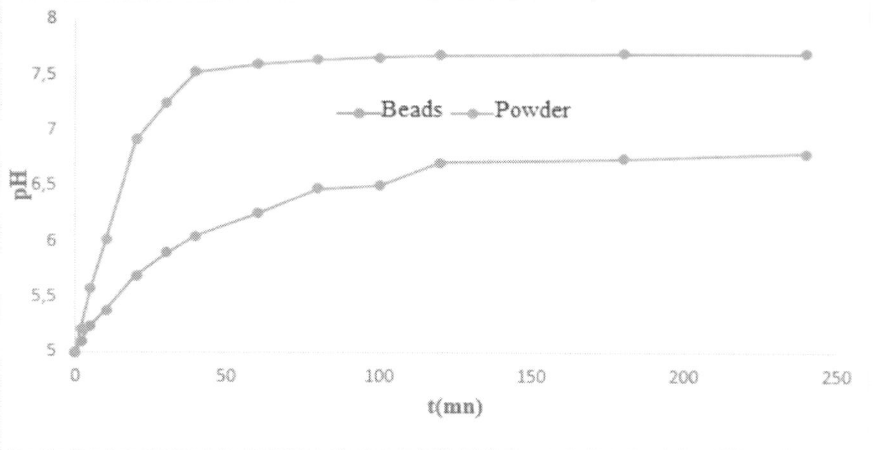

Fig.1. Equilibrium pH of PANa (beads and powder)

The evolution of the pH of PANa in contact with distilled water is shown in fig.1. The curve obtained shows an increase in pH values during the experiment. At (t = 0), we recorded a value of pH = 5, after this value the pH increases gradually until the equilibrium pH = 6.5 for beads and 7.7 for powder, which prove the acid-base rebalancing of the gel with the aqueous solution. It therefore appears obvious that the PANa gel equilibrates with the aqueous phase by consuming protons according to an exchange process of Na+ ions of PANa against protons of the aqueous solution according to:

$$\text{-}COONa + H^+ \rightleftharpoons \text{-}COOH + Na^+$$

3.2. Determination of pHpzc

The point of zero charge (pHpzc) of adsorbent was determined by the method described elsewhere [17]. In brief, 100 ml of 0.01M NaCl solutions were placed in various Erlenmeyer flasks. Their pH was adjusted to different values between 2 to 12 by the addition of 0.1 M HCl or NaOH solutions. Now the mass adsorbent beads and powder was added into each solution and the final pH of these solutions was recorded after 48 hours of agitation [18].

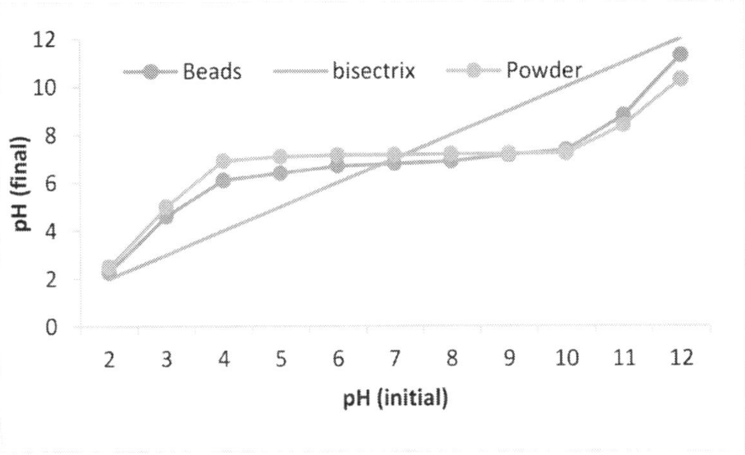

Fig.2. Point of zero charge

The pHpzc is the point where the curve pHfinal verses pHinitial intersects the straight line corresponding to: pH initial = pHfinal. The point of zero-charge, as shown in Fig.2, was evaluated to be equal to 7.12 for PANa beads and 6.8 powder form [19].

3.3. Adsorption of Chromium (III) by PANa

3.3.1. Effect of contact time

The study of adsorption kinetics has a great importance. In fact, it makes it possible to determine the adsorption process and subsequently to calculate the time required to eliminate the heavy metal ions. In figure (3(a) and 3(b)), the kinetics of adsorption of Cr3+ ions in the nitrate medium from an aqueous solution by a PANa in two forms (beads and powder) have been represented. The curves show the evolution of the concentration and the adsorption rate of chromium versus time of contact of the gel with the aqueous solution. It appears from the results a significant and rapid removal of the heavy metal ions after 20 minutes for the powder form and 40 minutes in the case of beads of PANa which can be attributed to the presence of a large contact surface of PANa powder in comparison with beads form. After equilibrium where the concentration of Cr3+ ions becomes independent of time, the contact time required to reach equilibrium is about 30 minutes and 85 minutes for PANa powder and beads respectively. This translates the extraction performance of chromium ions, related to the form of the gel. The evolution of the adsorption rate versus contact time with the heavy metal solution shows that more than 70% of Cr3+ is adsorbed after 30 minutes; R (%) reaches its limit value 95% at 45 minutes for the powder and about 89% for beads at 85 minutes.

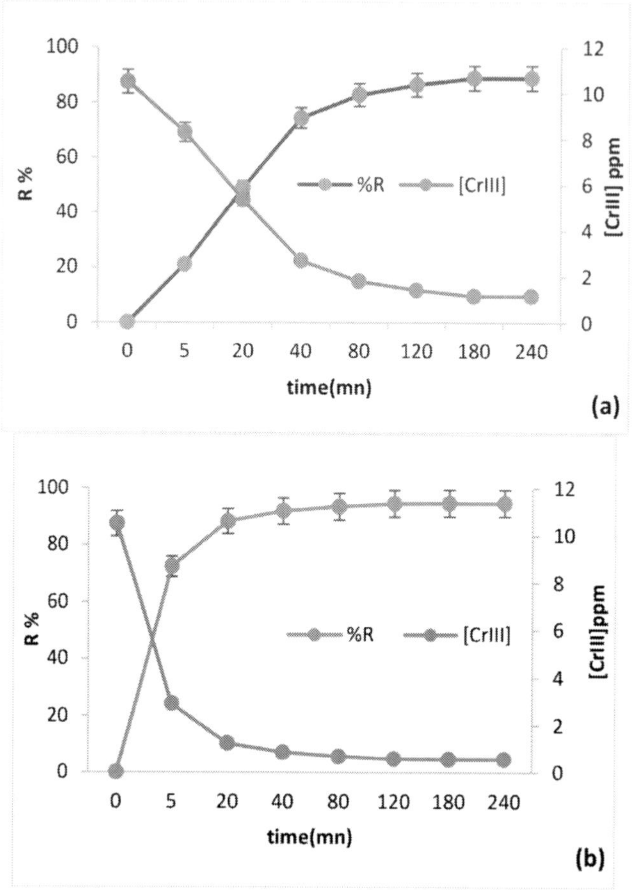

Fig.3. Evolution of adsorption rate (a) PANa beads and (b) PANa powder

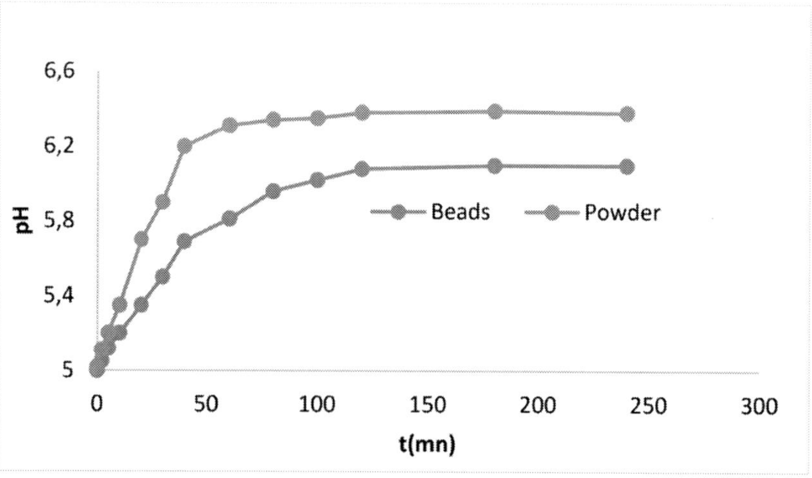

Fig.4. Evolution of pH of PANa in contact with Cr^{3+}

The decrease in metal concentration of aqueous solution in contact with PANa Beads form (from 10.5 to 1.16 ppm) and powder form (from 10.5 to 0.56 ppm) is accompanied by an increase in pH values, which shows that the hydrogel equilibrates with the metal solution by fixing H + protons and Cr3+ ions by releasing Na+ ions in the aqueous solution, according to an exchange process.

3.3.2. *Effect of stirring speed*

By the same experimental protocol previously described, we introduce volumes of 100 ml of 10 mg / l solution in chromium with 0.035 g of dry polymer at room temperature. The stirring speed is varied between 100 and 1000 rpm until equilibrium. The samples have been taken after 120 minutes of PANa contact with the metal solution and analyzed by ICP. The curve (fig.5) represents the evolution of removal rate versus the stirring speed.

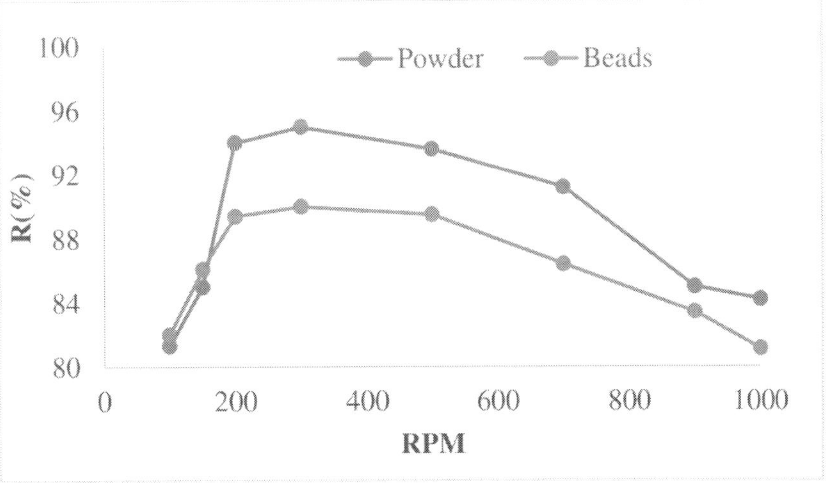

Fig.4. Evolution of adsorption rate versus stirring speed

Stirring of the system generally decreases the thickness of layer around the adsorbent particles that results from the increase in the speed of mixing. When the system is agitated, the suspensions move rapidly with the solution, thereby increasing the concentration of the metal near the surface of the extractant, probably to a value close to the concentration within the solution. The results reveal that the Cr (III) removal efficiency is influenced by the stirring speed. However, the speed corresponding to 300 rpm with both types of PANa gives better results, and it is found that at low stirring speed the retention is less favored. Beyond a stirring speed of 500 rpm there is an important rate in the removal efficiency of Cr (III) this may be due to additional energy sufficient to destroy the bonds possibly formed between the ions chromium and the surface of the PANa hydrogel[15].

4. Conclusion

The results obtained by equilibrium pH after contact of PANa with distilled water have shown an acid-base rebalancing of material with the double distilled water. The study of point of pH zero charge of each form of PANa indicated that the PANA beads acquires a positive charge below pH 7.12 while the powder form have a positive charge below pH

6.8. The study of the chromium ions Cr3+ extraction by PANa through following the evolution of the concentration and the rate of adsorption versus time has shown that the fixation of Cr3+ on the area of our material has been detected on a short time. The pH measurements have increased with time; this is the rebalancing of the acid - base of PANa (beads and powder) with the solution of the metal Cr3+. In brief, PANa can be an effective adsorbent potential of heavy metals contaminating wastewaters.

Acknowledgment

The authors are pleased to acknowledge:

- The 6th international congress "water, waste and environment", Sultan Moulay slimane University, Beni Mellal, Morocco for giving the opportunity to publish this work.
- The Ibn Tofail University, Kenitra, Morocco, for providing the facilities for the research. Also the Director of Center of Analysis, Expertise, Transfer of Technology and Incubation (CUAE2TI) "M. Ibn Touhami" is gratefully acknowledged for his kind permission to use their facilities for the analysis.

Reference

[1] A. Ouass, I. Ismi, H. Elaidi, A. Lebkiri, M. Cherkaoui, and E. H. Rifi, "Mathematical Modeling Of The Adsorption Of Trivalent Chromium By The Sodium Polyacrylate Beads," JMES, vol. 8, no. 10, pp. 3448–3456, 2017.

[2] M. Fabbricino and R. Gallo, "Chromium removal from tannery wastewater using ground shrimp shells," Desalination Water Treat., vol. 23, no. 1–3, pp. 194–198, Nov. 2010.

[3] H. Taouil, S. I. Ahmed, E. Rifi, and A. El Assyry, "Zinc extraction from dilute aqueous solutions, by hydrogels of polyacrylic acid-polyacrylate sodium (Extraction du zinc à partir des solutions aqueuses diluées, par d'hydrogels d'acide polyacrylique-polyacrylate de sodium)," JMES, vol. 5, no. 4, pp. 1069–1074.

[4] F. Fu and Q. Wang, "Removal of heavy metal ions from wastewaters: A review," J. Environ. Manage., vol. 92, no. 3, pp. 407–418, Mar. 2011.

[5] S. Afroze, T. K. Sen, and H. M. Ang, "Adsorption removal of zinc (II) from aqueous phase by raw and base modified Eucalyptus sheathiana bark: Kinetics, mechanism and equilibrium study," Process Saf. Environ. Prot., vol. 102, pp. 336–352, Jul. 2016.

[6] I. Ahamad, R. Prasad, and M. A. Quraishi, "Adsorption and inhibitive properties of some new Mannich bases of Isatin derivatives on corrosion of mild steel in acidic media," Corros. Sci., vol. 52, no. 4, pp. 1472–1481, Apr. 2010.

[7] A. Adamczuk and D. Kołodyńska, "Equilibrium, thermodynamic and kinetic studies on removal of chromium, copper, zinc and arsenic from aqueous solutions onto fly ash coated by chitosan," Chem. Eng. J., vol. 274, pp. 200–212, Aug. 2015.

[8] N. T. Abdel-Ghani and G. A. El-Chaghaby, "Biosorption for metal ions removal from aqueous solutions: a review of recent studies," Int J Latest Res Sci Technol, vol. 3, no. 1, pp. 24–42, 2014.

[9] Y.Essaadaoui, L.Kadiri, E.H.Rifi, and A.Lebkiri, "Microstructure characterization of the barks of eucalyptus "eucalyptus," IJSER, vol. 7, no. 11, pp. 1382–1387, Nov. 2016.

[10] S. Elanza, A. Lebkiri, S. Marzak, E. H. Rifi, M. Lebkiri, and C. Satif, "Removal of lead ions from aqueous solution by the sugarcane bagasse," J Mater Env. Sci, vol. 5, no. 5, pp. 1591–1598, 2014.
[11] Z. Mohamed, A. Abdelkarim, K. Ziat, and S. Mohamed, "Adsorption of Cu (II) onto natural clay: Equilibrium and thermodynamic studies," system, vol. 10, p. 11, 2016.
[12] S. Sahoo, Uma, S. Banerjee, and Y. C. Sharma, "Application of natural clay as a potential adsorbent for the removal of a toxic dye from aqueous solutions," Desalination Water Treat., vol. 52, no. 34–36, pp. 6703–6711, Oct. 2014.
[13] E.-H. Rifi, D. Lakkis, and J. F. M. Leroy, "Extraction de l'argent par des gels d'acide polyacrylique–polyacrylate de sodium. Application à l'élimination des ions chlorures," Comptes Rendus Chim., vol. 8, no. 5, pp. 917–921, May 2005.
[14] H. Bousfiha, E.-H. Rifi, A. Essamri, and Z. Hatim, "L'extraction liquide–solide du zinc par des supports à base de phosphate de calcium," Comptes Rendus Chim., vol. 8, no. 2, pp. 215–218, Feb. 2005.
[15] I. Ismi, E. H. Rifi, A. Lebkiri, and H. Oudda, "Spectral characterization of PA-Cu under two polymeric forms and their complex PA-Cu," 2015.
[16] I. Ismi, H. Elaidi, E. H. Rifi, A. Lebkiri, and A. Skalli, "Comportement de polyacrylate de sodium au contact des solutions aqueuses en absence et en presence d'ions métalliques (Ag, Cu 2, Cr 3)(Behaviour of sodium polyacrylate in contact with aqueous solutions in the absence and the presence of ions metal (Ag, Cu 2, Cr 3))."
[17] D. Schimmel, K. C. Fagnani, J. B. Santos, M. Barros, and E. Silva, "Adsorption of turquoise blue QG reactive bye commercial activated carbon in batch reactor: kinetic and equilibrium studies," Braz. J. Chem. Eng., vol. 27, no. 2, pp. 289–298, 2010.
[18] L. KADIRI et al., "Characterization of coriander seeds coriandrum sativum," Int. J. Sci. Eng. Res., vol. 8, no. 7, Jul. 2017.
[19] S. K. Bajpai and A. Jain, "Removal of copper (II) from aqueous solution using spent tea leaves (STL) as a potential sorbent," Water Sa, vol. 36, no. 3, pp. 221–228, 2010.

Adsorption of an anionic dye on a novel low-cost mesoporous adsorbent: kinetic, thermodynamic and isotherm studies

Asmaa MSAAD [1]*, *Mounir* BELBAHLOUL[1], and *Abdeljalil* ZOUHRI[1]

[1]Chemistry and Environment Laboratory, Faculty of Science and Technology, University Hassan 1, BP 577, Settat 26000, Morocco

Abstract. Our activated carbon was prepared successfully using phosphoric acid as an activated agent. The activated carbon was characterized by Scanning Electron Micrograph (SEM), Brunauer-Emmett-Teller (BET), Fourier transform infrared spectroscopy (FT-IR) and X-ray diffraction (XRD). The aim of our study is to evaluate the adsorption capacity of Methyl Orange (MO) on *Ziziphus lotus* activated carbon. Adsorption isotherms were studied according to Langmuir and Freundlich Model, and adsorption kinetics according to pseudo-first and second-order. Results show that the maximum adsorption was reached in the first 10min at ambient temperature with a yield of 96.31%. The Langmuir isotherm shows a correlation coefficient of 99.4 % higher than Freundlich model and the adsorption kinetic model follow a pseudo-second-order with a maximum adsorption capacity of 769.23 mg/g. FTIR and X-Ray spectroscopy indicate that our activated carbon has an amorphous structure with the presence of functional groups, where BET analysis revealed a high surface area of 553 mg/g, which facilitate the adsorption process.

Keywords: Adsorption, Methyl Orange, *Ziziphus lotus*.

***Corresponding author**: asmaa.msd@gmail.com;

1. Introduction

Numerous industries such as textile, leather, paper, rubber, paint and a plastic used in their application extensive compounds lead to discharges toxic colored effluent, contaminates surface and groundwater. Textile wastewater, generally, contains a big amount of pollutants materials like colored materials or dyes, organic compounds and heavy metals ions, these colored materials can affect the physicochemical and the biological properties of the sea, drinking water and globally the ecosystem. In addition to the undesirable colors of textile effluents, some dyes may degrade to produce carcinogens and toxic products **[1]**. Methyl orange (MO), an anionic dye belongs to the azo and the more toxic group of dyes. MO has been widely used in textile, paper, plastics, rubber, cosmetics, printing, dye manufacturing and pharmaceutical industries **[2]**.

The removal of industrial pollutants containing the aromatic band from water has become increasingly necessary in recent years and determining processes to achieve this is the focus of much scientific research. Researchers are indulging their interest in the wastewater treatment by various processes such as precipitation, ion exchange, reverse osmosis, and adsorption. Among these methods, Adsorption process is recognized as one of the most promising and effective approaches in wastewater treatment because of its low-cost ease of operation and high-efficiency **[3]**. There are currently a number of adsorbents including clay minerals, agricultural solid wastes are economical adsorbents, but they suffer from limited adsorption capacities of color removal.

This study highlights the adsorption activity of Methyl Orange (MO) from an aqueous solution onto *Ziziphus Lotus* activated carbon. MO is stable, shows solubility in water and low biodegradability; hence it is difficult to remove from aqueous solutions by common water purification or treatment methods **[4]**, several factors affected the biosorption were studied in this work as pH, adsorbent dose, contact time and temperature, to deduce the adsorption thermodynamic and kinetic behavior process. The experimental results revealed a high adsorption capacity of MO adsorption on our sorbent.

2. Material and methods

2.1. Material and reagents:

To test the capacity of *Ziziphus Lotus* sorption, diverse laboratories materials were used in this work: UV visible spectroscopy, multi-parameter Consort C3040 was used for pH measurements, the pH values were adjusted using HCl (0.1N) or NaOH (0.1N) buffer solutions, multi agitator. Methyl orange (MO) ($C_{14}H_{14}N_3NaO_3S$), obtained from Sigma Aldrich, with a molecular weight of 327.32 g/mol and λ max of 464 nm. Solutions were prepared by dilution with distilled water of the stock solution of MO with the initial concentration of 1000 ppm to reach the desired concentration.

2.2 Characterization of the activated carbon:

Ziziphus leaves are very abundant in Morocco and not valorized were collected from Settat-Casablanca, Morocco, washed several times to eliminate the impurities and then activated with the phosphoric acid to obtain the activated carbon. The physicochemical properties of our activated carbon were determined by several analysis methods such as the (BET)

Surface Area Analysis using a micrometric machine model 3 FLEX 3500, the surface morphology and the elemental composition was executed by SEM and Energy Dispersive X-Ray (EDX) using an FEI Quanta 450 FEG Environmental Scanning Electron Microscope (ESEM), the X-ray diffraction using a D2 PHASER-BREKUR and FT-IR spectroscopy analyses were performed using a Transform Spectrometer (SHIMADZU FTIR-8400S).

2.3 Adsorption experiments:

The adsorption of Methyl Orange on the activated carbon was investigated as an effect of pH, initial dye concentration, contact time, and temperature. Adsorption experiments were reached after stirring 100mL of MO solution with different adsorbent doses and contact time at ambient temperature and agitation speed of 300 Rpm. The suspensions were collected then centrifuged and MO equilibrium concentrations were determined by spectrophotometry UV-VIS at 464 nm. The amount of MO adsorbed at equilibrium noted q_e in $mg \cdot g^{-1}$ and R the yield of adsorption in (%), respectively, were calculated by the following equations (Eq.1 and Eq.2):

$$q_e = \frac{(C_0 - C_t)V}{m} \quad (1)$$

$$R\,(\%) = \frac{C_0 - C_t}{C_0} . 100 \quad (2)$$

Where:
C_0: the initial dye concentration ($mg \cdot L^{-1}$),
C_t: the equilibrium dye concentration ($mg \cdot L^{-1}$),
V: the volume of the solution,
m: the mass of the adsorbent (g).

Kinetic experiments were carried out by agitating 100 ml of MO solution at a constant dye concentration of 100ppm with 60 mg the activated carbon at fixed agitation speed, ambient temperature and pH = 2. Agitation was made from 5 to 120 min. Adsorption isotherm was obtained by varying the initial MO concentration between 10 to 100 mg/L to provide us several informations about the adsorption behavior using two famous models: Langmuir and Freundlich to fit the equilibrium experimental data of MO adsorption into our activated carbon. Thermodynamic behavior was also studied by varying the temperatures (20-60°C) using 60 mg of adsorbent mixed with 1000 mL of the MO dye solution at pH=2.

3. Results and discussion

3.1 Characterization of the biosorbent:

The total surface area (or BET surface area S_{BET}) is one of the most widely known parameters associated with activated carbons **[5]**. The Table 1 presents the textural properties of *Ziziphus* activated carbon. The S_{BET} of this low-cost activated carbon is 553 $m^2.g^{-1}$, this value are higher than 377 $m^2.g^{-1}$ reported by Wang et al., (2011) **[6]** using activated carbon from sewage sludge. A light difference is observed using commercial activated carbon with 564 $m^2.g^{-1}$ **[7]** However, **[8]** Anisuzzaman et al., (2015) found 966.74 $m^2.g^{-1}$ with activated carbon from *Typha orientalis* leaves. Generally, adsorption capacities increase in the growth of this parameter **[9, 10]**.While, there are various exceptions to other types of activated carbon **[10, 11]**.

Table 1: Chemical and textural properties of the activated carbon.

Properties	Value
Textural BET surface area (m^2/g)	553.39
Micropore volume (cm^3/g)	0.156
Micropore area (m^2/g)	307.25
Mesoporous volume (cm^3/g)	0.163
Total vol. BJH (cm^3/g)	0.319

The SEM analysis was also performed. The Fig. 1 shows the SEM images of the activated carbon at 2400 magnifications. This result confirms those obtained using BET analysis and indicated a large development of pores that are clearly found on the surface of the *Ziziphus Lotus* activated carbon, this could be thanks to the activation process using the phosphoric acid (H_3PO_4) as chemical activating agent **[12]**. Also, we notice the presence of heterogeneous pores which are essential for the adsorption of different types of pollutants **[13]**. Thus, it can be claimed that the surface of *Ziziphus* leaves activated carbon has nanoporous structure. Similar observations were reported by Ricou-Hoeffer et al., (2001) **[14]** and Hameed and Daud (2008) **[15]** using fly ash and agricultural waste activated carbon, respectively.

Fig.1. Scanning Electron Micrograph for the prepared activated carbon (a) × 2400.

FT-IR analysis of the bio-carbon was performed to determine the functional groups, and then evaluate the involvement of these functional groups in the adsorption of MO dye onto *Ziziphus Lotus* leaves activated carbon. The Fig.2 (a) display the spectra of the activated carbon, the band located between 3200 and 3600 cm^{-1} can be attributed to the hydroxyl groups –OH to lignin **[16]** and the hydrogen-bonded OH vibration of the cellulosic structure. The band observed around 1610 cm^{-1} could be attributed to the stretching of the aromatic rings (C=C). The band at 1223 cm^{-1} is partly associated with C-O stretching and O-H bending modes in the functional group, the band at around 1080 cm^{-1} can be attributed to the C–OH stretching of phenolic groups, also it can be seen a band around 640 cm^{-1} that justify the presence of the out-of-plane deformation mode of O-H of cellulose **[17]**. The presence of lignin and cellulose structure suggest a lignocellulosic structure of active and carbonized *Ziziphus Lotus*, this structure is also observed for other carbon, such as Brazilian coconut shell and Tunisian olive-waste cakes **[18]**. It is well acknowledged that the polyphenolic compounds like lignin groups present in lignocellulosic materials are the active sites for attachment of ions, the use of these materials in the natural state, without any treatment, can cause problems of lower durability, leaching of soluble organic components at strong acidic and basic media, and lower adsorption capacity.

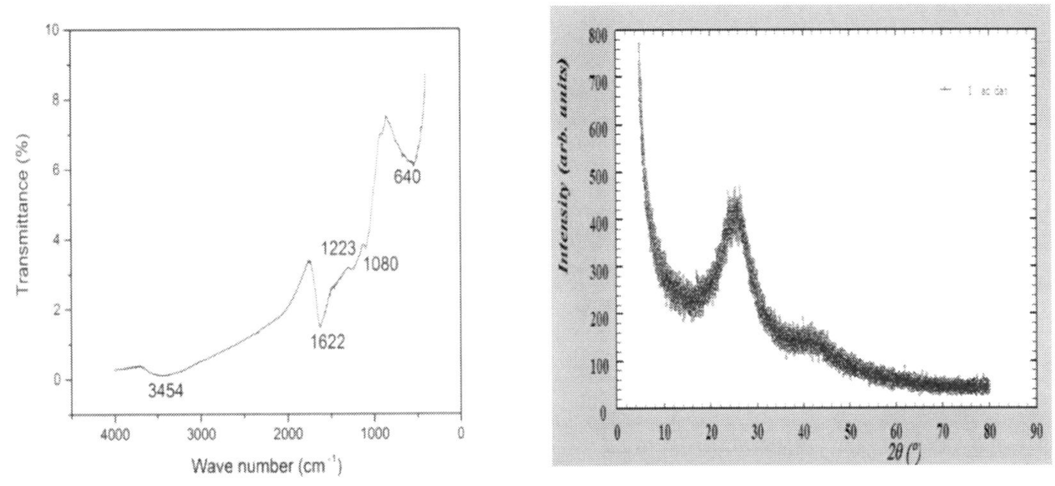

(a) (b)

Fig.2. **(a)** FTIR and **(b)** XRD spectra of *Ziziphus lotus* activated carbon

The XRD spectrum of activated carbon was also performed in the Fig.2 (b). It's indicates the presence of broad diffraction background and the absence of a sharp peak reveal a predominantly of amorphous structure **[19]**. The amorphous structure of the activated carbon was revealed by several products such as oil palm, empty fruit brunch and coconut shell **[20]**.

3.2 Kinetic studies:

The first and second order kinetic models were plotted (Fig. 3); Table 2 resumes the obtained results. The validity of each model was evaluated by the fitness of the straight line (R^2), from the results we observed that the adsorption of MO was better fitted with the pseudo-second order with a R^2=0.99, The Table.2 shows the kinetic models parameters; the high equilibrium adsorption q_e (769.23mg/g) at an equilibrium time of 10min indicates a high degree of affinity existing between MO and our activated carbon.

Table 2. Adsorption kinetic parameters of Methyl Orange on *Ziziphus leaves* adsorbent

Pseudo-second order			**Pseudo-first order**	
Qe (mg/g)	**K_2 (g /mg .min)**	**R^2**	**K_1 (min^{-1})**	**R^2**
769.23	3.67.10^{-4}	0.99	0.0175	0.90

These results indicate that the kinetic model of the adsorption is based on the assumption that the rate-limiting step is a chemical adsorption involving valance force through sharing or exchange of electrons between adsorbent and adsorbate. A similar result was also obtained by **[21]** Huang et al, (2013) using protonated cross-linked chitosan.

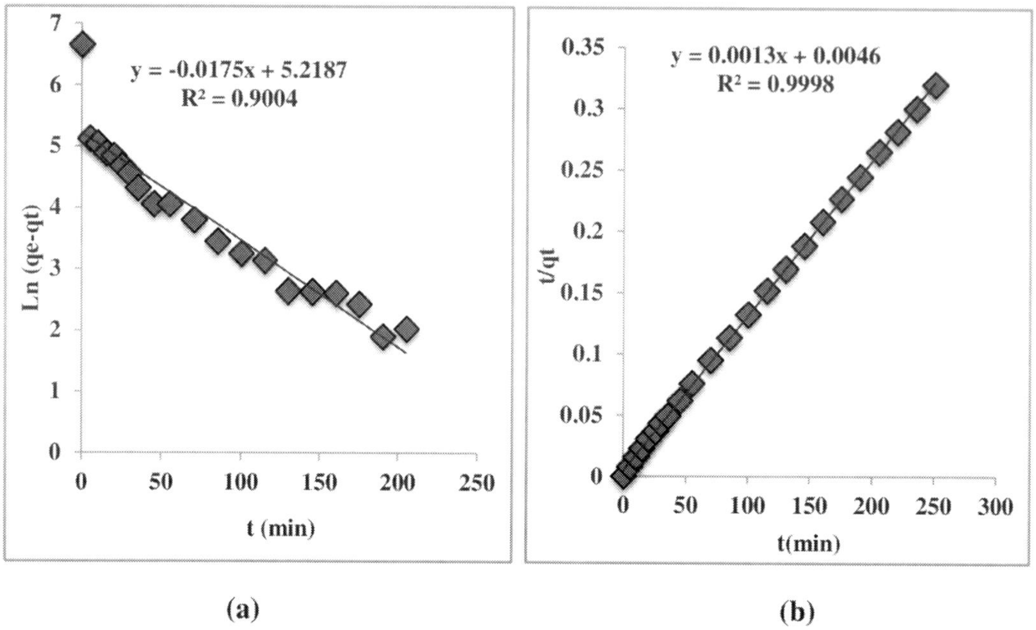

Figure.3. Adsorption kinetic models of MO on *Ziziphus* activated carbon (a) pseudo-first order, (b) pseudo-second order model.

3.3 Adsorption isotherm study:

Freundlich and Langmuir isotherm models explain the capacity of the adsorbent, the interaction between the solute-solution and the nature of adsorbed accumulation materials on the surface. The Fig. 4 shows the plot of Langmuir and Freundlich isotherm models. The results shown in Fig.4 revealed that the sorption of MO on *Ziziphus Lotus* adsorbent is the monolayer type since the Correlation coefficient (R^2=0.99) calculated from Langmuir isotherm, which indicates the applicability for Langmuir isotherm. The results are in accordance with the literature, where Chen et al., (2011) **[22]** were found the same behavior of MO on the surfactant-modified silkworm exuviae.

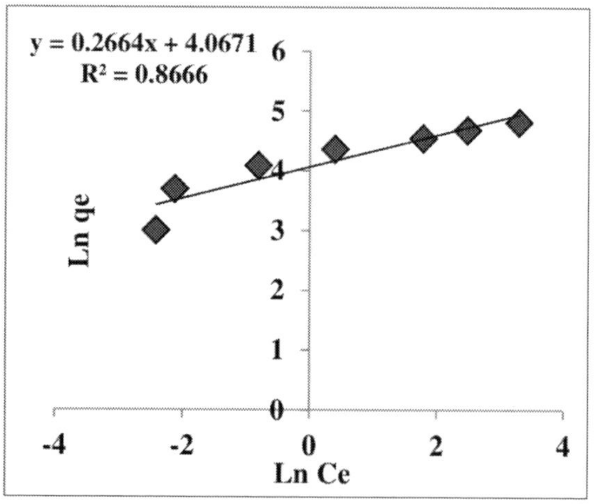

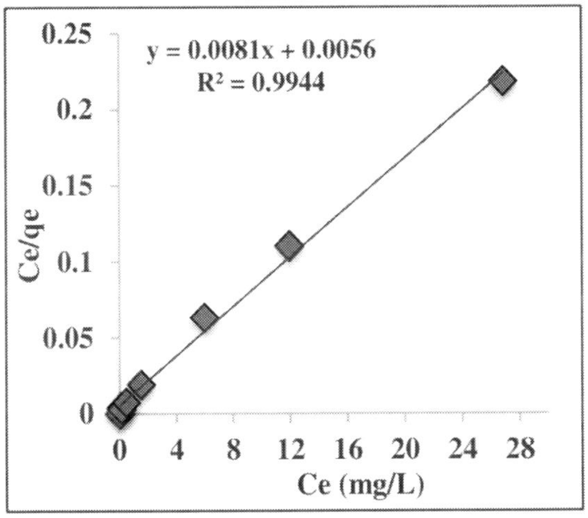

(a)

(b)

Figure.4. Adsorption isotherm of MO on *Ziziphus* activated carbon (a) Langmuir model, (b) Freundlich model

3.4 Thermodynamic studies:

The effect of temperature (20, 40, 50 and 60 °C) on the adsorption of MO by our activated carbon was studied. The values of ΔH^0 and ΔS^0 can be calculated from the slope and the intercept of the plot of ln K_L versus 1/T, respectively, when the values of ΔG^0 were calculated using the thermodynamic equations (Table 3), that indicates a physisorption type of the MO sorption. The negative values of overall free energy changes during the adsorption process indicate the spontaneous nature of the adsorption process. The positive value of ΔS^0 (72.74 J/mol.K) resulted from the increased randomness due to the adsorption of MO, similar thermodynamic behavior was obtained by Subbaiah and Kim, (2016) **[23]** using aminated pumpkin seed powder.

Table 3. Thermodynamic parameters of adsorption of MO onto *Ziziphus lotus* adsorbent

T (°K)	1/T	Ln K_d	R^2	ΔS° (J/mol.K)	ΔH° (j/mol)	ΔG° (J/mol)
333	0.0030	2.788				-7721.15
313	0.0031	2.390	0.991	72.74	16489.15	-6221.03
293	0.00347	1.930				-4703.41

Conclusion

The present work shows that a new activated carbon can be prepared from *Ziziphus* leaves through phosphoric acid activation. *Ziziphus* leaves were selected as a suitable agriculture product thanks to its abundance in Morocco and its disposal in the environment without any treatment. Several parameters were studied to deduce their effect on Methyl Orange adsorption. Adsorption of Methyl orange onto our activated carbon increases with decreasing temperature, with decreasing pH and with increasing concentration, as expected from the law of mass action, in the range studied. The results show that the pH =2, adsorbent dose of 60mg and very low contact time of 10min are sufficient to eliminate 97% of MO.

The adsorption isotherm was well fitted with Langmuir model and show that the adsorption is the monolayer type, the adsorption rate process is best described by the Lagergren pseudo-second-order rate model.

The kinetic study shows that the adsorption follows the pseudo-second-order model and the thermodynamic study revealed that the adsorption was the type spontaneous and endothermic. Finally, we can relate the *Ziziphus* lotus activated carbon efficiency to the nature and the chemical structure of the product found by BET, SEM-EDX, FT-IR and DRX analysis.

References

[1] R. Srivastava, D. C. Rupainwar. Indian J Chem Technol, **18**, 67-75(2011).
[2] V. K Gupta, A. Mittal, V. Gajbe, J. Mittal. Ind. Eng. Chem. Res, **45**, 1446–1453 (2006).
[3] C. Luo, Z. Tian, B. Yang, L. Zhang, S. Q. Yan. Chem. Eng. J, **234**, 256–265 (2013).
[4] N. A. Suciu, T. Ferrari, F. Ferrari, M. Trevisan, E. Capri. Environ. Sci. Pollut. Res Int, **19**, 1374–1383 (2011).
[5] S. J. Allen, L. Whitten, G. McKay. Dev. Chem. Eng. Mineral process. **6**,231–261 (1998).
[6] X. Wang, X. Liang, Y. Wang, Xi. Wang, M. Liu, D. Yin, S. Xia, J. Zhao, Y. Zhang. J. Desalin, **278**, 231–237(2011).
[7] C.M. Gonza, S. Roma, E. Sabio, F. Zamora, J.F Gonz. J. Appl. Surf. Sci, **252**, 6058–6063 (2006).
[8] S.M. Anisuzzaman, C.G. Joseph, W.L.W.D. Ashri, D. Krishnaiah, S.Y. Ho. Int. J. Ind. Chem, **6**, 9–21(2015).
[9] L. Bonomo, *Wastewater Treatment*, McGraw Hill Education, Italy, (2008).
[10] A. Pelaez-Cid, A. Herrera-gonz, M. Salazar-Villanueva, A. Bautista Hernandèz, J. Environ. Manage, **181**, 269-278(2016).
[11] M. Valix, W. H. Cheung, G. McKay. J. Chemosphere.**56**, 493-501(2004).
[12]A. M. M. Vargas, A. L. Cazetta, C. A. Garcia, J. C. G. Moraes, E. M. Nogami, E. Lenzi, W. F. Costa, V. C. Almeida. J. Environ. Manage, **92**, 178–184(2011).
[13] B. H. Hameed, M.I. El-khaiary, J. Hazard. Mater, **159**, 574–579(2008).
[14] P. Ricou-Hoeffer, I. Lecuyer, L. P. Cloirec. J. Water Res, **35**, 965-976 (2001).
[15] B. H Hameed, F. B. M Daud. J. Chem. Ing, **139**, 48–55(2008).
[16]M. Myglovets, O.I. Poddubnaya, O.Sevastyanova, M.E. Lindström, B. Gawdzik, M.So biesiak, M.M. Tsyba, V.I. Sapsay, D.O.Klymchuk, A.M. Puziy.Carbon, **80**,771–783(2014).
[17] J. Zheng, Q. Zhao, Z. Ye, J. Appl. Surf. Sci, **299**, 86-91(2014).
[18] R. Baccar, J. Bouzid, M. Feki, A. Montiel. J. Hazard. Mater, **162**, 1522–1529(2009).
[19] A. Omri, M. Benzina. J. Soc. Chim.Tunis. **14**, 175–183(2012).
[20] H.P Khalil, M. Jawaid, P. Firoozian, U. Rashid, J. Biobased. Mater. Bioenergy, **7**, 708-714(2013).
[21] R. Huang, Q. Liu, J. Huo, B. Yang, B. **10**, 24–32(2013).
[22] H. Chen, J. Zhao, J. Wu, G. Dai, G. J. Hazard Mater, **192**, 246-254(2011).
[23] M. V. Subbaiah, D.S. Kim. Ecotoxicol Environ Saf, **128**, 109-117(2016).

Assessment of leachates from uncontrolled landfill: Tangier case study

I. ELMAGHNOUGI*, *A.* AFILAL TRIBAK*, *M.*MAATOUK[1]

Abdelmalek Essaadi university, Science and Technical Faculty, Geology Department, Tangier, Morocco *

*Research Team of Marine Environment and Natural Risks, Faculty of Science and Technical Tangier

[1]Research Team of Geo-Information and Planning, Faculty of Science and Technical Tangier.

Abstract. Landfill site of Tangier City is non-engineered low lying open dump. It has neither bottom liner nor leachate collection and treatment system. Therefore, all the leachate generated finds its paths into the surrounding environment.

Leachate samples of landfill site were collected and analyzed to estimate its pollution potential. The analyzed samples contained a high concentration of organic and inorganic compounds, beyond the permissible limits.

Keywords: Landfill, Leachate, pollution

*Corresponding author: elmaghnougiikram@gmail.com

Introduction

Urban waste is one of the major challenges facing our societies. The combination of factors ranging from population growth to urban expansion, to the development of socio-economic and production activities, and to changes in lifestyles and consumption, leads to an increase and waste products

Landfill was originally the technique for the disposal of the most widely used waste in the world, easy to implement and relatively (Saadi et al., 2013). In developing countries, mainly uncontrolled and open pit landfills, where all types of waste (urban, industrial hospital and agricultural) are discarded in mixed. One of the major problems associated with this landfill is the production of liquid effluents rich in organic and mineral matter called leachate or commonly "discharge juice". They come from the percolation of rainwater through the deposits and the contained water in waste and their degradation (Hakkou et al., 2001).

Indeed, as soon as the deposition phase, the waste is subjected to degradation processes linked to complex biological and physico-chemical reactions. Water infiltrates and produces leachates loaded with organic or mineral substances which generate a pollution essentially of organic and inorganic type, in relation with the natural biodegradation of the confined waste. These leachates are a source of contamination of the surrounding environments if they are not collected and processed (Hakkou et al., 2001).

In Morocco, studies carried out by Hamid Chiguer et al in 2016 on the leachates from the Essouira landfill revealed high levels of biodegradable organic matter, heavy metals, minerals and suspended matter. Those of Saadi et al, 2013 on the Oujda dump and Rassam et al, 2012 on the Hoceima landfill showed similar results.

In Algeria, work carried out by Mokhtaria et al in 2007 on the Tiaret landfill showed that leachates contain many organic and metallic contaminants (Pb, Zn, Cd, Ni, Cr + 6) that exceed the accepted standards. Those of Bennama et al, in 2010 on the El Kerma dump showed similar results.

For this purpose, this study will use the physicochemical analysis methods to characterize the pollutant load of the leachates from the Tangier dump.

Study Area

Geographic location

The public dump of Tangier was put into operation in the early 1970s. It stretches over 30 ha 5 km from the city center on the road to Tetouan. The coordinates of the location are: latitude 35°44'35.32 "N and longitude 5°45'17.39" .The site belongs to the private domain of the State. It serves the following boroughs: Charf Mghogha, Souani, Tangier Médina and Beni Makada plus waste from industrial zones.

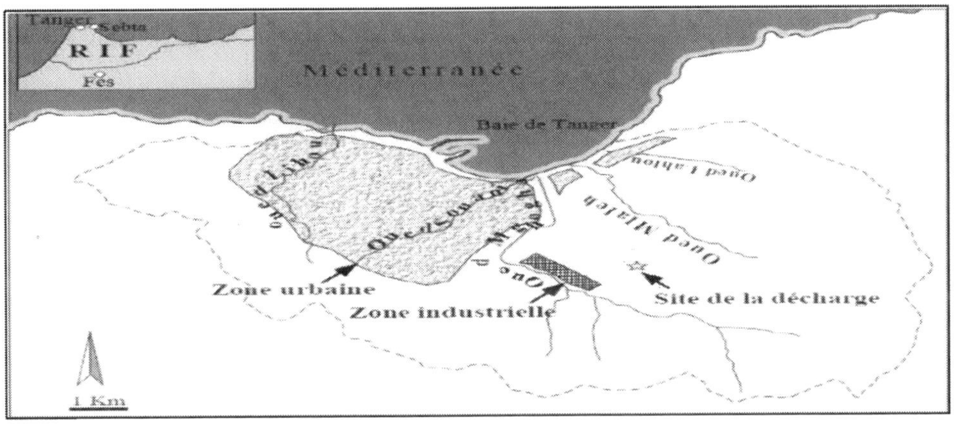

Fig.1. A map showing the location of the existing landfill of Tangier

This landfill is partially fenced and has two premises and a machine for the settlement of waste received daily.

At a distance of 100m, the landfill is limited to the south by a quarry of clay extraction and by the industrial district about 1km to the southwest. To the west of the dump, at a distance of about 200 meters, is a small residential area. To the west of the main entrance to the garbage dump is the shelters of ragpickers, which are built either of metal or wood. The perimeter of the landfill is only fenced to the south, on the other sides, the landfill is bounded by natural obstacles (valley floor to the north, cutting of a slope to the west), so it is easily accessible to the public.

Hydrographic network

The hydraulic network consists of two wadis around the landfill: Mghogha and M'laleh. The surface water and the leachate flowing over the area of the Tangier landfill flow into the Wadi Ayn At-Taleb Ahmed of Wadi Mghogha, which leads to the bay of Tangier in the immediate vicinity of the residence Sanaa Beach (Figure1).

Field geology

The unity is characterized by a vast schistous or clayey clay mass of the upper Cretaceous. The hills are surrounded by the Wadi Moghogha and Ghir Boudra valleys in the south and the wadis of Mlaleh and Khandak Bou Hajjar in the east. (GIZ Report, 2006).

Materials and Methods

Sampling of the leachates were carried out on April 2017, the samples were taken from polyethylene bottles, before the bottles were filled, the samples were washed with the sample to be taken, the bottles were filled edge and then the plugs aims to avoid any gaseous exhaust with the atmosphere. The leachate samples were stored in a cooling apparatus at 4°C during transport to the laboratory, they were analyzed immediately.

To avoid the evolution of the samples the conductivity, PH, dissolved oxygen and the temperature were measured in situ using a multi-parameter Brand HACH HQ40d. The physical parameters tested for included:

Color, Conductivity and Temperature. Chemical parameters analyzed were PH, Dissolved oxygen (DO) According to "NM ISO 10523" and ISO 5814 respectively , Nitrate, Nitrite, Chemical Oxygen Demand (COD), Ammonium Ions, Sulfures and Azote were analyzed using Quick method LCK,. HACH LANGE.

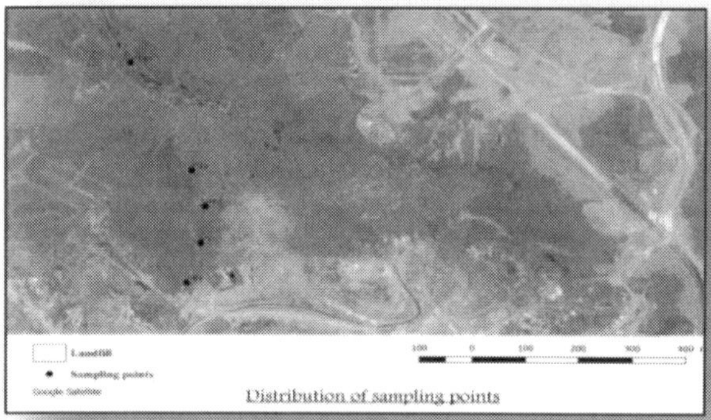

Fig.2. Distribution of sampling points

Fig.3. flow of leachate from Tangier discharge

Results and Discussion

In the landfill, through rainfall, rainwater is loaded with pathogenic germs, organic pollutants and minerals (Ozane, 1990, Martens et al., 1995).
The chemical composition of the leachates is specified at each discharge. Indeed, it varies closely with the nature and age of the landfill, the type of waste, degree of decomposition, method of landfilling, nature of landfill and climatic conditions (Navarro et al 1988, Matejka et al. al. 1994; Khattabi 2002; Aluko et al. 2003; Chofqi et al. 2004; Kurniawan et al. 2006; Renou et al. 2008).

Color and odor

The leachate from Tangier dump is blackish in color and has a fecal odor. The smell is due to the presence of organic acids from the decomposition of the highly concentrated organic matter in the waste. The dark color in landfill leachate is also associated with the high concentration of the organic substances (Aziz H.A et al., 2007).

PH

PH is an indicator of the different phases of waste degradation. During the various phases of the evolution of a landfill, the pH values obtained in the leachate could be related to the low concentration of volatile organic compounds. Indeed, during the acid fermentation phase of anaerobic decomposition of waste, young leachates are rich in volatile organic compounds. During this phase, the pH values are generally lower than 4 (Tchobanoglous et al., 1993). As they age of the landfill, leachate is depleted of volatile organic compounds. This will then cause a rise in pH = 7 or above (Kjeldsen et al., 2002). The pH varied according to the age of landfills (Kulikowska D et al., 2008).Generally, the pH of a stabilized leachate is higher than that of a young leachate (Poznyak T.G et al., 2008). In general, leachates have pH between 4.5 and 9 (Christensen T.H et al., 2001). The pH of young leachate is less than 6.5 whereas, pH of old landfill leachate has pH higher than 7.5 (Abbas A.A et al., 2009). Initial low pH is due to high concentration of volatile fatty acids

(Bohdziewicz J et al., 2008). Stabilized leachate shows fairly constant pH with little variations and it may range between 7.5 and 9.

The PH values of the leachate vary between 7,81and 8, 3 which is suitable for methanogenic bacteria. Similar results were obtained by El Khamlichi et al. (1997) on the landfill in Rabat (Morocco) and also Trankler et al. who found that leachate samples had a slightly high pH and remained in the range of 7.0–8.0 during the operations which indicates the short acidic phase and early methanogenic phase. On the other hand, Bahaa-eldin et al. found that the average value of pH was 6.7 for the municipal landfill leachate in Malaysia indicating the young leachate and the waste degradation was at its late stage of acidic phase.

Conductivity

The electrical conductivity reflects the mineralization of the sample analyzed. It assesses the overall mineralization and estimates the total water soluble salts, the values of the recorded electrical conductivity fluctuate between 20007-28002 µs / cm) with a mean value of 22072 This mean value is lower than that recorded in Malaysia discharge (31.68 lS/cm) (Bahaa-eldin et al., 2010) and also considerably lower than those recorded in the discharge of Agadir (64650µs / cm), (Commune urbaine d'agadir, 2010). The results obtained show an important mineralization of the leachate of the public dump in the city of Tangier.

Chemical Oxygen Demand COD

The COD represents the amount of oxygen consumed by the oxidizable materials chemically contained in the water. It is representative of the majority of organic compounds and oxidizable mineral salts (Makhoukh & al, 2001). In general, young landfill leachate (the acid-phase landfill, <5 years) contain large amounts of biodegradable organic matter. More than 95% of the dissolved organic carbon (DOC) consists of volatile fatty acids, and some high molecular weight compounds. In mature landfills (the methanogenic-phase landfill), the organic fraction in the leachate becomes dominated by refractory compounds, and the DOC content consists of high molecular weight compounds (Wang F et al.,2003; Wang F et al.,1997).

The high COD values obtained indicate a high organic load; The COD levels recorded in our study range from 1525-9800 mgO2 /L they are higher than those reported by (Kouadio & al,2000) in the Akouedo-Abidjan landfill (between 310 and 2495 mg / l) and by (Mekaikia & al, 2007) in the Tiaret landfill. Algeria (1048 mg / l), and also by (Commune urbaine d'agadir, 2010) in the Agadir landfill (6220 to 7640 mg / l).

On the other hand, they are much lower than those of the Mohammedia landfill (51456 mg/l), and also in Oujda dump (68036mg / l), and 20468 mg O2 / L at the discharge of Fes (Chtioui H et al., 2008). This difference could be related to the age, nature and quantity of the waste, as well as the various climatic factors such as rainfall, air humidity and temperature. Indeed, according to (Christensen et al. 2001), these various factors underlie the variability of pollutant loads.

Nitrate

Nitrates are the final stage of oxidation of nitrogen and represent the form of nitrogen with the highest degree of oxidation in water. The levels recorded ranged from 32,7to 73,1mg / L with an average grade of 53.14 mg / L. This value is considerably higher than the maximum concentration allowed by WHO (1980) which is 50 mg / L. Nitrites are derived either from nitrate reduction or incomplete oxidation of ammonium ions (Rodier, 1996).

The NO2 - leachate levels are low and range from 1,26to 4,24mg / L. These low levels are due to the low values of dissolved oxygen and the redox potential in the leachate.

In the leachate of the public dump in the town of Mohammedia, nitrates reached a concentration of 71 mg / L (Sabine, 2002). According to (Suabi et al., 2011), the leachate from a waste repository can reach a maximum value of 845 mg / L. High concentrations of nitrates in rivers lead to eutrophication, resulting in a significant development of microorganisms and algae, disrupting natural ecological equilibria (El Kharmouz et al., 2013).

Ions Amonium

In fact, due to the reducing conditions of the anaerobic reactor, only the reduced forms of nitrogen are encountered very high concentrations of ammonia nitrogen which is one of the delicate problems of leachate treatment.

Observed ammonia concentrations ranged from 875 to 1055 mg/l with a mean value of 990 mg/l. At this concentration the methanogenic is only slightly inhibited by ammonia, this chemical is the main reducing agent in landfill leachate and is a long-term pollutant (Christensen et al., 2001) lower mean values of ammonia concentrations (600 mg/l) than those reported in the present study were obtained by Hassan and Ramadan .

Total Kjeldahl Nitrogen and phosphate

In the present study, it is expected that the mean values of total Kjeldahl nitrogen (1060, 8 mg/l) and phosphates (19.19 mg/l) decrease during the stabilization process as found by Hassan and Ramadan (mean values of 973 mg/l for total nitrogen and 0.33 mg/l for total phosphate). This may be attributed to the compaction of the wastes in the landfill. In mature leachate ammonia-NH3/total Kjeldahl nitrogen ratio is usually greater than 70%. Which is the case in our study the ratio is about 93%.

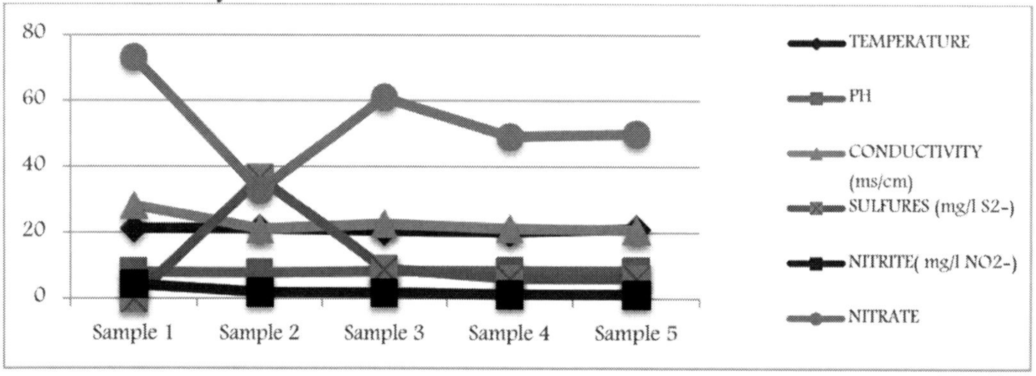

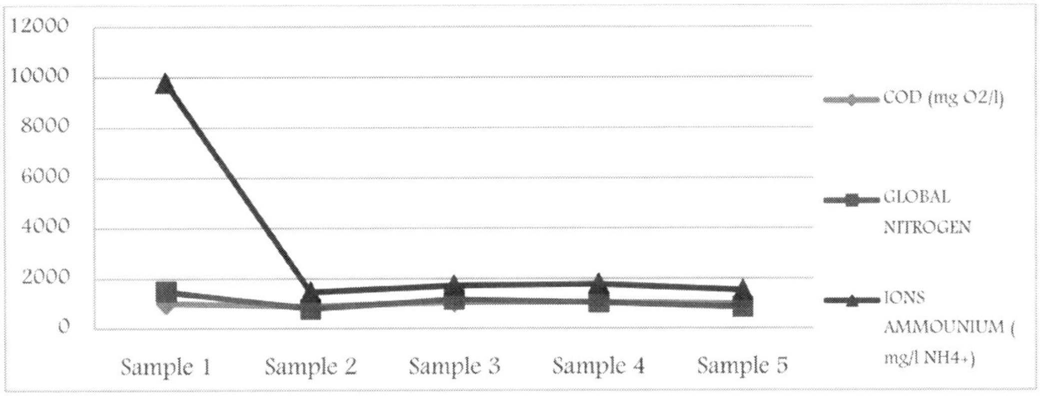

Fig.4. Results of physico-chemical analyzes

Table1. Physicochemical characteristics of leachate from various discharges in the world

Parameters	Unit	OUJDA (KHARMOUZ, 2013)	AGADIR (Y. JIROUI, 2014)	KERMA ALGERIE (TAHAR BENNAMA, 2011)	HOCEIMA (A. RASSAM, 2012)	AKOUEDO (ABIDJAN-CÔTE D'IVOIRE)(A.E. KOUASSI ET AL.2014)	RABAT (EL KHAMLICHI ET AL, 2007)	OUJDA (SAADI ET AL, 2013)	SOUIRA (HAMID CHIGUERI, 2016)	FES (F. KHALIL ET AL ,2015)	MOHAMEDIA (ABDELAZIZ MADINZI, 2014)	AGADIR (ELKADI ET AL 2015)	TANGER (ELKADI ET AL 2015)	ETUEFFONT 2009
PH	-	-	4.3	8.19	6	7,68	7,8	5,59-6,08	8,44	7,66	8.27	7,6-7,9	6,8-7,6	7.69
Temperature		17,7	19,2-20,8	23,7	-	29,8	-	14,9-15,7	-	-	-	20,9	27,6	18,3
conductivity	US/cm	170	14,6	120.3	75	78,05	23,13-74,6	4,8-3,7	39,9	-	-	58,65	33,23	30.06
COD	mg/l	-	72000	19333	76450	1306,25	260-50112	68036-89472	10361	5400	2473,9	38496	16810	355
BDO5	mg/l	13520	44000	3301	38200	588,08	5-10500	38188,7	632	1700	400,8	25395	2000-30000	36,7
Nitrite .	mg/l		0,125	43	-	0,69	-	-	-	-		-	-	9,58
Nitrate	mg/l	50	62	0,93	-	92,92	-	-	-	-	173,2	-	-	126
Phosphates	mg/l	-	188,4	1,84	-	128,74	-	-	-	-	-	-	-	
NH4	mg/l	-	0,1	-	-	388,4	-	-	-	-	723	-	-	65,4
Sulfates	mg/l	-	-	1717	-	-	-	-	0.73	-	156,04	-		260

All the results show that besides methanogenesis, is active in the discharge and that this discharge is an old discharge. The basic pH character of these leachates reflects this phase of biological evolution. The results of leachate analyzes have also shown that percolates are very poor in nitrite. On the other hand, the concentrations of COD, ammonia nitrogen and organic nitrogen (NTK) are relatively high. We can conclude that the leachate from the Tangier landfill is an old leachate, stabilized and characterized by low and difficult biodegradability and a very complex organic load.

Conclusion

The characterization of the leachates generated by the uncontrolled discharge and open pit of Tangier has shown that it is old and stabilized leachate, conveying a large mineral, organic, metallic pollutant load. These leachates with a high polluting charge risk contaminating the water table under medium permeable bedrock.

References

- Abbas A.A., Jingsong G., Ping. L. Z., Ya. P. Y., Al-Rekabi. W. S., J. Appl. *Review on LandWll Leachate Treatments* .Sci. Res. 5(5), 534-545.2009
- Aluko OO, Sridhar MKC, Oluwande PA. *Characterization of leachates from a municipal solid waste landfi ll site in Ibadan*, Nigeria. J. Environ. Health Res. 2(1):32–37. 2003
- A. Rassam, B. Bourkhiss, A. Chaouch et M. Bourkhiss. Caractérisation des lixiviats et évaluation de la pollution de la décharge de la ville d´al Hoceima / Maroc. Science Lib Editions Mersenne : Volume 4, N ° 120602. 2012
- Aziz H.A., Alias S., Adlan M. N., Asaari F. A. H., Zahari M. S. M., *Colour removal from landfill leachate by coagulation and flocculation processes*, J. Bioresource Technology 98 pp.218-220. 2007
- Bahaa-eldin EAR, Yusoff I, Samsudin AR, Yaacob WZW, Rafek AGM. *Deterioration of groundwater quality in the vicinity of an active open-tipping site in West Malaysia.* Hydrogeol J 2010; 18:997–1006.
- Bohdziewicz J., Kwarciak A., *Landfill leachate treatment by means of anaerobic membrane* Desalination. 222(1-3), (2008)
- Chofqi A, Younsi A, Lhadi E, Mania J, Mudry J, Veron A. *Environemental impact of an urban landfi ll on a coastal aquifer (El Jadida, Morocco).* J. Afr. Earth Sci. 39:509–516. 2004
- Chtioui H., Khalil F., Souabi S., Aboulhassan M. : *Evaluation de la pollution générée par les lixiviats de la décharge publique de la ville de Fés.* 2008
- Christensen T.H., Kjeldsen P., Bjerg P.L., Jensen D.L., Christensen J.B., Baum A., Albrechtsen H.J., Heron G. *Biogéochemistry of landfill leachate plumes*, Appl. Geochem., Vol 16, 659- 718. 2001
- COMMUNE URBAINE D'AGADIR ET LABORATOIRE INTERFACE MATERIAUX ENVIRONNEMENT (LIME) DE La Faculté Des Sciences Ain Chock - Casablanca, Maroc, Bulletin d'analyse de lixiviat a la décharge contrôlée grand Agadir – Maroc. 16-10-2010. 5p.
- El Khamlichi M.A., Lakrabni S., Kabbaj M., Jarby E., Kouhen M. *Etude d'impact de la décharge d'Akrach (Rabat, Maroc) sur la qualité des ressources en eau.* Revue Marocaine Civil, n°68, 17-31. 1997
- EL Kharmouz M, Sbaa M, Chafi A, Saadi S. *L'étude de l'impact des lixiviats de l'ancienne décharge publique de la ville d'Oujda (Maroc oriental) sur la qualité physico-chimique des eaux souterraines et superficielles.* Larhyss Journal, 16 : 105-119. 2013
- Elkadi Ahmed, Maatouk Mustapha, Tarik Chafik, Raissouni Mohammed. *Physico-chemical characterization of landfill leachates: a comparison case between Tangier and Agadir (Morocco).* J. Mater. Environ. Sci. 6 (11). 2016.
- Emilien Belle. *Evolution de l'impact environnemental de lixiviats d'ordures m´ménagères sur les ´ eaux superficielles et souterraines, approche hydrobiologies et hydrogéologique. Site d''étude : d´décharge d'Etueffont (Territoire de Belfort – France).* Hydrologie. Université de Franche- ´ Comté, Français. 2008
- F. Khalil, O. Bouaouine, H. Chtiouil, S. Souabi, M.A. Aboulhassan, A. Ouammou. *Traitement des lixiviats de décharge par coagulation-floculation (Treatment of Landfill Leachate by coagulation-flocculation).* J. Mater. Environ. Sci. 6 (5). 2015
- HAKKOU R. *La décharge publique de Marrakech: caractérisation des lixiviats, étude de leur impact sur les ressources en eau et essai de leur traitement.* Thèse de doctorat d'état, université de CAPI AYYAD, Maroc, 150 p. 2001
- Hamid Chiguer, Fatiha E Khayyat, Omar El Rhaouat, Redouane Rifki, Adnane Bensaid, Khadija El Kharrim,And Driss Belghyti. *Evaluation of the pollution load of leachates*

*from the landfill in the city Essaouira(MOROCCO).*International Journal of Innovation and Applied Studies ISSN 2028-9324 Vol. 14. 2016

- Hassan AH, Ramadan MH. *Assessment of sanitary landfill leachate characterizations and its impacts on groundwater at Alexandria. J Egypt Public Health* Assoc;80:27–49. 2005
- Khattabi H., *Intérêts de l'étude des paramètres hydrogéologiques et hydrobiologiques pour la compréhension du fonctionnement de la station de traitement des lixiviats de la décharge d'ordures ménagères d'Etueffont (Belfort, France).* Thèse de Doctorat, Université Franche Comté, France, 171p. 2002.
- KJELDSEN P., BARLAZ M.A., ROOKER A.P., BAUM A., LEDIN A., CHRISTENSEN T.A. *Present and long-term composition of MSW landfill leachate*, Crit. Rev., Environ. Sci. Technol., Vol. 32, n°4, 297-336. 2002.
- Kouassi A.E., Ahoussi K.E., Koffi Y.B., Kouame I.K.,Soro N., Biemi J. *Caractérisation Physico-Chimique Du Lixiviat D'une Decharge De L'afrique De L'ouest: Cas De La Décharge D'AKOUEDO (ABIDJAN-CÔTE D'IVOIRE).*Larhyss Journal, ISSN 1112-3680, n°19. 2014
- Kulikowska D., Klimiuk E., *The effect of landfill age on municipal leachate composition*. Bioresour Technol. 99, 5981-5985. 2008
- Kurniawan TA, Lo W, Chan GYS. *Physico-chemical treatments for removal of recalcitrant contaminants from landfi ll leachate.* J. Hazard. Mater. B129:80–100. 2006
- MAKHOUKH, M., SBAA, M., BERRAHOU, A., VAN CLOUSTER, M. *Contribution à l'étude physico-chimique des eaux superficieles de l'Oued Moulouya (Maroc oriental),* Larhys journal, ISN 112-3680, n° 09. pp.149-169. 2001
- Martens P., Bichelle B., Watel M. Wollast A. *Audit technique de caractérisation d'une ancienne décharge: composition d'un lixiviat particulier.* Techniques, sciences et méthodes l'Eau n°6. 1995
- Matejka P., *Pollution engendrée par un lixiviat de décharge d'ordures ménagères : bilan hydrique et caractérisation.* Environmental Technology, vol.15: 313□322. 1994
- MEKAIKIA, M, M; BELABBED, B, E; DJABRI, L; HANI, A; LAOUR, R. *Caractéristiques de la décharge publique de la ville de Tiaret et son impact sur la qualité des eaux souterraine*. Courrier du Savoir – N°08, Juin 2007. pp.93-99. 2007
- Navarro A., Bernard D., Millot N., *Les problèmes de pollution par les lixiviats de décharge. Techniques, Sciences et Méthodes-L'Eau* 3: 541–545. 1988
- Ozzane, F., *Les lixiviats de decharge, les points de connaissance*. TSM-Eau, juin 1990. pp.289-312.
- Poznyak T.G., Bautista L., Chairez I.R., Cordova. I., *Decomposition of toxic pollutants in landfill leachate by ozone after coagulation treatment* .Rios. E., J Hazard Mater. 152, 1108-1114. 2008
- Rapport final de l'étude de faisabilité relatif à l'amélioration de la gestion de la décharge publique de Tanger. Coopération Technique Allemande Programme de Gestion et de Protection de l'Environnement (PGPE).2006
- Renou S, Poulain S, Gagnaire J, Cadarache D, Marrot B, Moulin P. *Lixiviat de centre de stockage : déchet généré par des déchets. L'eau, l'industrie, les nuisances*, 310:37–43. 2008
- Rodier J. *L'Analyse de l'Eau: Eaux Naturelles, Eaux Résiduaires, Eau de Mer : Physico-chimie, Bactériologie et Biologie (edn).* The Dunod : Paris. 1996
- Saadi S., Sbaa M., El Kharmouz M. *Caractérisation physicochimique de lixiviats du centre d'enfouissement technique de la ville d'Oujda (Maroc oriental)*, Science Lib., Editions Mersenne, Vol. 5, n° 130517, ISSN 2111-4706, 1-12. 2013

- Sabine G., *Risques sanitaires liés aux fuites de lixiviats des centres de stockage de déchets ménagers et assimilés*. Mémoire de fin d'études, Formation des ingénieurs des génies sanitaires, 19. 2002
- Souabi S., Touzare K., Digua K., Chtioui H., Khalil. F., et Tahiri M. *Triage et valorisation des déchets solides à la décharge publique de la ville de Mohammedia. Les technologies de laboratoire*, 6 121-130. 2011
- Tahar Bennama, Abdelkader Younsi, Zoubir Derriche et Abdelkader Debab. *Characterization and physico-chemical treatment of El-Kerma (Algeria) landfi ll leachates by batch adsorption on untreated and chemically activated sawdust*. Water Qual. Res. J. Can. · Volume 45, No. 1, 81–90.2010
- TCHOBANOGLOUS G., THEISEN H.,VIGIL S. *Integrated Solid Waste Management*, McGraw-Hill International Edition, New York, 978 p.1993
- Trankler J, Visvanathan C, Kuruparan P, Tubtimthai O. *Influence of tropical seasonal variations on landfill leachate characteristics—results from lysimeter studies.* Waste Manage; 25:1013–20. 2005
- Wang F., Smith D.W., El-Din M.G., *Application of advanced oxidation methods for landfill leachate treatment-a review* .J. Environ. Eng. Sci. 2) 413-427. 2003
- Welander U., Henryson T., Welander T., *Nitrification of landfill leachate using suspended-carrier biofilm technology*.Water. RES. 31 2351-2355. 1997
- Y. JIROU, M.C. HARROUNI, M. BELATTAR, M. FATMI1 & S. DAOUD. *Traitement des lixiviats de la décharge contrôlée du Grand Agadir par aération intensive. Rev. Mar. Sci. Agron. Vét. 2 (2):59-69.*2001

Characterization of pyrolysis products derived from three biological wastes and their effect on plant growth and soil water retention

Laila Bouqbis[1], *Hans Werner* Koyro [2], *Claudia* Kammann[3], *Lalla Fatima Zohra* Ainlhout[4] *Laila* Boukhalef [5] and *Moulay Cherif* Harrouni[6]

(1, 4, 5 Polydisciplinary Faculty, Ibn Zohr University, Taroudant, Morocco)
(2, 3 Department of Plant Ecology, Justus-Liebig-University, Heinrich-Buff-Ring26-32, 35392, Giessen, Germany)
(6) Hassan II Agronomic and Veterinary Institute, Agadir, Morocco

Summary. Over two-thirds of Morocco can be classified as semiarid, arid and desert with low and variable rainfalls. While the country is subject to frequent drought, groundwater resources are predominantly consumed by irrigated agriculture leading to the depletion of water resources and degradation of soil quality. Application of bio-resources wastes to soils after pyrolysis process is well documented to help retain water and nutrients in soils. In this study, three bio-resources wastes derived from argan shells, wood chip, a blend of paper sludge and wheat husks are characterized for physical and chemical properties. To determine the potential impact of salt stress and toxic substances the second part of this study focused on the effect these bio-resources wastes have on germination of salad and barley respectively. The three bio-resources obtained from different biomass showed some unique properties compared to the soil, such as high electrical conductivity (EC), high content of K, Na and Mg, low content of heavy metals. Moreover, the water holding capacities increased with increasing application of bio-resources wastes. Concerning the phytotoxic tests, no negative effect was observed neither for salad (Lactuca sativa L.) nor for barley (Hordeum vulgare) indicating that the three bio-resources could be safely used for agriculture. Collectively, the use of these bio-resources wastes as a soil amendment is anticipated to increase both water and nutrient and could provide the potential for a better plant growth mainly in semiarid, arid and desert climatic conditions like the case of Morocco in which the agricultural practices reserve a majority of the water resources to be used for irrigation.

Key words: bio-resources wastes, water retention, phytotoxic tests, plant growth.

*Corresponding author : bouqbisl@yahoo.fr

1. Introduction

In arid and semi-arid regions, availability of fresh water is one of the important factors that affects biological systems worldwide (Koyro et al., 2011; Gallardo et al., 2015). This is especially of importance where low rainfall imposes constraints on plant growth and development (Schlesinger et al., 2015, Abideen et al., 2017). The social demand on agriculture is not only to produce high-quality food for a growing world population but farming also has to deal with careful use of resources and the mitigation of climate change (von Glisczynski et al., 2016). The use of biochar (charcoal produced from biomass by pyrolysis in a low-oxygen environment for use as a soil amendment or carbon sequestration) as a soil amendment could help to achieve all these requirements. Biochar is attracting recently growing interest for its potential use as a carbon sequestration agent and as a soil amendment for improved agricultural productivity (Lehmann et al., 2006; Shackley et al., 2013; Sorrenti et Toselli., 2016; Forján et al., 2017). The ability of biochar to retain soil water is a function of the combination of its porosity and surface functionality (Suliman et al., 2017). Previous studies showed that soil properties such as water holding capacity (WHC), cation exchange capacity (CEC) and nutrient retention application of biochar can be affected positively by biochar treatment (Liu et al., 2012; Ventura et al., 2013, Forján et al., 2017).

However, the use of biochar is not without its critics. Research shows that biochar can contain dangerous inorganic contaminants and organic ones as well as dioxins and furans (Hale et al., 2012; Oleszczuk et al., 2013; Buss et Mazek., 2014; Kołtowski and Oleszczuk., 2015; Domene et al., 2015). In the case of high levels of contaminants there is a risk of their uptake by plants or migration down the soil profile to ground-waters. This may have negative effects for humans, for the environment and for living organisms. Thus, biochar applied to soils should be free of toxic substances before any future large-scale application.

In this study, we determined the chemical and physical properties of two biochars obtained from commercial vendors in Germany from different wastes: Biochar I (BC I) derived from wood chip and biochar II (BC II) derived from a blend of paper sludge and wheat husks and their properties are compared to those of biochar produced in Morocco from argan shells (ASBC). Therefore, we investigated the effects these three biochars have on (i) germination of salad (test for salt stress) and (ii) barley (test for toxic substances) and the findings are compared to those obtained from argan shells biochar (test procedures are adapted from Busch et al., 2012).

2. Materials and Methods

2.1. Soil, Peat and Biochar analysis

All physical and chemical analysis were performed in the Hassan II Agronomic and Veterinary Institute (IAV) soil science laboratory. The pH and electrical conductivity (EC) were both measured in water extracts with standard electrodes. For chemical analysis, the total concentration of Na, K, Ca and Mg were determined by flame emission spectrophotometer and Fe, Mn, Zn and Cu by atomic absorption spectrophotometer and colorimetrically for NaNO3 and KH2PO4.

2.2. Water holding capacity

Six biochar-fine sand mixtures have been prepared. We have used small amounts of biochar that equal 0, 0.5, 1, 2, 4 and 8%, which means that 1% describes a mixture of 1 g dry biochar with 99 g dry sand. Around 40 g of fresh weight of each mixture were filled in small gray PE-tubes (height: 5,5 cm; inner diameter: 3,6 cm) that were closed at the bottom with gaze and filter paper. The filled tubes were placed in a plastic box that was then filled with tap water. The box was covered with aluminum foil. After 24 h the tubes were placed on test tube racks to let water drip out. 24 h later, the weight of the tubes were determined and the water content was calculated (knowing in advance the dry weight of the substrate). Then, the maximum water holding capacity can be determined.

2.3. Salad germination test

The biochar was mixed with an inert, fine-sand medium and used in the salad germination test based on ISO 17126 (Busch et al., 2012), where the germination of a salt-sensitive species (Lactuca sativa L.) was evaluated. Several different proportions of biochar and fine sand were mixed following geometrical dilution. In accordance with the ISO test, the factor 2 is the maximal allowed, so the series used was: 0% (control), 0.5%, 1%, 2%, 4% and 8%. For one replicate, an amount of fresh weight equal to 100 g of dry weight was taken out of the whole mixture and filled into a Petri dish. Tap water was added to set the water content to 85% of the maximum WHC. Forty seeds of Lactuca sativa L. were evenly sowed, leaving a free space of around 1 cm to the border of the Petri dishes. The seeds were pressed softly into the substrate. Afterward, 90 g (dry weight) of coarse sand was distributed on top. The prepared Petri dishes were placed open in a bloated, zippered plastic bag and positioned in the green house. For the first 48 h, black plastic foil covered the dishes. After 5 d incubation, harvesting commenced and the number of germinated seedlings and the fresh and dry weights of the above ground biomass were determined. The Petri dishes, which still contained the soil mixture were mixed and 25 g of dry weight were taken. Around 100 ml of water were added. After shaking for 1 h at 150 rpm, the substrate was allowed to settle for 30 min and then the pH and electrical conductivity (EC) were measured. All determinations were replicated three times.

2.4. Barley germination test

Five biochar-peat mixtures were prepared, using small amounts of biochar (0%, 1%, 2.5%, 5% and 10%) where 1% describes a mixture of 1 g dry biochar with 99 g dry peat. Then, the dry weight and water holding capacity (WHC) of the substrates were determined. To prepare 1000 ml of dry weight for every mixture, the amount of fresh weight of biochar and peat and the amount of water needed to set the mixture to 60% of the maximum water holding capacity (WHCmax) were calculated. Afterward, some textile/filter paper was placed into the bottom of the plant pots and the mixtures were split into four replicates while leaving sufficient to cover the seeds. After sowing 20 seeds of barely in every pot, the rest of the mixture was distributed over the seeds and the initial weight of the whole construction (pot + mixture with 60% of WHCmax + seeds) was recorded. The plots were then randomly placed in the greenhouse. The weight of each pot was recorded daily and the difference from the initial weight was made up by adding tap water. After 9 d, the germination rate, biomass fresh weight and dry weight were determined.

2.5. Statistical analysis

For the barley and salad germination tests, the effects of different biochar additions on all replicated measurements were tested using one-way analysis of variance. The significance of differences among treatment groups was determined using the Tukey test. A result was considered significant at $p < 0.05$. All statistical tests were performed using the SigmaPlot 11 software (Systat Inc.; Chicago, IL, USA).

3. Results and discussion

3.1. Soil, Peat and Biochar properties

Several studies demonstrated that the environmental function of biochar in soil vary depending on the physical and chemical characteristics of the biochar which depend on feedstock used in the production of biochar and production conditions (Smider and Singh,

2014; Sun et al., 2014; Nansubuga et al., 2015). The physical and chemical properties of three biochars compared to other media are presented in table 1.

Table 1. Properties of different biochars used in this experiment

	ASBC	BC I	BC II	Sand	Peat
pH	10,7	9,8	8,25	7,93	5,81
EC (milliS cm^{-1})	4,83	1,25	1,33	0,48	0,94
K (ppm)	1906,25	521,73	421,87	7,11	78,26
Na (ppm)	339,2	46,54	242,85	3,47	257,94
Ca (ppm)	4,8	0,61	0,6	34,09	190,9
Mg (ppm)	1258,92	1303,73	1299,1	222,89	1191,58
Fe (ppm)	0	0	0	9,07	0
Mn (ppm)	0,04	0,083	0,02	7,7	0,14
Cu (ppm)	0	0	0,062	1,3	0,06
Zn (ppm)	0,04	0,0042	0,04	0,42	0,04
$NaNO_3$ (ppm)	100	100	2666,66	100	133,33
KH_2PO_4 (ppm)	0,33	0,46	1,93	1	2,33

The pH ranged from 5,81 to 10,7 and was higher in argan shells biochar. Depending on the pyrolysis conditions and the nature of the feedstock, different pH values ranging in pH from 4 to 13 can be reached (Chan et al., 2008; Sigua et al., 2016). In this sense, BC I, BC II and argan shells biochar resulted in the same pH range. The alkaline nature of all biochars has been reported to be useful to increase the pH of acidic soils.
The EC value was lower for sandy soil than for the other media, and was higher for argan shells biochar. Previous research reported that added biochar with high EC value into soil with low EC value (indicating its low salinity) increased EC of soil (Chan et al., 2008; Sigua et al., 2016). Thus, we would expect that the EC of the soil (0,48mS/cm) would increase with biochar application.

Mineral composition differed among the substrates. The cation concentrations except Ca were the highest with different biochars and the lowest with sandy soil while Ca concentration was the highest with argan shells biochar and the lowest with biochar I and biochar II. The Mg content was relatively higher in the three biochars than Ca and Na. It is reported that the chemical constituents of biochar are directly affected by different parameters: the temperature, the time and the heating rate during pyrolysis (Lima and Marshall, 2005) which is in line with our result. Furthermore, ASBC, BC I and soil showed the same value of NaNO3; whereas BC II had the higher value (Table 1). For KH2PO4, BC II and peat showed the higher value. Biochar, as reported by Coates, 2000, contain little or no NO3−, which is an essential nutrient for plant growth while the PO43- was characteristic for each feedstock (Uras et al., 2012). Given the high nutrient content of the three biochars, a significant positive effect on plant growth especially in the sandy soil was expected. However, previous studies showed that biochar can contain dangerous inorganic contaminants and organic ones (Hale et al., 2012; Oleszczuk et al., 2013; Buss et Mazek., 2014; Kołtowski and Oleszczuk., 2015). Heavy metal contents were characteristic for each feedstock, but generally increased after pyrolysis and with pyrolysis temperature. This was not the case in our study where heavy metal contents were higher with soil and lower with biochars. The three biochars used in this study contain more major nutrients and less heavy metals than the soil. These properties of biochars make them more attractive as a soil amendment agent.

3.2. Water holding capacities

For argan shells biochar, the water holding capacities were 0.36, 0.39 and 0.43 gH2O g−1 soil (dry weight) in 0%, 0.5% and 8% biochar-sand mixtures, respectively. The biochar application increased the WHC by 7,69% and 16,28 % compared to the control. For BC I, the water holding capacities were 0.30, 0.33 and 0.51 gH2O g−1 soil (dry weight) in 0%, 0.5% and 8% biochar-sand mixtures, respectively. The biochar application increased the WHC by 10% and 70% compared to the control. For BC II, the water holding capacities were 0.30, 0.32 and 0.42 gH2O g−1 soil (dry weight) in 0%, 0.5% and 8% biochar-sand mixtures, respectively. The biochar application increased the WHC by 9% and 43 % compared to the control. The biochar application increased the WHC by 70%, 43% and 16,28 for BC I, BC II and argan shells biochar, respectively indicating that the most important value was observed for BC I. An increase in water holding capacity has also been reported for wet conditions with biochar application (Abel et al., 2013).

3.3. Salad Germination Test

Our results, as shown in Fig. 1A, revealed a positive effect of ASBC, BC I and BC II on germination rate of salad in all of the biochar treatments except for the mixture with 4% of biochar I in which a non-significant decrease was observed. In other words, with the highest application, biochar increases the germination rate by 50%, 21% and 17% for ASBC, BC II and BC I respectively indicating that the most important value was observed for ASBC. A positive effect was also observed for the more sensitive parameter to biotoxic substances in biochar: fresh weight of seedlings (Fig. 1B and Fig. 1C). Even when the biochar contributed 8% of the volume of the germination mixture, no negative effect was observed. In other words, the biochar application increases the fresh weight per Petri dish (Fig. 1B) by 145%, 98% and 76% for BC II, BC I and ASBC respectively indicating that the most important value was observed for BC II. Moreover, the biochar application increases the fresh weight per plant (Fig. 1C) by 101%, 70% and 40% for BC II, BC I and ASBC respectively indicating that the most important value was observed for BC II. Collectively, our results indicate that biochar feedstock materials vary in their characteristics (e.g., pH, EC, nutrient levels) which also influence application rate and germination results. Thus before any large scale application, it is necessary to determine which biochar materials are best suited for application and at which rates to specific soils.

https://doi.org/10.1051/e3sconf/20183702004

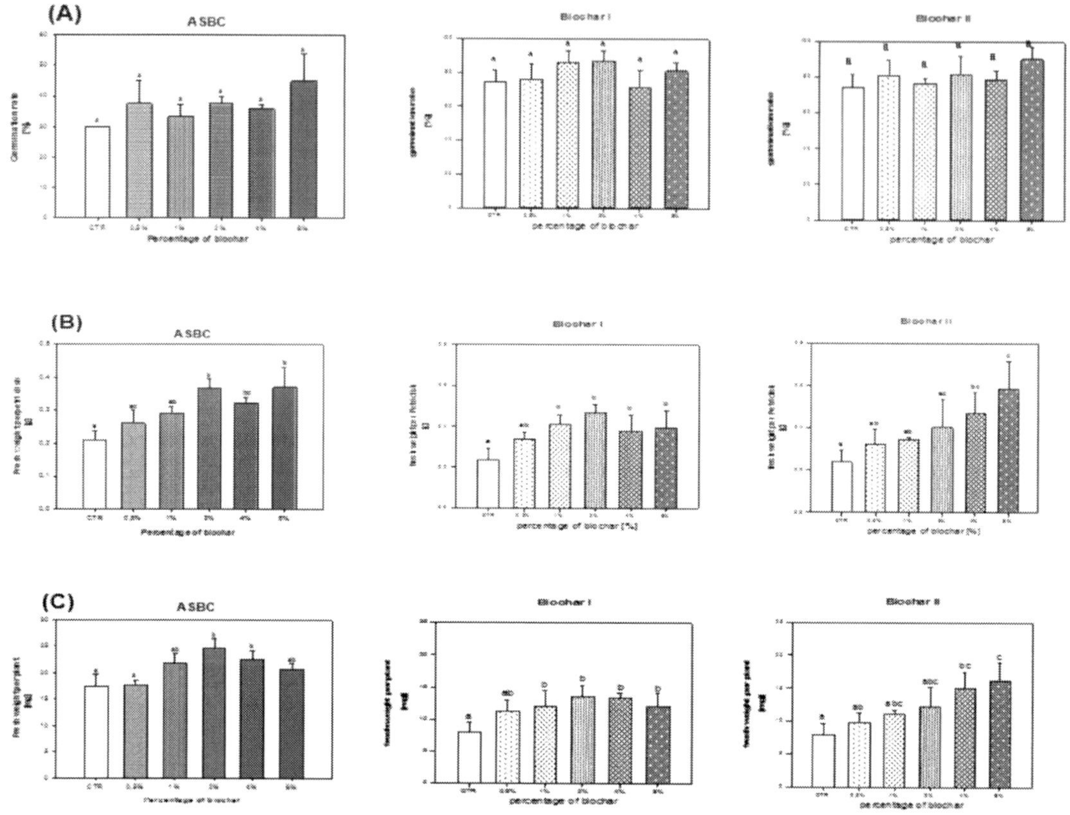

Fig.1. Salad germination test results: (A): germination rate, (B): fresh weight per Petri dish, (C): fresh weight per plant of three repeated test runs. The graphs present mean values and error bars present the standard deviation of the mean (n=3). Different letters indicate significant differences (one-way analysis of variance, ANOVA) between different mixtures with biochar and the control

3.4. Barley Germination Test

In all biochar treatments (1, 2.5, 5% and 10% biochar-peat mixture), no negative effect of the three biochars on germination rate of barley has been revealed (Fig. 2A). The same result was observed for the more sensitive parameters to biotoxic substances in biochar: fresh and dry weight of seedlings (Fig. 2B and Fig. 2C). In other words, biochar application increases the fresh weight by 14%, 7% and 7% for ASBC, BC I and BC II respectively indicating that the most important value was observed for argan shells biochar. Furthermore, biochar application increases the dry weight by 13%, 6% and 3% for ASBC, BC I and BC II respectively indicating that the most important value was observed for argan shells biochar. The low increases of biomass with biochar application could be due to much shorter test (9 days). Regarding the chemical properties of the ASBC, BC I and BC II compared to the peat (high content of K, Na, and Mg, low content of heavy metals as presented in Table 1), the observed increases, even low, in fresh and dry weight of biomass could be the results of high level of water soluble nutrients in mixture with biochar application. Thus the three biochars could provide the potential for a better plant growth.

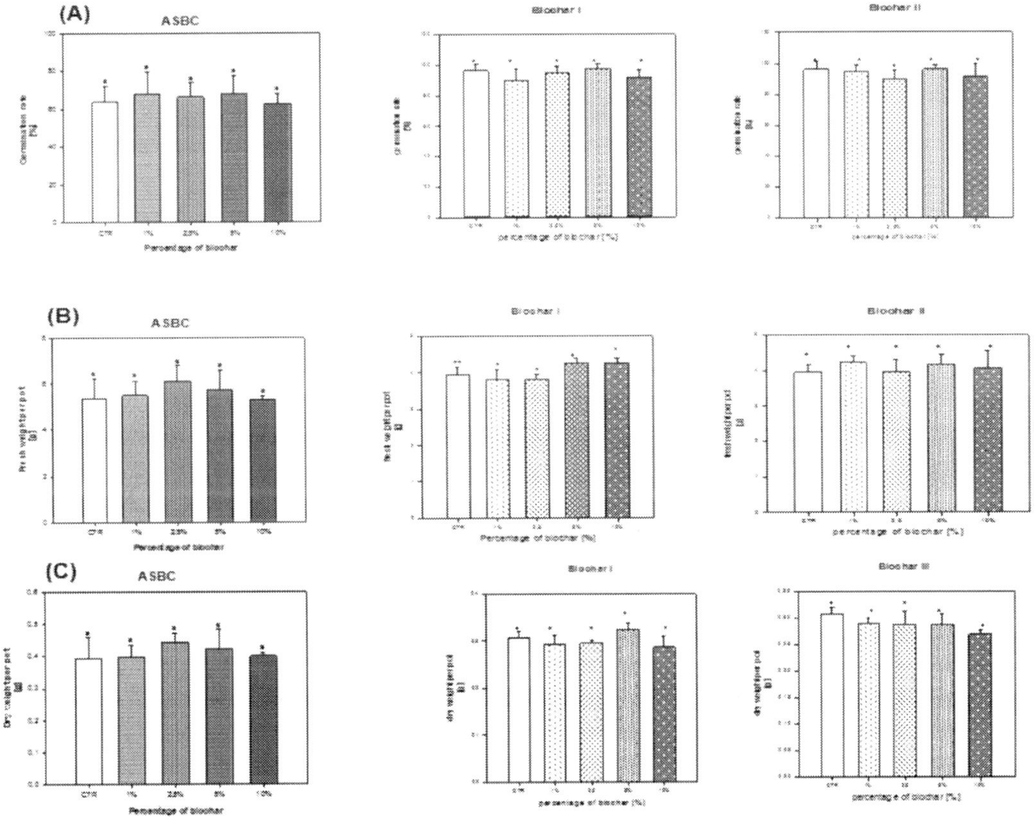

Fig.2. Effect of three biochars on germination of barley. (A): germination rate, (B): fresh weight per plant and (C): dry weight per plant for four repeated test runs. The graphs present mean values and error bars present the standard deviation of the mean (n=4). Different letters indicate significant differences (one-way analysis of variance, ANOVA) between different mixtures with biochar and the control

4. Conclusion

The major reason of degradation in existing agricultural land is due to various abiotic factors including drought. The use of organic amendments like biochar has the advantage to increase soil organic carbon, soil aggregation and water retention capacity. In this study, the three biochars obtained from different biomass showed some unique properties compared to the soil, such as high EC, high content of K, Na and Mg, low content of heavy metals. In addition, the results obtained from the two phytotoxicity test (test for salt stress and test for toxic substances) revealed no negative effect of the three biochars which provide a preliminary indication that they could be safely used for agriculture. When the three biochars are used in mixture with sandy soil, they increase the water holding capacities and the most important value was observed for biochar I. Thus, the use of agran shells biochar, Biochar I and biochar II as a soil amendment will be anticipated to increase both nutrient and water retention.

References

H. W. KOYRO, M. A. KHAN and H. LIETH (2011): Halophytic crops: a resource for the future to reduce the water crisis?. Em J Food Agric. 23: 1-16

B. GALLARDO, A. ZIERITZ, D. C. ALDRIDGE (2015): The importance of the human footprint in shaping the global distribution of terrestrial, freshwater and marine invaders. PloS one. 10: 1-17

W. H. SCHLESINGER, M. C. DIETZE, R. B. JACKSON, R. P. PHILLIPS, C. C. RHOADES, L. E. RUSTAD, J. M. VOSE (2015): Forest biogeochemistry in response to drought. Glob Change Biol. 22: 2318-2328

Z. ABIDEEN, H.W. KOYRO, B. HUCHZERMEYER, B. GUL AND M.A. KHAN (2017): Impact of a Biochar or a Compost-Biochar Mixture on Water relation, Nutrient uptake and Photosynthesis of Phragmites karka. Pedosphere. doi:10.1016/S1002-0160(17)60362-X

F. VON GLISCZYNSKI, A. SANDHAGE-HOFMANNA, W. AMELUNGA, R. PUDE (2016): Biochar-compost substrates do not promote growth and fruit quality of a replanted German apple orchard with fertile Haplic Luvicsol soils. Scientia Horticulturae 213: 110-114

G. SORRENTI, M. TOSELLI (2016): Soil leaching as affected by the amendment with biochar and compost. Agriculture, Ecosystems and Environment 226: 56–64

R. FORJÁN, A. RODRÍGUEZ-VILA, B. CERQUEIRA, E.F. COVELO (2017): Comparison of the effects of compost versus compost and biochar on the recovery of a mine soil by improving the nutrient content. Journal of Geochemical Exploration. 183: 46-57

W. SULIMAN, J.B. HARSH, N.I. ABU-LAIL, A.M. FORTUNA, I. DALLMEYER AND M. GARCIA-PÉREZ (2017): The role of biochar porosity and surface functionality in augmenting hydrologic properties of a sandy soil. Sci. Total Environ. 574: 139-147

J. LIU, H. SCHULZ, S. BRANDL, H. MIEHTKE, B. HUWE AND B. GLASER (2012): Short-term effect of biochar and compost on soil fertility and water status of a Dystric Cambisol in NE Germany under field conditions. J. Plant Nutr. Soil Sci. 175: 698-707

M. VENTURA, G. SORRENTI, P. PANZACCHI, E. GEORGE AND G. TONON (2013) : Biochar reduces short-term nitrate leaching from a horizon in an apple orchard. J. Environ. Qual. 42: 76-82

S.E HALE, J. LEHMANN, D. RUTHERFORD, A.R. ZIMMERMAN, R.T. BACHMANN, V. SHITUMBANUMA, A. O'TOOLE, K.L. SUNDQVIST, H.P.H. ARP AND G. CORNELISSEN, G. (2012): Quantifying the total and bioavailable polycyclic aromatic hydrocarbons and dioxins in biochars. Environ. Sci. Technol. 46 :2830-2838

P. OLESZCZUK, I. JO_SKO AND M. KU_SMIERZ (2013): Biochar properties regarding to contaminants content and ecotoxicological assessment. J. Hazard. Mater. 260: 375-382

W. BUSS AND O. MASEK (2014): Mobile organic compounds in biochar : A potential source of contamination-Phytotoxic effects on cress seed (Lepidium sativum) germination. Journal of Environmental Management 137: 111-119

M. KOŁTOWSKI AND P. OLESZCZUK (2015): Toxicity of biochars after polycyclic aromatic hydrocarbons removal by thermal treatment. Ecological Engineering 75: 79-85

X. DOMENE, A. ENDERS, K. HANLEY, J. LEHMANN (2015): Ecotoxicological characterization of biochars: Role of feedstock and pyrolysis temperature. Science of the Total Environment 512–513: 552-561

D. BUSCH, C. KAMMANN, L. GRÜNHAGE AND CHRISTOPH MÜLLER (2012): Simple Biotoxicity Tests for Evaluation of Carbonaceous Soil Additives: Establishment and Reproducibility of Four Test Procedures. Journal of Environmental Quality 41:1023-1032

B. SMIDER and B. SINGH (2014): Agronomic performance of a high ash biochar in two contrasting soils," *Agriculture Ecosystems and Environment* 191: 99-107

Y. SUN, B. GAO, Y. YAO, J. FANG, M. ZHANG, Y. ZHOU, H. CHEN, L. YANG (2014): Effects of feedstock type, production method, and pyrolysis temperature on biochar and hydrochar properties," Chemical Engineering Journal 240: 574-578

I. NANSUBUGA, N. BANADDA, F. RONSSE, W. VERSTRAETE, K. RABAEY (2015): Digestion of high rate activated sludge coupled to biochar formation for soil improvement in the tropics, " Water Research 81: 216-222

K.Y. CHAN, L. VAN ZWIETEN, I. MESZAROS, A. DOWNIE, S. JOSEPH (2008): Using poultry litter biochars as soil amendments," *Australian Journal of Soil Research* 46: 437-444

G.C. SIGUA, J.M. NOVAK, D.W. WATTS, M.G. JOHNSON, K. SPOKAS (2016): Efficacies of designer biochars in improving biomass and nutrient uptake of winter wheat grown in a hard setting subsoil layer," *Chemosphere* 142: 176-183

I.M LIMA AND W.E. MARSHALL (2005): Granular activated carbons from broiler manure: physical,chemical and adsorptive properties. Biores. Techno 96: 699-706

Ü. URAS, M. CARRIER, A. G. HARDIE, J. H. KNOETZE (2012): Physico-chemical characterization of biochars from vacuum pyrolysis of South African agricultural wastes for application as soil amendments. Journal of Analytical and Applied Pyrolysis 98: 207-213

S. ABEL, A. PETERS, S. TRINKS, H. SCHONSKY, M. FACKLAM, G. WESSOLEK (2013): Impact of biochar and hydrochar addition on water retention and water repellency of sandy soil. Geoderma 202:183-191

 https://doi.org/10.1051/e3sconf/20183702005

Kinetic studies of adsorption of Cu (II) from aqueous solution by coriander seeds (Coriandrum Sativum)

L. KADIRI*, *A.* LEBKIRI, *E.H.* RIFI, *A.* OUASS, *Y.* ESSAADAOUI, *I.* LEBKIRI, *H.* HAMAD

Laboratory of Organic Synthesis and Extraction Processes - Department of Chemistry
University Ibn Tofail, Kenitra - Faculty of Sciences, PO Box. 133 – 14000- Morocco

Abstract. The adsorption of copper ions Cu2+ by Coriandrum Sativum seeds (CSS) from aqueous solution was studied in order to highlight the importance of coriander seeds as a potential tool in the treatment of wastewaters containing heavy metals. The kinetic studies of adsorption of Cu (II) were discussed using the spectroscopic technique "Inducting Coupled Plasma" (ICP). The effects of initial copper ion concentration and contact time were determined. All results show that coriander seeds have, over their culinary and medicinal benefits, a significant adsorbent power of copper ions.
Keywords: coriander seeds, adsorption, Cu (II), ICP.

**E-mail:* kadiri.lamya@gmail.com

1. Introduction

Technological applications of heavy metals (lead, copper, zinc, cadmium...) in many industrial processes (electroplating, pigments, paper pulp, and textiles, etc.) [1], have produced contaminated wastewaters that are one of the main sources of environmental pollution [2], [3]. Heavy metals is an important issue and have serious ecological and human health consequences, mainly because the heavy metals cannot be destroyed or degraded, have a toxic effect on most life forms, and also tend to accumulate in the environment [4]. Therefore, it is necessary to find some effective methods for removal of heavy metals from industrial wastewaters [5], in order to reduce their discharge into environment [6]. Adsorption of heavy metals from aqueous solutions is a relatively new process that has been confirmed a very promising process in the removal of heavy metal contaminants. The major advantages of adsorption are its high effectiveness in reducing the heavy metal ions and the use of inexpensive adsorbents [7], [8]. Adsorption processes are particularly suitable to treat dilute heavy metal wastewater.

Different forms of plant material such as *iris pseudacorus* [9], sawdust [10], black gram husk [11], eggshell [12], seed shells [13], orange peel [14], sugar-beet pectin gels [15], coffee husks [16], barks of *eucalyptus camaldulensis* [17], sugarcane bagasse [18] etc., have been widely investigated as potential adsorbents for heavy metals.
several depollution techniques such as membrane systems, precipitation [19], ion-exchange resins, electro dialysis are still very expensive [20]. Recently, the fight against metallic pollution has stimulated the search for ecofriendly technologies.

Coriander (Coriandrum Sativum.L) is a popular spice [21] and finely ground seed is a major ingredient of curry powder, an effective antioxidant [22], [23] and a drug for indigestion[24], against worms [25], rheumatism [26], [27]... Coriander seeds have also several properties such as antimicrobial activity [28], insecticidal effect, hypolipidemic activity [29], hypoglycemic action [30]...
The objective of the present work is to investigate the possible use of Coriandrum Sativum seeds as a removal of copper ions from contaminated solutions since its adsorption potential has not been previously reported. The adsorption parameters, contact time and initial ions concentration, were investigated using ICP to evaluate the possible adsorbent-metal ion interactions.

2. Material and methods

2.1. Preparation of adsorbent

The coriander seeds (Coriandrum Sativum) were collected from the west of morocco (Kenitra), washed with double distilled water to remove impurities such as sand and dust, dried in air at 50 °C for 24 hours, ground using mortar, and then sieved, until the grain-size of particles was lower than 250 μm. The obtained powder was dissolved in double distilled water and filtrated several times in order to obtain a clean powder. The resulted material was stored in a desiccator to prevent it from humidity and was used as an adsorbent.

2.2. Preparation of aqueous solution

Aqueous solutions were prepared from copper sulfate salt (purchased from Sigma Aldrich). The stock solutions were then used to obtain the working solutions through dilution with double distilled water. Fresh dilutions were prepared and used for each experiment. 10-1 mol /L HNO3 or NaOH solutions were used to adjust the initial pH of working solutions.

2.3. Adsorption experiments

The extraction experiments were realized by adding the given quantity of adsorbent with 100 mL of aqueous solution with known heavy metal ion concentration, in 200 mL conical flask. The flask was intermittently stirred for an adequate period of time, depending on the purpose of the experiment. All experiments were studied at room temperature (20°C). After adsorption, samples were analyzed using the Inductively Coupled Plasma (ICP) in the University Center for Analysis, Expertise, Transfer of Technology and Incubation, Ibn Tofail University, Kenitra, Morocco.

2.4. Data evaluation

The adsorption process of copper ions by coriander seeds was evaluated using the following parameters, which were calculated from experimental results:
Rate of copper ions removed (%):

$$R = \frac{C0 - Ce}{C0} * 100$$

amount of copper ions retained on mass unit of coriander seeds (mg/g):

$$Qe = \frac{C0 - Ce}{m} * V$$

C0 is the initial concentration of copper ions in the solution (ppm).
Ce is the equilibrium concentration of heavy metal ions in the solution (ppm).
V is volume of solution (L).
m is the mass of coriander seeds (g).
All the data are the mean values of three replicate measurements, and the standard deviation calculated in each case was lower than ± 1.0%.

3. Results and discussion

3.1. Effect of contact time

The influence of contact time on Cu (II) ions adsorption onto coriander seeds, at room temperature (20°C), constant adsorbent mass (0.5 g), initial solution pH of 5.5 and initial heavy metal ions concentration 10 ppm, is illustrated in (Fig.1). The experimental data indicates that the adsorption efficiency of the considered heavy metal ions onto coriander seeds increases with the increase of contact time (ranging from 10 to 240 minutes), and reaches a maximum after 30 min. The increase of pH solution shows that the coriander seeds balance with the metal solution by consuming H+ protons. There is competition between Cu2+ and the protons to be fixed to coriander seeds surface.

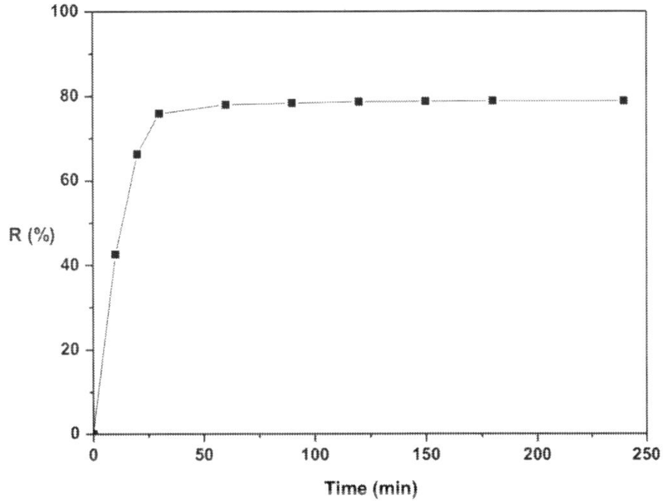

Fig.1. Effect of contact time on the adsorption rate

3.2. Effect of initial metal ion concentration

The effect of initial copper ions concentrations on the adsorption performances of coriander seeds CSS, under optimum experimental conditions (0.5 g of adsorbent; initial solution pH of 5.5 and 4 hours of contact time) is illustrated in (Fig.2. and Fig.3). The study is carried out at different concentrations ranging from 6.5 to 150 ppm .It can be seen that the adsorption capacity of coriander seeds increases with the increase of initial copper ions concentration from 1.03 to 5.58. These significant increases of the adsorption capacity values are expected because high initial copper ions concentrations correlate with a higher probability of collisions between Cu^{2+} and superficial functional groups of the adsorbent.

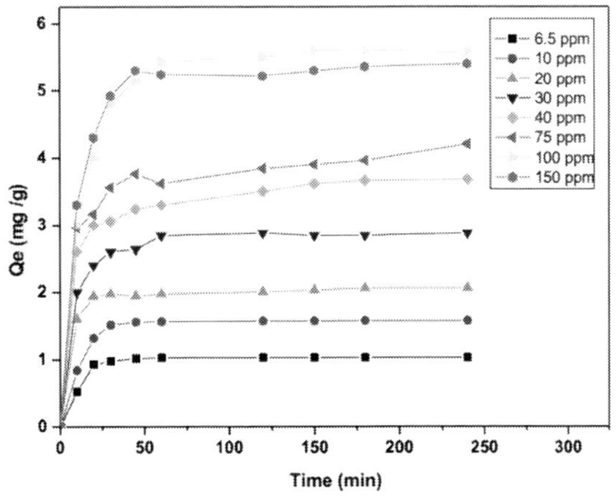

Fig.2. Effect of initial concentration of Cu^{2+} on the adsorption kinetic by CSS

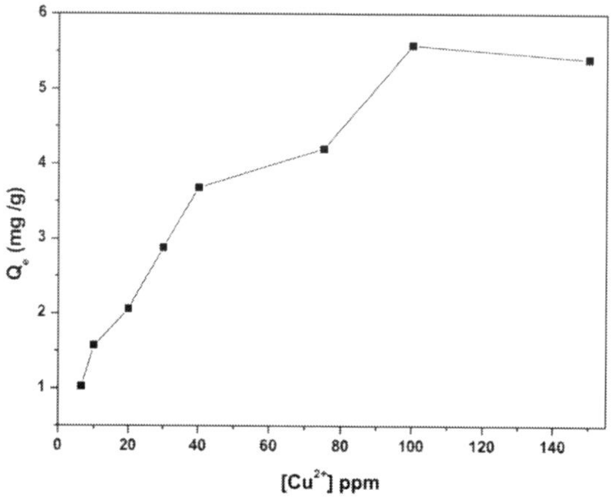

Fig.3. Effect of initial concentration of copper ions on the amount of adsorption

4. Conclusion:

The present investigation emphasizes adsorption potential of Coriandrum Sativum seeds (CSS) as an ecofriendly adsorbent of copper ions from wastewaters.
The following conclusion can be made from this study:
The equilibrium of system metallic solution- CSS was observed after 30 minutes of contact with a low quantity of adsorbent.
Therefore it can be concluded that CSS fulfills the criteria of an effective adsorbent for copper adsorption.

5. Acknowledgment:

The authors are pleased to acknowledge:
- The 6th international congress "water, waste and environment", Sultan Moulay slimane University, Beni Mellal, Morocco for giving the opportunity to publish this work.
- The Ibn Tofail University, Kenitra, Morocco, for providing the facilities for the research. Also the Director of Center of Analysis, Expertise, Transfer of Technology and Incubation (CUAE2TI) "M. Ibn Touhami" is gratefully acknowledged for his kind permission to use their facilities for the analysis.

References:

[1] F. Fu and Q. Wang, "Removal of heavy metal ions from wastewaters: A review," *J. Environ. Manage.*, vol. 92, no. 3, pp. 407–418, Mar. 2011.
[2] R. Han, H. Li, Y. Li, J. Zhang, H. Xiao, and J. Shi, "Biosorption of copper and lead ions by waste beer yeast," *J. Hazard. Mater.*, vol. 137, no. 3, pp. 1569–1576, Oct. 2006.

[3] N. T. Abdel-Ghani and G. A. El-Chaghaby, "Biosorption for metal ions removal from aqueous solutions: a review of recent studies," *Int J Latest Res Sci Technol*, vol. 3, no. 1, pp. 24–42, 2014.

[4] A. Malik, "Metal bioremediation through growing cells," *Environ. Int.*, vol. 30, no. 2, pp. 261–278, Apr. 2004.

[5] Z. Mohamed, A. Abdelkarim, K. Ziat, and S. Mohamed, "Adsorption of Cu (II) onto natural clay: Equilibrium and thermodynamic studies," *system*, vol. 10, p. 11, 2016.

[6] H. Bousfiha, E.-H. Rifi, A. Essamri, and Z. Hatim, "L'extraction liquide–solide du zinc par des supports à base de phosphate de calcium," *Comptes Rendus Chim.*, vol. 8, no. 2, pp. 215–218, Feb. 2005.

[7] A. Ouass, I. Ismi, H. Elaidi, A. Lebkiri, M. Cherkaoui, and E. H. Rifi, "Mathematical Modeling Of The Adsorption Of Trivalent Chromium By The Sodium Polyacrylate Beads," *JMES*, vol. 8, no. 10, pp. 3448–3456, 2017.

[8] H. Taouil, S. I. Ahmed, E. Rifi, and A. El Assyry, "Zinc extraction from dilute aqueous solutions, by hydrogels of polyacrylic acid-polyacrylate sodium (Extraction du zinc à partir des solutions aqueuses diluées, par d'hydrogels d'acide polyacrylique-polyacrylate de sodium)," *JMES*, vol. 5, no. 4, pp. 1069–1074.

[9] E. Barbolani, M. Clauser, F. Pantani, and R. Gellini, "Residual heavy metal (Cu and Cd) removal byIris Pseudacorus," *Water. Air. Soil Pollut.*, vol. 28, no. 3, pp. 277–282, 1986.

[10] F. Kaczala, M. Marques, and W. Hogland, "Lead and vanadium removal from a real industrial wastewater by gravitational settling/sedimentation and sorption onto Pinus sylvestris sawdust," *Bioresour. Technol.*, vol. 100, no. 1, pp. 235–243, Jan. 2009.

[11] A. Saeed, M. Iqbal, and M. W. Akhtar, "Removal and recovery of lead(II) from single and multimetal (Cd, Cu, Ni, Zn) solutions by crop milling waste (black gram husk)," *J. Hazard. Mater.*, vol. 117, no. 1, pp. 65–73, Jan. 2005.

[12] H. J. Park, S. W. Jeong, J. K. Yang, B. G. Kim, and S. M. LEE, "Removal of heavy metals using waste eggshell," *J. Environ. Sci.*, vol. 19, no. 12, pp. 1436–1441, 2007.

[13] O. S. Amuda, F. E. Adelowo, and M. O. Ologunde, "Kinetics and equilibrium studies of adsorption of chromium(VI) ion from industrial wastewater using Chrysophyllum albidum (Sapotaceae) seed shells," *Colloids Surf. B Biointerfaces*, vol. 68, no. 2, pp. 184–192, Feb. 2009.

[14] N. Feng, X. Guo, S. Liang, Y. Zhu, and J. Liu, "Biosorption of heavy metals from aqueous solutions by chemically modified orange peel," *J. Hazard. Mater.*, vol. 185, no. 1, pp. 49–54, Jan. 2011.

[15] Y. N. Mata, M. L. Blàzquez, A. Ballester, F. Gonzàlez, and J. A. Munoz, "Sugar-beet pulp pectin gels as biosorbent for heavy metals: Preparation and determination of biosorption and desorption characteristics," *Chem. Eng. J.*, vol. 150, no. 2–3, pp. 289–301, Aug. 2009.

[16] W. E. Oliveira, A. S. Franca, L. S. Oliveira, and S. D. Rocha, "Untreated coffee husks as biosorbents for the removal of heavy metals from aqueous solutions," *J. Hazard. Mater.*, vol. 152, no. 3, pp. 1073–1081, Apr. 2008.

[17] Y.Essaadaoui, L.Kadiri, E.H.Rifi, and A.Lebkiri, "Microstructure characterization of the barks of eucalyptus "eucalyptus," *IJSER*, vol. 7, no. 11, pp. 1382–1387, Nov. 2016.

[18] S. Elanza, A. Lebkiri, S. Marzak, E. H. Rifi, M. Lebkiri, and C. Satif, "Removal of lead ions from aqueous solution by the sugarcane bagasse," *J Mater Env. Sci*, vol. 5, no. 5, pp. 1591–1598, 2014.
[19] T. C. Wang, J. C. Weissman, G. Ramesh, R. Varadarajan, and J. R. Benemann, "Heavy metal binding and removal by Phormidium," *Bull. Environ. Contam. Toxicol.*, vol. 60, no. 5, pp. 739–744, 1998.
[20] I. Kim, M. Lee, and S. Wang, "Heavy metal removal in groundwater originating from acid mine drainage using dead Bacillus drentensis sp. immobilized in polysulfone polymer," *J. Environ. Manage.*, vol. 146, pp. 568–574, Dec. 2014.
[21] A. Sarimeseli, "Microwave drying characteristics of coriander (Coriandrum sativum L.) leaves," *Energy Convers. Manag.*, vol. 52, no. 2, pp. 1449–1453, Feb. 2011.
[22] M. K. Shahwar, A. H. El-Ghorab, F. M. Anjum, M. S. Butt, S. Hussain, and M. Nadeem, "Characterization of Coriander (*Coriandrum sativum* L.) Seeds and Leaves: Volatile and Non Volatile Extracts," *Int. J. Food Prop.*, vol. 15, no. 4, pp. 736–747, Jul. 2012.
[23] Z. Zeković *et al.*, "Chemical characterization of polyphenols and volatile fraction of coriander (Coriandrum sativum L.) extracts obtained by subcritical water extraction," *Ind. Crops Prod.*, vol. 87, pp. 54–63, Sep. 2016.
[24] K. Ghedira and P. Goetz, "Coriandrum sativum L. (Apiaceae) : Coriandre," *Phytothérapie*, vol. 13, no. 2, pp. 130–134, Apr. 2015.
[25] B. Laribi, K. Kouki, M. M'Hamdi, and T. Bettaieb, "Coriander (Coriandrum sativum L.) and its bioactive constituents," *Fitoterapia*, vol. 103, pp. 9–26, Jun. 2015.
[26] K. Msaada *et al.*, "Antioxidant activity of methanolic extracts from three coriander (Coriandrum sativum L.) fruit varieties," *Arab. J. Chem.*, Dec. 2013.
[27] H. Wangensteen, A. B. Samuelsen, and K. E. Malterud, "Antioxidant activity in extracts from coriander," *Food Chem.*, vol. 88, no. 2, pp. 293–297, Nov. 2004.
[28] P. J. Delaquis, K. Stanich, B. Girard, and G. Mazza, "Antimicrobial activity of individual and mixed fractions of dill, cilantro, coriander and eucalyptus essential oils," *Int. J. Food Microbiol.*, vol. 74, no. 1, pp. 101–109, 2002.
[29] S. Mandal and M. Mandal, "Coriander (Coriandrum sativum L.) essential oil: Chemistry and biological activity," *Asian Pac. J. Trop. Biomed.*, vol. 5, no. 6, pp. 421–428, Jun. 2015.
[30] V. Chithra and S. Leelamma, "Coriandrum sativum—mechanism of hypoglycemic action," *Food Chem.*, vol. 67, no. 3, pp. 229–231, 1999.

Optimization and characterization of gelatin and chitosan extracted from fish and shrimp waste

AIT BOULAHSEN *M*[1*], CHAIRI *H*[1], LAGLAOUI *A*[2], ARAKRAK *A*[2], ZANTAR *S*[3], BAKKALI *M*[2], HASSANI *M*[1]

1. Department of Biology, Polydisciplinary Faculty of Larache, ABDELMALEK ESSAADI University, Tangier, Morocco
2. Department of Biology, Faculty of Science and Technology, ABDELMALEK ESSAADI University, Tangier, Morocco
3. National Institute of Agricultural Research, Tangier, Morocco

Abstract. Fish and seafood processing industries generate large quantities of waste which are at the origin of several environmental, economic and social problems. However fish waste could contain high value-added substances such as biopolymers. This work focuses on optimizing the gelatin and chitosan extraction from tilapia fish skins and shrimp shells respectively. The gelatin extraction process was optimized using alkali acid treatment prior to thermal hydrolysis. Three different acids were tested at different concentrations. Chitosan was obtained after acid demineralization followed by simultaneous hydrothermal deproteinization and deacetylation by an alkali treatment with different concentrations of HCl and NaOH. The extracted gelatin and chitosan with the highest yield were characterized by determining their main physicochemical properties (Degree of deacetylation, viscosity, pH, moisture and ash content). Results show a significant influence of the acid type and concentration on the extraction yield of gelatin and chitosan, with an average yield of 12.24% and 3.85% respectively. Furthermore, the obtained physicochemical properties of both extracted gelatin and chitosan were within the recommended standard values of the commercial ones used in the industry.

Key words: Waste, extraction, characterization, chitosan, gelatin

* Corresponding author: maitboulahsen@uae.ac.ma

1. Introduction

Fish and seafood processing industries occupy a strategic place in the Moroccan economy. They ensure 50% of the country's agri-food exports for a value of 14 billion dirhams. Among several activities, there are about 20 shrimp shelling and semi-preserved smoked fish companies processing more than 20 000 tons of fish [1]. They generate more than 14 000 tons of waste that are thrown directly into landfills causing a serious environmental problems.

However, these wastes are composed of skins and shells that may contain high value-added substances. Extraction of biopolymers like gelatin and chitosan is one of the most important ways for waste valorization. Gelatin, a polypeptide extracted from fish skins, offers various functional properties such as film-forming, gelling and emulsifying properties and can be applied in several fields like food, pharmaceutical and cosmetic industry [2].On the other hand, chitosan, a polysaccharide derived from chitin can be extracted from crustaceans shells, especially shrimps, is widely studied for its excellent functional and biological properties, and has various applications in several fields such as wastewater treatment, food, pharmaceutical, textile and cosmetic industry [3].Therefore, the valorization of these by-products can represent an important eco-friendly alternative that allow, in addition to reducing the environmental impact, the enhancement of creativity, competitiveness and productivity of companies. Despite that extraction and characterization of these two biopolymers have been already studied by several authors all over the world [4–6], to the best of our knowledge, studies at national level are inexistent. Moreover, data obtained from other authors cannot be extrapolated due to the specificity and characteristics of each raw material. Thus, the main objective of this work is to optimize the extraction and characterization of gelatin and chitosan from tilapia skins (aquaculture fish) and shrimp shells generated by agribusiness units in the northern region of Morocco.

2. Material and methods

2.1. Raw material and chemical reagents

Shrimp waste and tilapia skins have been used as a source of chitosan and gelatin extraction respectively. They have been provided by two companies, located in the north of Morocco. All the chemicals reagents used were of analytical grade.

2.2. Chitosan extraction

Chitosan was obtained after the two following steps: acid demineralization followed by simultaneous deproteinization and deacetylation by hydrothermal-chemical method in an alkali environment, with a solid-liquid ratio of 1:10 w/v (weight of dry shell / volume of diluted solution), according to the method described by Truong et al. (2007) [7] with some modifications.

Shrimp shells was first pretreated prior to their use, washed with excess water to remove organic residues (tissues), and then dried for 48 hours. Once dried, they were reduced to powder using a mixer in order to have particles with a uniform granulometry of less than 0.5mm.

Demineralization step: 20 g of dried powder of shrimp shells were added to HCl solution(1/10 w/v) at 3 different concentrations (0.5 M, 2 M and 3.5 M), the solutions were stirred for 150 min at 50°C, then washed with distilled water until a neutral pH of 7 was obtained.

Deproteinization and deacetylation steps: These two steps were performed simultaneously in an alkali environment. The product obtained from the previous step was added to NaOH solution at 2 different concentrations (7.5 and 10 M) for 120 min at 100°C. The resulted product was then filtered and washed several times with distilled water. At about neutral pH, the substrate was dried in an oven at 80 ° C for 24 hours.

2.3. Gelatin extraction

The extraction method of gelatin was carried out as described by Niu et al. (2013) [8]. First, the tilapia skins (30 g) were soaked in water (1: 6 w/v) for 10 min, drained into a filter cloth and then washed twice to remove the residues. The washing step consists of stirring the mixture for 4 minutes at 180 rpm and then draining the skins in a filter cloth. After washing, the excess water was removed, and then the skins were immersed in 0.3M NaOH solution (1: 6 w/v) for 1 h. Then, the skins were drained for 5 minutes, washed five times and dipped in 3 different acid solutions with 2 different concentrations: HCl at (0.05 and 0.07 M), acetic acid at (0.18 and 0.20 M) and citric acid at (0.03 and 0.05 M) for 1 h. The skins were drained and washed five times as described above. After the acid treatment, they were soaked in distilled water (1: 4 w/v) at 50 ° C in a water bath for 3 h and then filtered to recover the gelatin-containing solution. The solution was dried in an oven until a constant weight was obtained.

2.4. Extraction Yield

The extraction yield of gelatin and chitosan was calculated according to the following formula:

Extraction Yield (%) = (dry weight after extraction *(g)/ weight of raw material (g))*100*

2.5. Physicochemical characteristics

Physicochemical properties of extracted gelatin and chitosan with the highest yield were determined. Moisture content (%), ash content (%), and pH were measured according to AOAC official methods (2000) [9]. The viscosity of gelatin solution was determined with a falling ball viscometer; a 6.67% (w/v) gelatin solution was obtained by dissolving gelatin in 45 mL of distilled water and maintained at 60°C for 30 minutes, after that, the gelatin solution was poured into the measuring tube, then the Ball (diameter =15.60 mm, ρ = 2.235 g/cm^3) was placed into the tube, the falling time (s) of the ball moving from the Upper ring to the lower ring of the tube was determined by using a stopwatch, the dynamic viscosity n (in mPa*s) was calculated using the following equation:

$$\eta = K*(\rho_1 - \rho_2)*t$$

Where, **K** = ball constant $mPa*s*cm^3/g*s$, ρ_1 = density of the Ball in g/cm^3, ρ_2 = density of the sample g/cm^3, and **t** = falling time of the ball in seconds.

The degree of deacetylation (DD) of chitosan was determined by potentiometric titration as described by Dima et al. (2017) [10] with a slight modifications; a solution of 100 mg of chitosan, 3 mL of 0.3M HCl and 40 mL of distilled water were prepared and stirred overnight, then the solution was titrated with NaOH 0.1 M, pH meter was used for pH measurements under continuous stirring, The titrant was added until the pH value reached 4.5, the volume of NaOH was recorded as V1, NaOH was then added until a pH of 8.5 was obtained and the added volume was recorded as V2. The DD was obtained by the following equation:

$$DD\ (\%) = 203M_{eq} / (1+42M_{eq})$$

Where, M_{eq} = (N*ΔV)/w, ΔV = V2-V1=volume added in mL of NaOH between pH 4.5 and pH 8.5., N=normality of NaOH Solution, w= dry weight of chitosan in the test sample, in mg, **203** is the molar mass of glucosamine and **42** is the molar mass of the acetyl group.

2.6. Statistical analysis

All measurements were done in triplicate. Data analysis was carried out by ANOVA using IBM SPSS STATISTICS 20. Differences between pairs of means were evaluated on the basis of 95% confidence intervals. The level of significance was $p \leq 0.05$.

3. Results and discussions

3.1. Yield of chitosan

Figure 1 shows the effect of NaOH (7.5, 10 M) and HCl (0.5, 2, 3.5 M) concentrations on the chitosan yield extraction. The yield varies between 2.1% and 4.4% for all treatments. The highest yield (4.4%) was obtained by a demineralization using 0.5M HCl, followed by a simultaneous deproteinization and deacetylation with 7.5 M NaOH. Acid concentration has a significant effect ($p \leq 0.05$) on the yield extraction, in fact, the yield decreased as the acid concentration increased. It can be explained by an extensive demineralization step with higher acid concentrations, that causes an important loss in the weight of the final product [11]. No significant influence of the NaOH concentration on the yield extraction has been shown ($p \geq 0.05$). Our results are similar to those obtained by Ahing et al. (2016) [12] and slightly lower than those obtained by Nouri et al. [4]. Differences in yield extraction can be due to the difference of the extraction conditions, precisely, time and temperature of deacetylation [6].

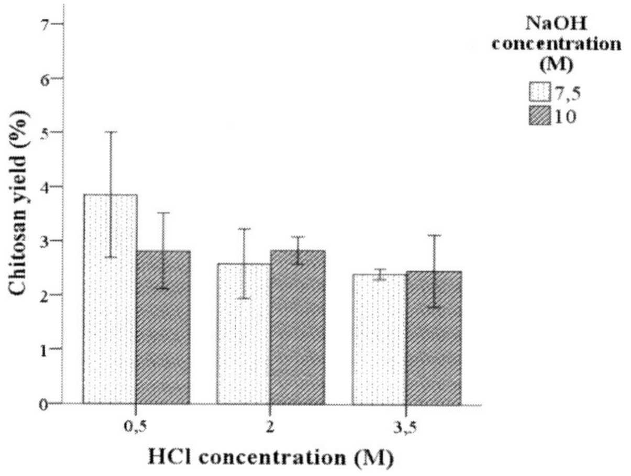

Fig.1. Effect of NaOH and HCl concentration on chitosan yield. Bars represent the standard deviation (n = 3)

3.2. Physicochemical characteristics of extracted chitosan

Table 1 represents the main physicochemical characteristics obtained for the extracted chitosan with the highest yield. The degree of deacetylation (DD%), described as the crucial factor determining the physicochemical behavior and biological functionality of chitosan [13], reached a value of 67.03 %. Our results are similar to those obtained by Tokatlı et al. (2017) [14], but higher than those presented by Kumari et al. (2015) [15], and lower than those obtained by Sagheer et al. (2009) [16]. These variations may be affected by the different nature of raw material and the parameters of deacetylation process (time and temperature). Truong et al. (2007) [7] and Kumar (2000) [17] reported that the DD%

necessary to obtain a chitosan product is 65% or higher. Ash content indicates the effectiveness of the demineralization step, No et al. (1995) [18] reported that chitosan of a good quality must have less than 1% of ash content, which is in agreement with the results obtained in this study (0.94%). Moisture content, on the other hand, reached an average percentage of 8.71, similar to the results obtained by Lertsutthiwong et al. (2002) [19] who showed that moisture content varies from 6 to 12% depending on the different chemical treatments, Alishahi et al. (2011) [20] also suggested that a suitable chitosan must contain lower than 10% of moisture content.

Table 1. Physicochemical characteristics of extracted chitosan

	Yield (%)	Moisture content (%)	Ash content (%)	Degree of desacetylation (%)
Chitosan (HCl 0.5 M, NaOH 7.5 M)	3,85±0,576	8,71±0,412	0,94±0,026	67,03±1,743

***Values were means ± standard deviation from triplicate determinations.**

3.3. Yield of gelatin

Figure 2 shows the effect of different acid treatments HCl (0.03 and 0.05 M), acetic acid (0.07 and 0.18 M) and citric acid (0.05 and 0.20 M) on the gelatin yield extraction. Results show that the yield varies between 6.26 % and 13.16 %, depending on the type and acid concentration used. Average yield was 10.53 %, 11.55 % and 7.55% for HCl, acetic acid and citric acid respectively. Our results are higher than those obtained by Jamilah and Harvinder (2002) [21] (7.81%) and in agreement with those reported by Jamilah et al. (2011) [5] (12.92%). Acid treatments eliminate certain proteins and other undesirable components, disrupts collagen molecule linkages and allows the skin to swell, which is effective for gelatin extraction [22]. In terms of the influence of acid concentration on the yield, Niu et al. (2013) [8] reported that the acid concentration should be high enough for optimal extraction, but should not be used in excess, to avoid over-hydrolysis of the collagen molecules and in consequence losses in yield. Our results show that the yield increases as the acid concentration increases for all the acid treatments, yet with no significant influence ($p \geq 0.05$) on the extraction yield, which is in agreement with the results obtained by Zeng et al. (2010) [23]

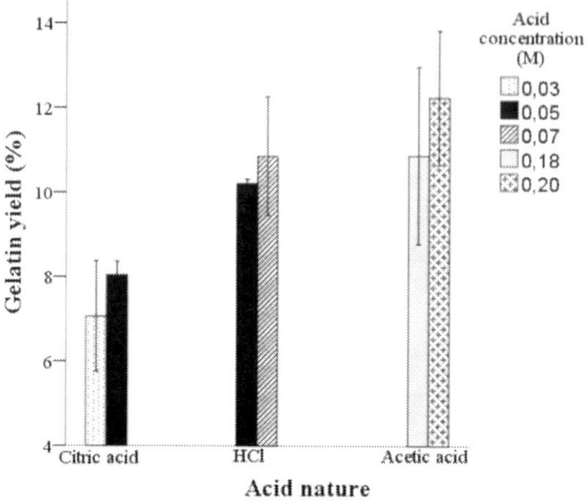

Fig.2. Effect of acid treatment on gelatin yield. Bars represent the standard deviation (n = 3)

3.4. Physicochemical characteristics of extracted gelatin

The extracted gelatin with the highest yield (0.20 M acetic acid treatment), was characterized, to determine its main physicochemical properties as presented in **table 2.** pH obtained was 5.62, which is similar to the results obtained by Jamilah et al. (2011) [5] for red tilapia, however pH values obtained by Choi et al. (2000) [24] ranged from 4.2 to 6.5, Alfaro et al. (2015) [25] reported that the variation of pH values is due mostly to the extraction process, they also reported that pH value affect directly the viscosity of gelatin. In our case, the viscosity obtained was 5.19 mPa*s at 60°C, which is within the range of the values for commercial gelatins [26]. The values obtained for moisture and ash content were 8.51 % and 0.67 % respectively, which are in agreement with the results obtained by Jamilah et al. (2011) [5] and Zeng et al. (2010) [23], and within the standard values of commercial gelatins, reported by Alfaro et al. (2015) [25]: 2.6 % as maximum ash content, and less than 14 % for the moisture content.

Table 2. Physicochemical characteristics of extracted gelatin

	Yield (%)	Moisture content (%)	Ash content (%)	Viscosity (mPa*s)	pH
Gelatin (0.20 M acetic acid)	12.24±0.888	8.51±0.122	0.67±0.090	5.19±0.025	5.62±0.091

***Values were means ± standard deviation from triplicate determinations.**

4. Conclusions

The methods used in this study for gelatin and chitosan biopolymers allowed an optimization of the extraction yield and a determination of the optimal conditions necessary to obtain the best yield. Gelatin extracted from tilapia skins, has a maximum average extraction yield of 11.51% obtained by 0.20 M acetic acid treatment, allowed an optimization of +3% of the yield compared to the other acids used in this study, with interesting values of physicochemical properties such as viscosity, pH, moisture and ash content, that range within the recommended grades for commercial gelatins. On the other hand, the chitosan obtained by the hydrothermal-chemical method has a maximum yield of 4%, a degree of deacetylation of 67%, allows the reduction of production costs and extraction time and can be an alternative to the classical techniques used. These findings could be used to help the fish and seafood industries to reuse and valorize the waste they generate and produce high value-added biopolymers in an economic, environmental and efficient manner. Furthermore, the use of fish gelatin and chitosan could be a potential alternative tool for innovative technologies to improve environmental quality and sustainability.

5. Acknowledgements

The authors thank CNRST (PPR/2015/10) for financial support.

References

[1] Département de la pêche maritime, Ministère de l'agriculture et de la pêche maritime, "Activités de valorisations du Secteur de l'industrie de la pêche," 2014.
[2] A. A. Karim and R. Bhat, "Fish gelatin: properties, challenges, and prospects as an alternative to mammalian gelatins," Food Hydrocoll., vol. 23, no. 3, pp. 563–576, May 2009.
[3] I. Hamed, F. Özogul, and J. M. Regenstein, "Industrial applications of crustacean by-products (chitin, chitosan, and chitooligosaccharides): A review," Trends Food Sci. Technol., vol. 48, pp. 40–50, Feb. 2016.
[4] M. Nouri, F. Khodaiyan, S. H. Razavi, and M. Mousavi, "Improvement of chitosan production from Persian Gulf shrimp waste by response surface methodology," Food Hydrocoll., vol. 59, pp. 50–58, Aug. 2016.
[5] B. Jamilah, K. W. Tan, M. R. Umi Hartina, and A. Azizah, "Gelatins from three cultured freshwater fish skins obtained by liming process," Food Hydrocoll., vol. 25, no. 5, pp. 1256–1260, Jul. 2011.
[6] A. Patria, "Production and characterization of Chitosan from shrimp shells waste," AACL Bioflux, vol. 6, no. 4, pp. 339–344, 2013.
[7] T. O. Truong, R. Hausler, F. Monette, and P. Niquette, "Valorisation des résidus industriels de pêches pour la transformation de chitosane par technique hydrothermo-chimique," Rev. Sci. Eau, vol. 20, no. 3, pp. 253–262, 2007.
[8] L. Niu, X. Zhou, C. Yuan, Y. Bai, K. Lai, F. Yang, Y. Huang, "Characterization of tilapia (Oreochromis niloticus) skin gelatin extracted with alkaline and different acid pretreatments," Food Hydrocoll., vol. 33, no. 2, pp. 336–341, Dec. 2013.
[9] W. Horwitz, Official methods of analysis of AOAC International. Gaithersburg, Md.: AOAC International, 2000.
[10] J. B. Dima, C. Sequeiros, and N. Zaritzky, "Chitosan from Marine Crustaceans: Production, Characterization and Applications," in Biological Activities and Application of Marine Polysaccharides, E. A. Shalaby, Ed. InTech, 2017.
[11] Naznin, Rokshana, "Extraction of Chitin and Chitosane from Shrimp (Metapenaeus monoceros) Shell by chemical method," Pak. J. Biol. Sci., vol. 8, no. 7, pp. 1051–1054, 2005.
[12] F. A. Ahing and N. Wid, "Extraction and Characterization of Chitosan from Shrimp Shell Waste in Sabah," Trans. Sci. Technol., vol. 3, no. 1–2, pp. 227–237, 2016.

[13] E. Szymańska and K. Winnicka, "Stability of Chitosan—A Challenge for Pharmaceutical and Biomedical Applications," Mar. Drugs, vol. 13, no. 4, pp. 1819–1846, Apr. 2015.
[14] K. Tokatlı and A. Demirdöven, "Optimization of chitin and chitosan production from shrimp wastes and characterization," J. Food Process. Preserv., p. e13494, Sep. 2017.
[15] S. Kumari, P. Rath, A. Sri Hari Kumar, and T. N. Tiwari, "Extraction and characterization of chitin and chitosan from fishery waste by chemical method," Environ. Technol. Innov., vol. 3, pp. 77–85, Apr. 2015.
[16] F. A. A. Sagheer, M. A. Al-Sughayer, S. Muslim, and M. Z. Elsabee, "Extraction and characterization of chitin and chitosan from marine sources in Arabian Gulf," Carbohydr. Polym., vol. 77, no. 2, pp. 410–419, Jun. 2009.
[17] M. N. R. Kumar, "A review of chitin and chitosan applications," React. Funct. Polym., vol. 46, no. 1, pp. 1–27, 2000.
[18] H. K. No and S. P. Meyers, "Preparation and Characterization of Chitin and Chitosan—A Review," J. Aquat. Food Prod. Technol., vol. 4, no. 2, pp. 27–52, Oct. 1995.
[19] P. Lertsutthiwong, N. C. How, S. Chandrkrachang, and W. F. Stevens, "Effect of Chemical Treatment on the Characteristics of Shrimp Chitosan.," J. Met. Mater. Miner., vol. 12, no. 1, pp. 11–18, 2002.
[20] A. Alishahi et al., "Enhancement and Characterization of Chitosan Extraction from the Wastes of Shrimp Packaging Plants," J. Polym. Environ., vol. 19, no. 3, pp. 776–783, Sep. 2011.
[21] B. Jamilah and K. G. Harvinder, "Properties of gelatins from skins of fish—black tilapia (Oreochromis mossambicus) and red tilapia (Oreochromis nilotica)," Food Chem., vol. 77, no. 1, pp. 81–84, 2002.
[22] M. Ahmad and S. Benjakul, "Characteristics of gelatin from the skin of unicorn leatherjacket (Aluterus monoceros) as influenced by acid pretreatment and extraction time," Food Hydrocoll., vol. 25, no. 3, pp. 381–388, May 2011.
[23] S. Zeng, X. Yan, W. Cao, P. Hong, C. Zhang, and L. Li, "Original article: Optimisation of extraction conditions and characteristics of skin gelatin from Nile tilapia (Oreochromis niloticus): Optimisation of extraction conditions and characteristics of skin gelatin," Int. J. Food Sci. Technol., vol. 45, no. 9, pp. 1807–1813, Sep. 2010.
[24] S.-S. Choi and J. M. Regenstein, "Physicochemical and sensory characteristics of fish gelatin," J. Food Sci., vol. 65, no. 2, pp. 194–199, 2000.
[25] A. da T. Alfaro, E. Balbinot, C. I. Weber, I. B. Tonial, and A. Machado-Lunkes, "Fish Gelatin: Characteristics, Functional Properties, Applications and Future Potentials," Food Eng. Rev., vol. 7, no. 1, pp. 33–44, Mar. 2015.
[26] P. Songchotikunpan, J. Tattiyakul, and P. Supaphol, "Extraction and electrospinning of gelatin from fish skin," Int. J. Biol. Macromol., vol. 42, no. 3, pp. 247–255, Apr. 2008.

E3S Web of Conferences 37, 02007 (2018) https://doi.org/10.1051/e3sconf/20183702007
EDE6-2017

Physicochemical and bacteriological characterization of raw sewage from the Meknes region Morocco

Ouallal *Hassan* [1, 4]*, El Hamzaoui *Najia*[2], Sbiti *Mohammed*[3], Machkor *Mohammed*[4], Messaoudi Lahcen4, Azrour *Mohamed*[1].

[1]Moulay Ismaïl University, Faculty of Sciences and Techniques, Departement of Chemistry fundamental and applied, laboratory of physic-chemistry Materiqls, PO Bosc 509 Irrachidia, Morocco

[2]Medical microbiology laboratory, Mohamed V Hospital, Meknes, Morocco

[3]Laboratory of bacteriology, Moulay Ismail Military Hospital Meknes Morocco.

[4]Moulay Ismaïl University, Faculty of Sciences, Dépertement of Chemistry,Materials, Membranes and Processes of Separation, PO Bosc 11201, Zitoune, Meknes, Morocco

Abstract. The region of Meknes suffers from a significant delay in the water sector, including sewage treatment. The Urban waste water in the region originates from various activities; households, industry, agriculture and hospital. These waters are released in to nature without any treatment, which constitutes a serious risk for the environment. The objective of this research is to characterize the physicochemical and bacteriological pollution of the three sewage collectors of the Meknes region (Aïn Choubbik, Oued Ouislane and Aïn Taoujdate). Indeed, the pollution parameters; biological and chemical oxygen demand are abnormally high, which indicates clearly that waste waters is rich of reducing or oxidizing organic materials. Ammonium, nitrogen, phosphorus and heavy metals concentrations exceed the limit values. So these waters are classified as bad or very bad quality. On the one hand, the presence of a large bacterial load indicates fecal contamination. The total coliform, fecal coliform and fecal streptococci levels are high exceeding exceed the values specified the WHO guidelines and the Moroccan standards. This situation is likely to cause serious and irreparable damage to the environment groundwater. To counter this, make the necessary investment in sewage treatment and other remedial measures is key to preventing pollution and its spread.

Keyword: waste, water, bacteriological, physicochemical,

*Corresponding author: hassanouallalaghbalou@gmail.com

 https://doi.org/10.1051/e3sconf/20183702007

1. Introduction

In Morocco, water scarcity, waste and pollution are increasingly becoming a concern. Indeed, population growth and droughts since the 1970s are affecting water availability. The gap between the needs and availability of water increase might increase the size with the increasing of water-consuming sectors of all water-consuming sectors of the country. In order to address this situation, promoting the re-use of treated wastewater in agriculture becomes essential. It's is a very useful and very ancient practice across the globe [1]. At present, there is an area of about 7000 hectares directly irrigated with raw wastewater discharged by towns (about 70 million m^3 / year) with no precautionary measures. Meknes Region is the largest surface of wastewater irrigated crops. It is estimated about 1400 ha. It generally corresponds to it is of the order of 1400 ha and it concerns the cereals, the market gardens and the arboriculture [2]. The waste water reuse in agriculture constitutes a potential risk to human health and the environment. This study aims to evaluate the physicochemical and bacteriological parameters of the three sewage collectors of the Meknes region (Aïn Choubbik, Oued Ouislane and Aïn Taoujdate).This is intended to demonstrate the need for the treatment of waste water before being used in agriculture.

2. Materials and Methods

2.1. Sampling sites and sampling method

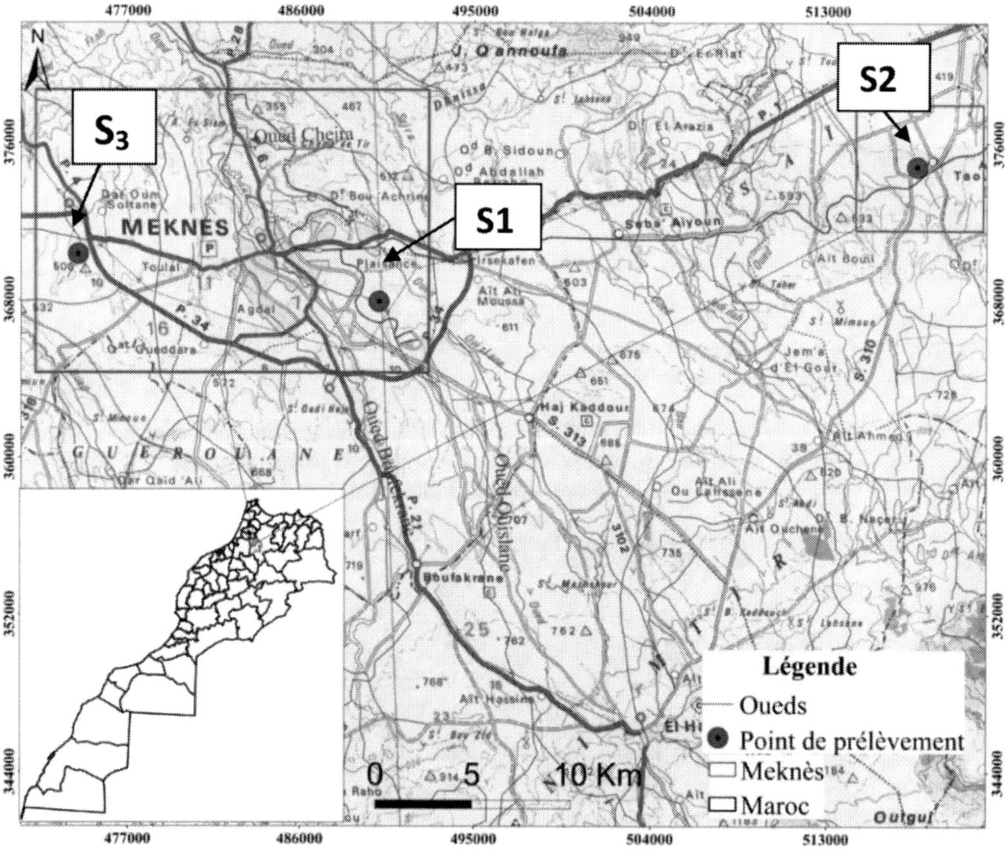

Fig1. Map of sampling points

The choice of sampling sites is based on their environmental features and on their use in agriculture.

The selected sites are;
Site (S1): Oued Ouislane belongs to the superficial hydrographic network of the region of Meknes, originates at the south-east of the village of Boufekrane. The sewage of Oued Ouislane is reused by the farmers.
Site (S2): Aïn Taoujdate is known for its agricultural and industrial activities (30 Km from Meknes). The residual urban water includes the flow from the households, industry, and discharges from the municipal slaughterhouse. This wastewater is reused for agriculture.
Site (S3): Aïn Choubbiik is the sanitation network, it receives the effluent from the Mohamed V hospital without treatment, and this wastewater will be directed towards Oued Bourouh located downstream. This water is reused in agriculture and for livestock watering.
Sampling is done in the morning between 10am and 1pm (time of rejects) in the summer 2017. Glass vials with screw caps are used and they are sterilized beforehand in the laboratory according to the protocol mentioned in Table 1.
Few drops of sodium thiosulphate are added to vials for bacteriological analysis sampling, they are added in the vials before sterilization to destroy the chlorine which may exert its action on the bacteria during transport. Some physico-chemical parameters as temperature and pH are measured in situ.

Table1. Sampling Protocol

Analysis	Amount
Physico-chemical	1 liter
Bactériological	250 ml sterile
Bactériological: search for salmonella	5 liter sterile

2.2. Methods of analysis of the different samples

All physicochemical and bacteriological analysis is made according to the techniques recommended by Rodier [3 and 4]: Tables 2 and 3

Table2. Methodology of physicochemical parameters

Parametres	Méthodes d'analyses
Temperature	Mercury thermometer
Conductivity	Conductivity meter of the type (meaning ion™EC71, HACH)
pH	pH-mètre
Oxygène dissous	Winkler
Nitrates	Molecular absorption spectrometry
Nitrites	Molecular absorption spectrometry
Sulfates	Nephelometric
Chlorures	Mohr

Table 3. Bacteriological analysis protocol

Germs	Volume	Culture media	Incubation	Techniques
Total coliforms	100 ml	TTC and Tergitol lactose agar	37 ° C during 24 hours	0.45µm filter membrane filtration technique
Fecal coliforms	100 ml	TTC and Tergitol lactose agar	44 ° C for 18 to 24 hours	0.45µm filter membrane filtration technique
Fecal streptococci	100 ml	Slanetz-Bartley	37 ° C / 48 hours	0.45µm filter membrane filtration technique
Total germs	1ml	PCA	37 ° C / 48 hours	Incorporation en gélose
staphylococci	100 ml	Chapman	37 ° C / 24-48 hours	0.45µm filter membrane filtration technique
Pseudomonas	100 ml	Cetrimide agar	42°C / 24hours	0.45µm filter membrane filtration technique
salmonella	5 Litres	sodium selenite then selective medium	Incubate 24h at 42 ° C then 24 hours at 37 ° C	Pre-enrichment during 24 hours then seeding and incubation

3. Results and discussion

3.1. Physicochemical characterization of analyzed waters

The results of physicochemical characterization of raw sewage in the region of Meknes are summarized in Table 4.

Table4. Results of physicochemical characterization of raw waste water from the studied sites during the period of April-August 2017.

Sites / Parametres	S1	S2	S3
Temperature °C	19,43	21	21
pH	8,72	9,23	7,67
DBO5 (mg O_2/l)	625	700	425
DCO (mg O_2/l)	1210	1530	970
conductivity (µs/cm)	1745	1895	1810
NH^{4+} (mg N/l)	66,8	74,3	66.7
Orthophosphate (mg P/l)	20,5	32,4	35.1
Chlorure (mg /l)	175,4	175,7	131.5
Sulfate (mg /l)	64,5	100,1	56.7
Cu (µg/l)	18,7	19,2	12.8
Cr (µg/l)	35,9	69,8	29.7
Cd (µg/l)	0,48	0,37	0.25
Fe (µg/l)	178	200,5	129
Mn (µg/l)	35	55	27
Mg (µg/l)	47	46	24
Pb (µg/l)	5,8	9,2	5.1
Zn (µg/l)	212	245	149
Ni (µg/l)	0	0	0
dissolved oxygen (mg /l)	4,5	0.23	0,97

3.1.1. Température

The average temperature of raw sewage in the region of Meknes is between 19.43 - 21 ° C. The temperatures recorded are included in the range of the direct discharge limit values in the receiving environment [5] and in the range of Moroccan standards for water quality for irrigation [6].

3.1.2.pH

The pH values of the raw sewage show that these waters are slightly alkaline and fall within the range of Moroccan standards for water quality for irrigation [6] and within the range of direct discharge limits [5].

3.1.3. Electrical conductivity

The values of the conductivity recorded at the level of the raw wastewater of the different sites vary between 1745- 1895µs / cm. This reflects an important mineralization (waters rich in salts and minerals). Electrical conductivity indicates the degree of overall mineralization; it provides information on the level of salinity. The electrical conductivity values meet the international regulatory standards for direct discharges and for irrigation of the crops [5, 6].

3.1.4. The chemical oxygen demand COD

Chemical Oxygen Demand is an important water quality parameter because, it provides an index to assess the effect discharged wastewater will have on the receiving environment. Higher COD levels mean a greater amount of oxidizable organic material in the sample, which will reduce dissolved oxygen (DO) levels[7].The average values recorded in the sites are very high (Table 4) and exceed the limit values of direct discharges [5], therefore, these waters are of very poor quality.

3.1.5. Biological oxygen demand DBO5

The BOD (Biochemical Oxygen Demand) is a measurement that aims to give the amount of organic material that can be broken down by micro-organisms. It is a widely-used parameter, which can be applied to all water types, especially sewage and purification processes [7].The average BOD5 concentrations of the three study sites ranges from 425 mg O_2 / L, 700 mg O_2 / L in urban waters of Ain Shubbiik and Ain Taoujdate respectively, these values are greater than limit values for direct discharges (100 mg O_2 / l) [5], therefore, these waters are of very poor quality.

3.1.6. Ratio DCO/DBO5

The COD / BOD5 ratio provides a deduction of the origin of wastewater pollution. In fact, a low value of the COD / BOD5 ratio indicates the presence of a large proportion of biodegradable materials and a biological treatment is envisaged. Inversely, a high value of this ratio indicates that a large part of the organic matter is not biodegradable therefore, physicochemical treatment is envisaged.

The average values of the ratio COD / BOD5 are 1.9, 2.18, and 2.28 in the waters of Oued ouislane, Ain Taoujdate and Aïn choubiik respectively. Therefore, it can be concluded that the wastewater in our study is easily biodegradable In spite of its high organic load. In addition, the values of the ratios are significantly lower than 2.5. On this basis, it can be concluded the dominance of domestic-type effluents [8].

3.1.7. Orthophosphate

The concentrations of orthophostate in the studied waste water are 20.5, 32.5 and 35.1 mgP /l for Oued Ouislane, Ain Taoujdate and Aïn choubiik respectively. These high values could be explained by wastewater discharges [8, 9].

3.1.8. Nitrogen

The concentrations of ammonium are high at different sites. A high level of nitrogen in the ammonium form is not toxic, but in alkaline waters (pH greater than 7.5), as was the case in our water analysis, a fraction could be transformed into ammonia form (ammonia gas) which is toxic.

3.1.8. Sulphate

Sulphate concentrations meet the limits values of direct discharges of wastewater and the ones of crop irrigation [5, 6].

3.1.9. Heavy metals (Cu, Cr, Fe, Mg, Mn, Pb, Zn, Ni)

Analysis of heavy metals have shown the presence of Cu, Cr, Fe, Mg, Mn, P and Zn with lower concentrations which meet the required limits values. These results were significantly different from those obtained from the sewage analysis of the Region of Oujda [10].

3.2. Bacteriological analyses

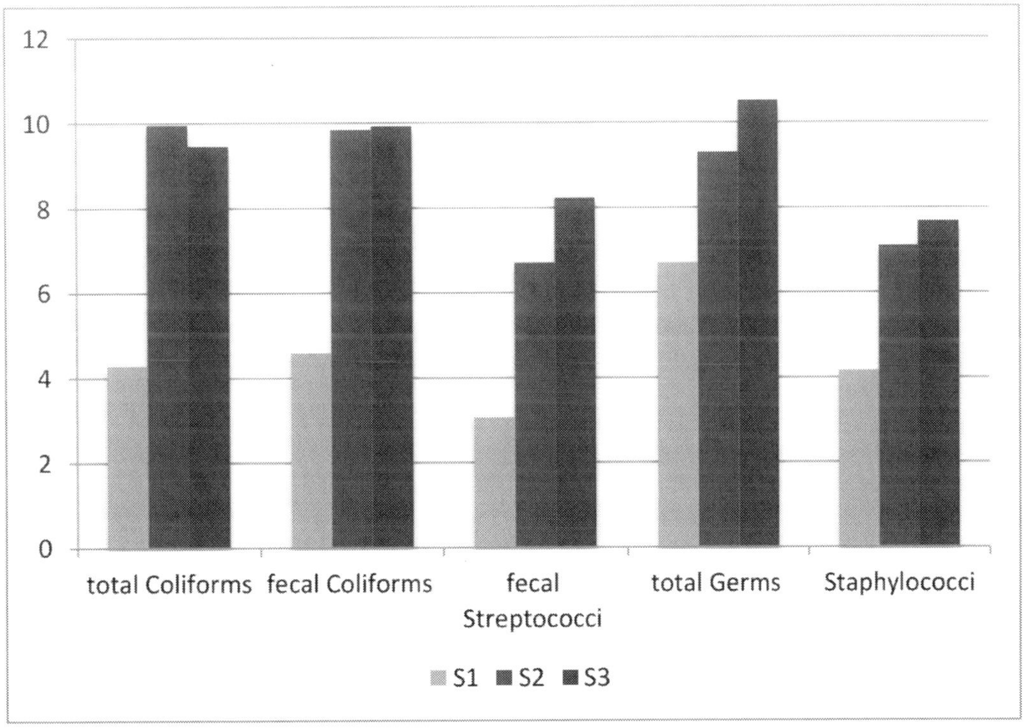

Fig2. Results of the bacteriological analysis of the three sites

Bacteriological analyses of different sites indicate the presence of germs that are indicators of fecal contamination (Figure 2). Some pathogenic germs are also present.
The average loads of total coliform (TC) are $1, 910^{4}$; $2, 910^{9}$ and 910^{9} germs / 100ml for Oued Ouislane, Ain Taoujdate, and for Aïn Choubbik respectively. This shows direct contamination by the fecal matter. On the other hand, the average values of fecal coliforms are $1, 210^{3}$; 710^{6} and $8, 410^{9}$ germs / 100 ml for oued ouislane, Aïn Choubbik and for Aïn Taoujdate respectively. Those values exceed the limit values recommended by the WHO [13].The high levels of fecal coliforms are attributable to the high content of organic matter (BOD5) and the lack of precipitation.
Results of the Oued Ouislane have been compared over the years [11] and allowed to conclude that the bacteriological quality has been deteriorating over the years. On, makes it possible to conclude that of the water has deteriorated over the years. The magnitude of fecal bacterial contamination of rivers, especially during the summer months, may be an imminent health risk [11].
The average value of the *fecal streptococci* (SF) are $1, 710^{8}$; 510^{6}; and $1, 210^{3}$ of Ain taoudjdate, Aïn Choubbik, Ouislane respectively.
Aïn choubbik site receives hospital Mohamed V effluents, which explains that Pseudomonas *aeruginosa*, were found only in this site. These waters are used without treatment in the irrigation, which constitutes health risk for major consumers.

The pathogenic germs such as genus *Salmonella* have been systematically sought for the three sites. None of the sites contained *Salmonella* bacteria. These results are consistent with previous study done on oued Ouislane [1, 12].

4. Conclusion

The physicochemical analysis of wastewater in the Meknes region has showed that these waters have a COD / BOD5 ratio less than 2.5, which indicates clearly that they are biodegradable. On the other hand, the analysis reveals the presence of heavy metals.
However, microbiological analysis revealed very high concentrations of germs in these waters, which constitutes a health risk. To address this issue, Wastewaters must, before being discharged into a natural receiving body, be treated. (The wastewater collected must flows through a number of preliminary treatment installations) (Case of Oued Ouislane and Mohamed V Hospital effluents).
To sum up, the studied wastewater should not be reused directly in agriculture or other activities as is no the case
To protect health or environmental quality, adequate treatment of wastewater discharges must be considered before any use. The quality of treated wastewater released must meet established standards.

References

1- CHAHLAOUI A. (1996). Etude hydro biologique de l'oued Boufekrane (Meknès), Impact sur l'environnement et la santé. Thèse d'état. Fac. Meknès. 234 p.
2- M. Bzioui : Rapport national sur les ressources en eau au Maroc.U.N. Water-Africa, 44p, 2004
3- Rodier, J., Bazin, C., Broutin, J.P., Chambon, P., Champsaur, H. et Rodi, L. 1996. L'analyse de l'eau. Eaux naturelles, eaux résiduaires, eaux de mer. Dunod, 8ème édition, Paris, pp. 1384
4- J. Rodier : L'analyse de l'eau. Dunod (Ed), 8ème Ed, 1384 p, 1996. B. Miller, Proc. 6 th Australian Electrochem. Conf., Geelong, Vic., 19-24 Feb., 1984; J. Electroanal. Chem., 168 (1984) 91
5- Valeurs limites des rejets directs et indirects du Maroc.
6- Ministère de l'Environnement du Maroc « Normes marocaines, Bulletin officiel du Maroc », N° 5062 du 30 ramadan.
7- Ouazzani N, Bousseltaj K & Abbas Y (1996) Reuse of waste water treated by infiltration percolation. WastSci Tech 33 (10/11): 401-408
8- Bechac JP, Boutin P, Mercier B & Nuer P. Traitement des eaux usées, Ed. Eyrolles, (Paris). 280 p.1987
9- NEAL C., JARVIE H. P., HOWARTH S. M., WHITEHEAD P. G., WILLIAMS R. J., NEAL M., HARROW M., WICKHAM H. The water quality of the River Kennet: initial observations on a lowland chalk stream impacted by sewage inputs and phosphorus remediation. Sci. Total Environ. 251-252: 477-495. 2000.
10- Aboueloufa M., El Halouani H., Kharboua M. &Berrichi A. Caractérisation physicochimique et bactériologique des eaux usées brutes de la ville d'Oujda : canal principal et oued Bounaîm. Actes Inst. Agronom. &Vet., Rabat, 22, 3, 143-150, 2002.
11- OMS (1989) Utilisation des eaux usées en agriculture et en aquaculture. Série de rapport technique 778

12- El Ouali Lalami1, A. Zanibou, K. Bekhti, F. Zerrouq, M. Merzouki. Contrôle de la qualité microbiologique des eaux usées domestiques et industrielles de la ville de Fès au Maroc. JMESCN J. Mater. 5 : 2325-2332.2014.

INTEGRATING THE ENVIRONMENT IN LOCAL STRATEGIC PLANNING : GUIDELINES (CASE OF MOROCCO)

Hafsa BENBRAHIM*

Phd Researcher
Laboratory Environment, Oceanology and Natural Resources (LEORN),
Faculté des Sciences et Techniques de Tanger-Université Abdelmalek Essaâdi,
BP. 416, Tangier, Morocco.

Abstract. Since 2010, an advanced regionalization project has been initiated by Morocco, which plans to consolidate the processes of decentralization and deconcentration by extending the powers of the regions and other local authorities. This project, institutionalized in the 2011 Constitution, defines the territorial organization of the Kingdom and reinforces decentralization according to a model of advanced regionalization. Through advanced regionalization, Morocco aims at integrated and sustainable development in economic, social, cultural and environmental terms, through the development of the potential and resources of each region. However, in order to honor this commitment of advanced regionalization, local authorities must be assisted in adopting a local strategic planning approach, allowing them to develop territorial plans for sustainable development in accordance with the national legal framework, specifically the Framework law 99-12, and international commitments in terms of environmental protection.

This research deals with the issue of environmental governance in relation to the role and duties of local authorities. Thus, the main goal of our study is to present the guidelines to be followed by the local authorities to improve the quality of the environment integration process in the local strategic planning with the aim of putting it in a perspective of sustainable development.

Key words : Sustainable Development, Local Environmental Integration, Territorial Strategic Planning.

*Corresponding author: hafsabenbrahim@yahoo.com

1. Introduction

The role of territorial communities in the process of implementing sustainable development was affirmed at the United Nations Conference on Environment and Development in Rio (1992): «It is local communities that build, monitor the planning processes, set local environmental guidelines and regulations, and support the implementation of environmental policies at the national or sub-national level. They play at the administrative level closest to the population, a key role in education, mobilization and taking into account public views for sustainable development» [1].

Agenda 21, for this purpose, suggests in Chapter 28[1] that local authorities also need to prepare action's strategies that respond, on their own scale, to the concern for sustainable development. These initiatives can be considered as an action plan resulting from the reflections made around the postulate of sustainable development and which the purpose is the integration of this vision in the management of local communities [1].

However, implementing sustainable development as a local public policy is problematic: Local and regional authorities are key players, but they intervene in increasingly large fields and face a growing complexity of the policies for which they are responsible. As a result, they do not always have the appropriate knowledge, forecasting and action tools to implement it. Thus, the integration of sustainable development requires a gradual and continuous learning and improvement process, under which sustainable development integrates, simultaneously and transversally, the economic, environmental and social dimensions into any development project, as well as their interrelations in order to find the best balance or the best possible compromise.

Sustainable development introduces new logics of transversality, long-term and participation, particularly in the objectives and practices of public action [2]. While the very definition of sustainable development is debated, a consensus seems to have emerged on the need to act locally to solve a global problem.

The territorial approach is gradually becoming an essential component of the implementation of sustainable development, especially since the diversity of its stakes can hardly be taken into account on a global scale. It is a multidimensional concept, based on the reconciliation of economic, social and environmental issues that can be contradictory. And since the operational translation of these dimensions leaves large place for interpretation, one can think that there is no sustainable development but rather sustainable developments that respond to the specificities of each territory. This territorialisation reinforces, moreover, one of the founding principles of sustainable development, that of the participation of actors in development policy. Various forms of participation have been highlighted in the work of REY-VALETTE et al. [2] and ANDRES and FARACO [3]. While at a global scale, the sustainable development process is based on a normative approach based on values accepted by all (convergence of discourses), its implementation at the lower levels is closely linked to the individual and collective visions of the decision-making actors (experts, politicians, citizens, etc.) and the specificities of the territory (geographic situation, economic and social specificities, etc.).

[1] http://www.un.org/french/ga/special/sids/agenda21/action28.htm

For DECAMPS and VICARD [4] , the application of the principles of sustainable development at the national and regional scales follows a procedural approach, its dimensions being then declined according to more precise orientations, and defined by actors pursuing a common general interest. Similarly, the implementation of sub-regional sustainable development is a collaborative, compromise-based approach in response to the particular interests of stakeholders. The issues of sustainable development are then broken down into consensual actions, adapted to the specificities of their territory and to the objectives and practices of local actors. Although the interest of a territorialisation of sustainable development is no longer demonstrated, it nonetheless leads to obvious complexification. Sustainable development is still a difficult concept to make operational for local authorities, which results in a great heterogeneity of implementation methodologies.

As a pillar of sustainable development, the environment has been in fact for several decades, and still remains today, a "laboratory" where new forms of governance are constantly being invented: democratic consultation procedures, flexible forms of coordination, modes of decentralized management. Various examples attest to this effort to modernize public action: "Agendas 21", "impact studies", "voluntary agreements", "rights to pollute markets", "independent authorities", "conferences" consensus ","precautionary principle "," sustainable development "," negotiated standards or agreements "..., etc. This trend obviously accelerated at the turn of the 90s, in a context marked by globalization, the emergence of major economic regions such as Europe, and the growing influence of liberal ideas, to the point that one can speak retrospectively of a true "silent revolution" in the design and implementation of environmental policies [5] .

If we take the case of Morocco, which since 2010 has experienced an unprecedented dynamism especially in terms of governance, many reform projects are indeed committed to endorse the process of democratization of society and guide the search for a sustainable development model at the national; regional and local level. Under these reforms, and in accordance with the Article 2 of the Framework law 99-12 [6], bearing the Charter of the Environment and Sustainable Development (CNEDD), the local authorities are called to take into account the objectives of State's action in terms of environmental protection and sustainable development in their strategies for developing and implementing development plans. This implies that the communal planning process needs to be revised to align more closely with the fundamental principles, rights and duties of sustainable development in accordance with Article 1 of the Framework law 99-12 [6]. Thus, territorial authorities have to watch quite particularly to adopt a global, intersectorial and transverse approach allowing to take into account environmental stakes in the elaboration of the visions and the plans of local development.

The integration of the environment into municipal planning is now recognized as an essential requirement to put local Action in the path of sustainable development. The main national references in terms of environmental integration and sustainable development are respectively the provisions contained in the 2011 Constitution [7], and in particular the Framework law 99-12, bearing Charter of Environment and Sustainable Development (CNEDD) [6]. In response to these legal references, it is fundamental to renew the tools and methods adopted so far in the context of municipal planning, to base this process on sustainable long-term visions and on strategic choices that are aligned perfectly for the purposes of sustainable development on a local scale.

The promulgation of the Framework law 99-12 represented a real achievement for the Moroccan legal framework in terms of protection of the environment and sustainable development at the national level. It constitutes a true legal reference for the decision-makers, planners and executors of public policies whose main aim is to decline and operationalize the National Charter of the Environment and Sustainable Development (CNEDD), by giving a legal basis to its content. This framework law establishes the principles, the rights, the duties and the commitments, proclaimed by the CNEDD, which aims at inscribing the processes of elaboration, implementation and monitoring-assessment of the strategies, as well as sectoral policies, programs and action plans in strict compliance with the requirements of environmental protection and sustainable development.

It should be noted that one of the important aspects of this framework law is that it defines the responsibilities and commitments that all stakeholders (State, local authorities, public institutions, private companies, civil society associations and citizens) must respect in terms of environmental protection and sustainable development.

2. Methodology :

Our approach of integration of the environment in the local strategic planning, is based on a qualitative analysis which takes as frame of reference:

- The National Charter of the Environment and Sustainable Development [6] which becomes the reference framework for the development of any territorial project;
- The new Constitution [7] and the process of advanced regionalization, which require a strong commitment of local authorities that allows them to dispose consequently of a broader decision-making and executive powers that should enable them to perform their duties and to honor their new responsibilities;
- The National Initiative for Human Development [8], which aims to struggle against poverty, exclusion and territorial disparities, and which intervenes mainly in the most vulnerable areas in both urban and rural areas;
- And finally, the broad project of support and accompaniment of the communes in the elaboration of their Communal Development Plan (PCD) [9]: "The commune at the 2020 horizon" which aims to make local authorities a locomotive of development at the local level, provincial and regional.

Our aim is to bring together the findings, to establish a reading grid on the integration of the 'environment' component in strategic planning at the local level under the territorial variation of the national environmental policy in a strategic and prospective framework.

3. Results and discussion

Our research aims to highlight guidelines for integrating the environment into local strategic planning, which challenges consequently the concept of environmental governance in relation to the role and duties of territorial communities.

3.1 Fundamental principles of environmental integration

Seven fundamental principles, in terms of environmental integration and sustainable development, are set out in Article 2 of the framework law 99-12, bearing Charter of Environment and Sustainable Development (CNEDD). These principles, considered as essential elements of framing that must be respected by local authorities and other levels of decision-making (national and regional) in any medium and long-term development action, are:

1. The principle of integration: it implies the need to adopt a global and transversal approach when drawing up and implementing development plans in the medium and long term;
2. The principle of territoriality: it imposes the taking into account of the territorial dimension, in particular the regional one, in order to ensure a better articulation of the measures initiated by the different levels of territorial decisions and to encourage the mobilization of the actors in favor of human development, sustainable and balanced territories;
3. The principle of solidarity: it allows to increase the capacity of the territory to face the vulnerabilities and to favor a rational, economical and balanced use of natural resources and spaces;
4. The precautionary principle: it recommends that adequate, economically viable and acceptable measures be taken to deal with environmental damage or potential risks, even in the absence of absolute scientific certainty about their real impacts;
5. The principle of prevention: it invites to provide tools for regular assessment and assessment of the impacts of activities likely to harm the environment, to recommend and implement concrete measures to eliminate these impacts, or at least reduce their negative effects;
6. The principle of responsibility: it specifies the obligation of any person, physical or moral, public or private, to repair the damage caused to the environment;
7. The principle of participation: it encourages the active participation of companies, civil society associations and the public in the process of developing and implementing plans for environmental protection and sustainable development.of participation.

The adoption of these fundamental principles of the Framework law 99-12 is aimed primarily at bridging existing legal gaps in the integration of the environment into strategies, policies, programs and action plans. At local authority levels, it lays the foundations for a new generation of communal development plans, based on a global, cross-sectoral and transversal approach, enabling to achieve sustainable territorial development in the long term.

3.2 Foundations of the integration approach of Environment and Sustainable Development at the local scale

According to Article 20 of Framework law 99-12, regions and other local authorities must ensure that the principles and objectives of environmental protection and sustainable development are integrated into planning tools and programs related to their respective territories.

On this basis, the Local Environment Integration (IEL) and Sustainable Development approach sets itself as key objectives to help local authorities respectively to:

- ➢ Base the communal planning process on the fundamental principles and objectives of environmental protection and sustainable development stated in the Framework law 99-12 (CNEDD);
- ➢ Basing communal planning on a vision shared by the populations, in the long term, which reconciles the economic, social and cultural objectives of the development activities and the issues of environmental protection of the territory;
- ➢ Articulate municipal planning with national strategies, policies and conventions related to human development and environmental protection;
- ➢ Integrate in the communal planning, at a time environmental protection and the valorization of environmental services, according to an ecosystem approach, while promoting the effects of synergy, subsidiarity and pooling of intercommunal efforts.

These key-objectives form the framework's elements subtending the guidelines to regulate and govern the communal planning process, but also permits to integrate systematically the environmental dimension and to suscribe the communal planning in a local sustainable development perspective. These elements of framing require that the new generation of Communal Development Plans (PCD) adopts respectively:

- a global, intersectoral and transversal approach;
- an active participation process;
- a relevant spatial scale;
- inter-municipal and regional solidarity;
- a high level commitment;
- a simple and practical environmental assessment tool;
- a strategy to mitigate potential risks.

3.3 Constitutive phases of the Integration approach of Environment at the local scale

The environmental integration approach (IEL) in municipal planning has four main stages, which are organized, according to the communal planning process, recommended by the General Directorate of Local Communities's guide (DGCL guide) [10], as follows:

- ➢ Stage 1: Integration of the environment into the preparation stage and into the launch of the Communal Development Plans (PCD);
- ➢ Stage 2: Integration of the environment into the participatory diagnosis stage;
- ➢ Stage 3: Integration of the environment into the planning, programming and putting in coherence stage;
- ➢ Stage 4: Integration of the environment in the Communal Information System (SIC).

In terms of discussion, and after having put forward the principles and foundations of the environmental integration approach in local strategic planning (IEL), it results that its application to the municipal planning process aims at three essential objectives:

1. Identify, understand and prioritize the environmental issues that are decisive for the future of the territory, and identify their interdependencies with the rest of the economic, social and cultural components;
2. Integrate, according to a participatory and transversal approach, the priority environmental issues of the territory, in the construction of the vision of development and in the programming of the actions;
3. Monitor progress made in integrating environmental issues into municipal planning, based on territorialized sustainable development indicators.

According to these objectives, it appears that the notion of environmental issue is at the heart of the "IEL" approach. And the environmental issue implies, at this level, any change in the state of the environment that significantly influences the future of the territory, and which often involves an ecosystem service, useful for the current and future social well-being, that it is not acceptable to see the disappearance or degradation, from the point of view of natural resources, public health or quality of life.

4. Conclusion :

The construction of a territorial project for sustainable development in Morocco requires the systematic consideration of interdependencies between the economic, social, cultural and environmental dimensions. The Local Environment Integration and Sustainable Development approach (IEL approach) should engage local and regional authorities in a process of continuous improvement, allowing to respond simultaneously to at least four types of essential needs:

1. Reconcile the short-term economic, social and cultural objectives with the decisive environmental issues of the territory and with the perspective of sustainable development in the long term;
2. Promote the active participation, commitment and responsiveness of local stakeholders and their partners in the territory, throughout the process of building and implementing the vision of sustainable development;
3. Use the relevant environmental information to forecast, anticipate and take the appropriate measures that are aligned with and fully meet the objectives of sustainable territorial development;
4. To dispose of monitoring tool which allow to report on local authorities efforts undertaken in terms of environmental protection and valorization of environmental services.

References :

- [1] UNCED, Agenda 21, The United Nations Program of Action from Rio, New York, United Nations Development Program (1992) : https://sustainabledevelopment.un.org/content/documents/Agenda21.pdf

- [2] Rey-Valette H., Laloë Francis, Le Fur Jean, Roussel S. (2006). Usages des indicateurs de développement durable: entre offre et demande d'indicateurs. Saint-Quentin-en-Yvelines: IRD-UVSQ, 15 p. multigr. Colloque GECOREV.Gestion Concertée des Ressources Naturelles et de l'Environnement, du Local au Mondial, Saint Quentin en Yvelines (FRA), 2006/06/26-28.

- [3] L. Andres, B. Faraco, *La territorialisation des normes du développement durable. Agenda 21 locaux : vers un modèle explicatif des différenciations »,* in A. FAURE et E. NEGRIER, *Les politiques publiques à l'épreuve de l'action locale, Critiques de la territorialisation*, L'Harmattan, Paris, (2007).

- [4] M. Decamps, F. VICARD, *Mesurer le développement durable: jeux d'indicateurs et enjeux locaux. Revue d'Économie Régionale & Urbaine,* 4 (2010).

- [5] W.D. SUNDERLIN, *Global environmental change, sociology and paradigm isolation*. Global environmental change, 5 (1995).

- [6] Loi n° 99-12 portant Charte de l'Environnement et du Développement Durable http://www.environnement.gov.ma/PDFs/loi_cadre_fr.pdf

- [7] Secrétariat Général du Gouvernement, *La Constitution, édition 2011,* Série Documentation juridique marocaine, (2011).

- [8] Conseil Économique, *Social et Environnemental, Initiative Nationale pour le Développement Humain: analyse et recommandations.* Saisine, 2 (2013).

- [9] Ministère de l'intérieur, *Projet de guide pour l'élaboration du Plan Communal de Développement (PCD) selon l'approche de planification strategique participative,* Direction Générale des Collectivités Locales, (2008).

- [10] Ministère Délégué auprès du Ministère de l'Energie, des Mines, de l'Eau et de l'Environnement, chargé de l'Environnement, *Guide pour l'intégration de l'environnement dans la planification communal,* (2014).

Complementary references:

- A. Boutaud, *Le développement durable : penser le changement ou changer le pansement ? Bilan et analyse des outils d'évaluation des politiques publiques locales en France : de l'émergence d'un changement dans les modes de faire au défi d'un changement dans les modes de penser,* Thèse de doctorat, Science et génie de l'environnement, Ecole des Mines de St Etienne-Université J. Monnet, St Etienne, (2005).

- A.Fouquet, J. Perriault, *Évaluation, politiques publiques, politique de recherche.* Communication et Organisation, 38 (2010).

- F. Bertrand, *Planification et développement durable : vers de nouvelles pratiques d'aménagement régional ? L'exemple de deux Régions françaises, Nord-Pas-de-Calais et Midi-Pyrénées*. Géographie. Université François Rabelais-Tours, (2004). [manuscrit disponible en ligne : http://tel.ccsd.cnrs.fr/tel-00012142].

- F. Rudolf, J. kosman, *Le développement durable entre programme d'action et applications* ? , Ecologie et politique, 29 (2004).

- G.H. Bruntland, *Notre avenir à tous (Our Common Future), Rapport de la Commission des Nations-Unies pour l'Environnement et le Développement (CNUED)*, Editions du Fleuve, Québec, (1987).

- J. Theys, *La Gouvernance, entre innovation et impuissance*, Développement durable et territoires 2 (2003). URL : http://developpementdurable.revues.org/1523; DOI : 10.4000/developpementdurable.1523.

- MEDD/PUCA programme « Politiques territoriales et développement durable » [http://citeres.univ-tours.fr//p_vst/contrats/D2RTrapportfinal.pdf].

E3S Web of Conferences **37**, 02009 (2018) https://doi.org/10.1051/e3sconf/20183702009
EDE6-2017

From the promotion of biodiversity to the Recovery of organic waste

Mustapha Jammoukh[1,*], *Khalifa* Mansouri[1] and *Bachir* Salhi[2]

[1]Laboratory of Signals, Distributed systems and Artificial Intelligence (SSDIA), Hassan II University of Casablanca, Higher Normal School of Technical Education, P.O. Box 159 28 820 Mohammedia, Morocco.
[2]Laboratory of Applied Mechanics and Technology (LAMAT), Mohammed V University of Rabat, Higher Normal School of Technical Education P.O. Box 6207, 10 100 Rabat, , Morocco.

Abstract. This article presents an empirical research to classify a new renewable resource material, as opposed to eco-composites, it has been neglected by the materials specialist. This classification is based on the typology of elastic behavior demonstrated by tensile tests. In addition, some identifying criterions of the usefulness of this material were examined. To justify the relevance of this classification, curves from the extension of tests focusing on the virgin material, illustrate significant results of the review. Obtained from waste, having a significant recycling possibilities and potential from renewable resources, bio-mechanically characterized loads will be injected into polymeric materials of different categories. All in the perspective of promoting changes in thermomechanical properties, whether static or dynamic; such as resistance to corrosion, heat, wear... They result in functional changes such as security, relief, coatings and stability...

* Corresponding Author : JAMOUKH@Yahoo.fr

 https://doi.org/10.1051/e3sconf/20183702009

1. Introduction

Since 1992, Kohmei Halada has defined eco-materials [1, 2, 3] as materials contributing to the reduction of the environmental impact associated with human activities. This concept takes its meaning in the context of sustainable development and eco-design. These can be innovative materials or original combinations of known materials: materials containing fewer hazardous substances, materials from renewable resources and some industrial coproducts [4].

The latest advances in organic or inorganic materials, plant, animal or hybrid origin are in vogue in the development of " Sustainable Materials ". It will be necessary to consider the use of natural and renewable resources both at the level of matrix and loads (natural loads and plants) for composite applications by focusing on the physico-chemical properties [5].

Based on these recommendations, we hope, through our research, to improve these eco-composites by reinforcing their thermo-mechanical qualities by a load of animal origin.

By general name of charge, we mean any inert substance, added to a basic polymer, makes it possible to modify the mechanical, electrical or thermal properties in a sensitive manner, to improve the surface appearance or simply to reduce the price of returns from the processed material [6].

For a given polymer, the choice of a load is determined based on the changes you are looking for the finished object. However, in general, substances used as loads of plastics will first have to meet several requirements: compatibility with basic resin, wettability, uniformity of quality and particle size, low abrasive action, low cost... [7, 8 ,9].
This work essentially attempts to unveil the physical performance of the virgin biomaterial found naturally so that other materials can benefit from: Hardness, Toughness, Resilience, Heat Resistance ...
[10].

This animal-based bio-charge comes in the form of a regular thickness sheath and generally smooth surface. It is finer at its base than at its extremity. The lower part is rough on the surface and has streaks at fairly regular intervals (Figure 1). Although the "Death" this fabric is produced by the living. It is composed of proteins rich in sulfur, forming fibers. It is present on the surface of the skin in the form of scales. Once it was a valuable material, ideal for making lots of items. In mythology, this matter of abundance symbolizes perpetual profligacy, it produces much from little [11].

After its preparation in powder or in cuttings, this substance will be injected into a basic composite material and then tested to evaluate its contribution to the reinforcement of the thermo-mechanical characteristics sought on the base material after summing up and also optimizing its work [12]. To develop this new material, a set of actions will be discussed and can be categorized into four phases (Figure 1).

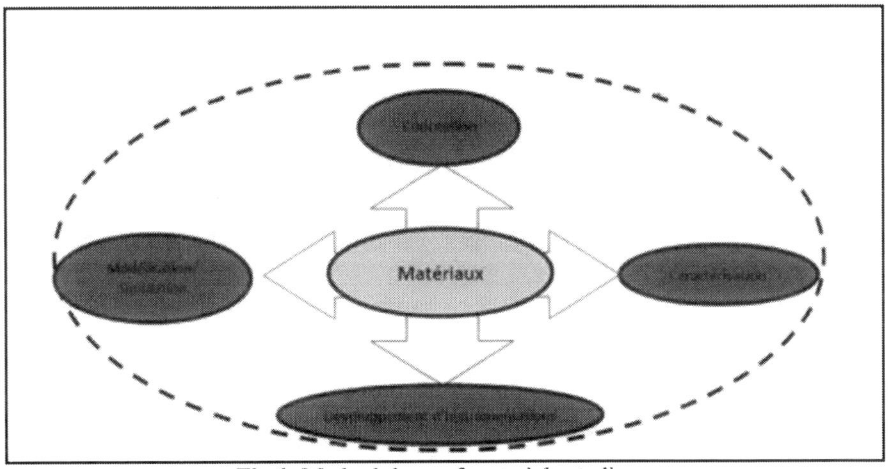

Fig.1. Methodology of materials studies.

2. Characterization Strategy of the Virgin material

2.1 process for the preparation of tensile specimens

The transformation of the sheath into bio-charge is obtained by separating it from the cornillon forming the selected substance. This material is very easy to work with, once heated and softened, it can be cracked, compressed, sawed, pierced or rotated. We can even solder parts to make them placquers [11]. This step is used to obtain specimens sample for mechanical tests by punching or cutting, particularly those for tensile tests. It can also be turned into cutters and powder to be eventually used as loads in plastic materials [6].

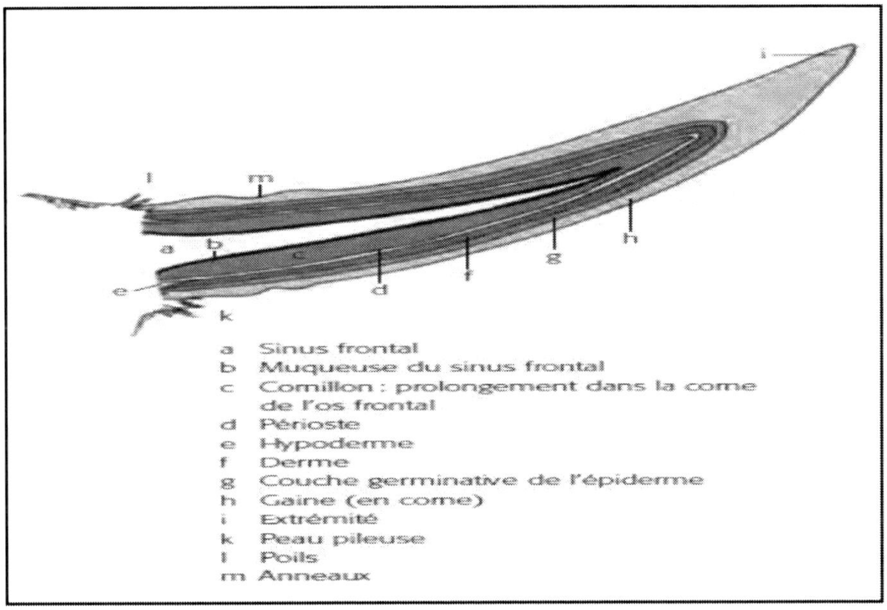

Fig. 2. Longitudinal sections of the corne.

Once heated and softened, the sheath is compressed for a quarter of an hour in a press. This phase is followed by the milling process for finishing the surface of the plate, according to ISO 527-2 (Figure 3).

 https://doi.org/10.1051/e3sconf/20183702009

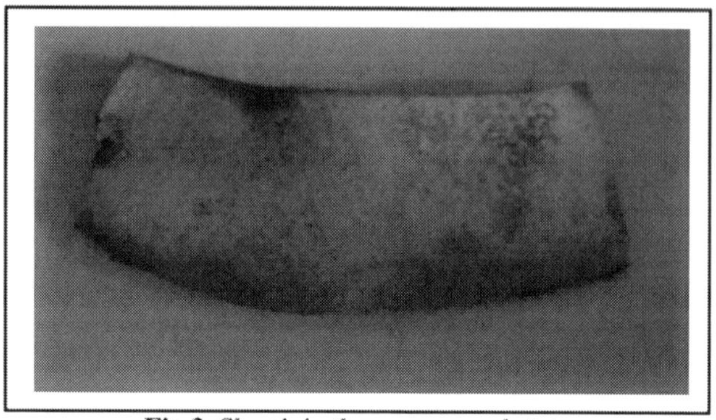

Fig.3. Sheath in the compressed state.

The plate is then punched or cut to have lots of test pieces in accordance with the current standard (Figure 4).

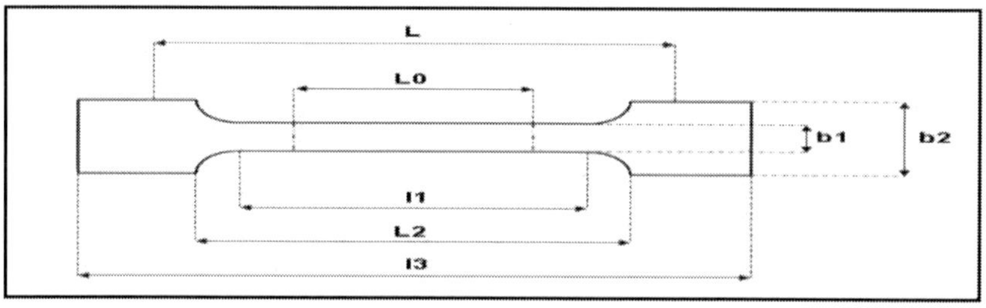

Scheme.1. Standardized specimen.

Table 1. Standardized values of the dimensions of the test specimen.

ISO 527-2	1BA	l3 > 75	l1 = $30^{\pm}$ 0.5	b_2= $10^{\pm}$ 0.5	b_1= $5^{\pm}$ 0.5	h> 2	L_0= 25 $^{\pm}$ 0.5	L_2 + 2; L_2= $58^{\pm}$	Dumbbell

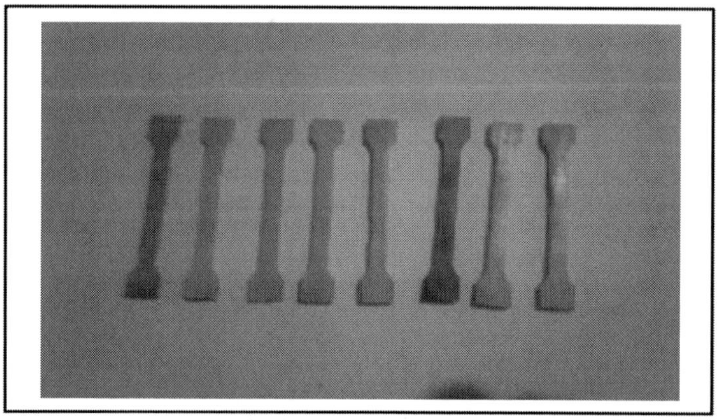

Fig.4. Tubes of traction.

The objective of our first experiment is to verify the advanced assumptions and to acquire positive classification and explore criteria in mechanical characteristics such as the maximum breaking stress, the relative strain and Young's modulus of virgin material [13]. To this end, the flank is cut out according to 45 ° angle of fibers direction in four test

 https://doi.org/10.1051/e3sconf/20183702009

specimens in ISO 527-2Standards, type 1BA and form "HALTERE "(table 1) and des traction tests are made on these specimens to test the reactivity of the virgin material to the solicitations exercised.

2.2 Process of characterization

The test bench used is a machine which is adapted both to quality control and to research and develop for various applications. Its modular system provides an optimum solution for demanding testing requests.

Fig.5. Traction Machine "AllroundLine"

Regardless of the load applied to the test piece, this traction machine has a test speed of 0.00005 to 3000 mm/min. The speed of the test machine is independent of the test force.
The traction machine rise test loads up to 250 kN. The test space height is 1030 mm to 2560 Mr.
Not requiring the addition of any extra cards, it can be used with a commercial PC or laptop.

https://doi.org/10.1051/e3sconf/20183702009

2.3 Experimental Results

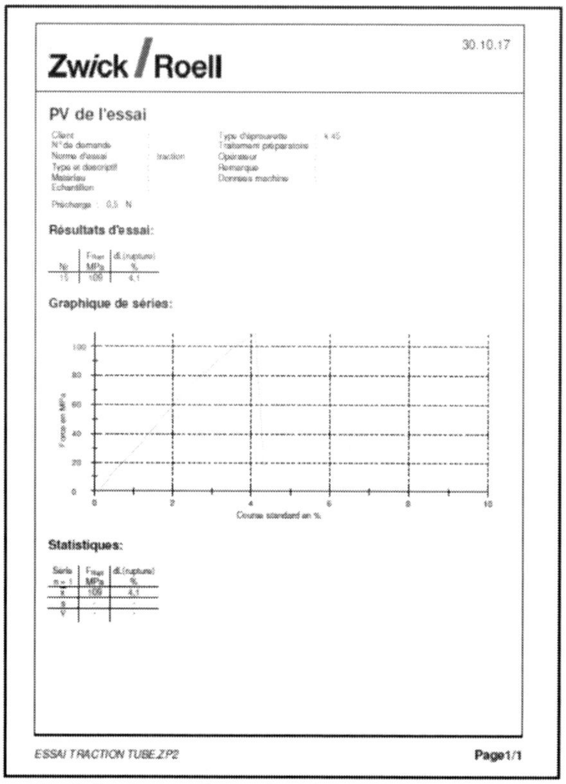

Fig.6-A. Stress-elongation diagram "EProof No. 1"

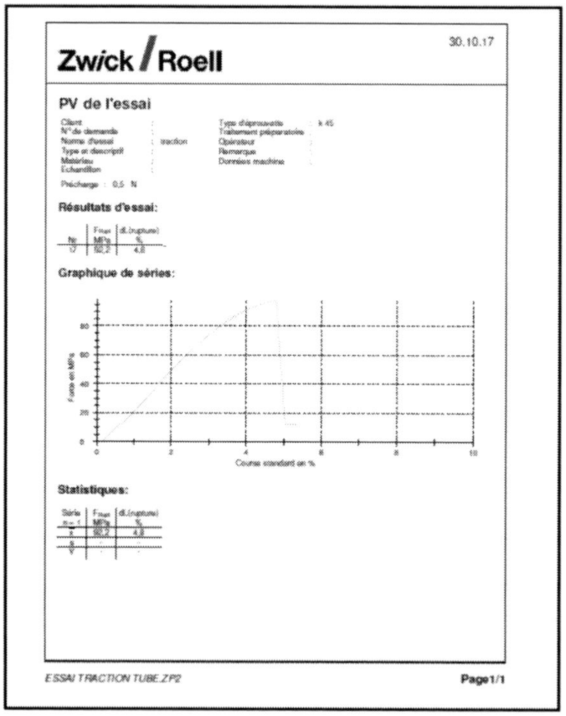

Fig.6-B. Stress Diagram-Elongation "EProof No. 2"

https://doi.org/10.1051/e3sconf/20183702009

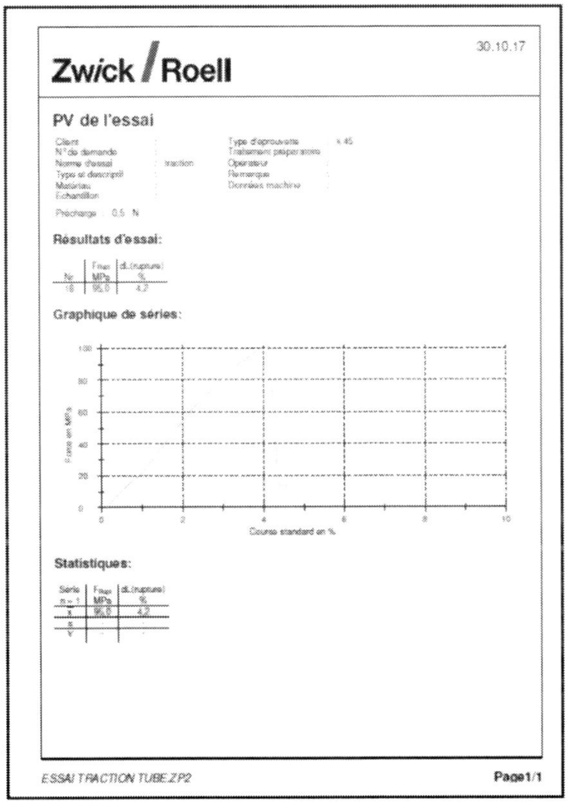

Fig.6-C. Stress-elongation diagram "EProve No. 3"

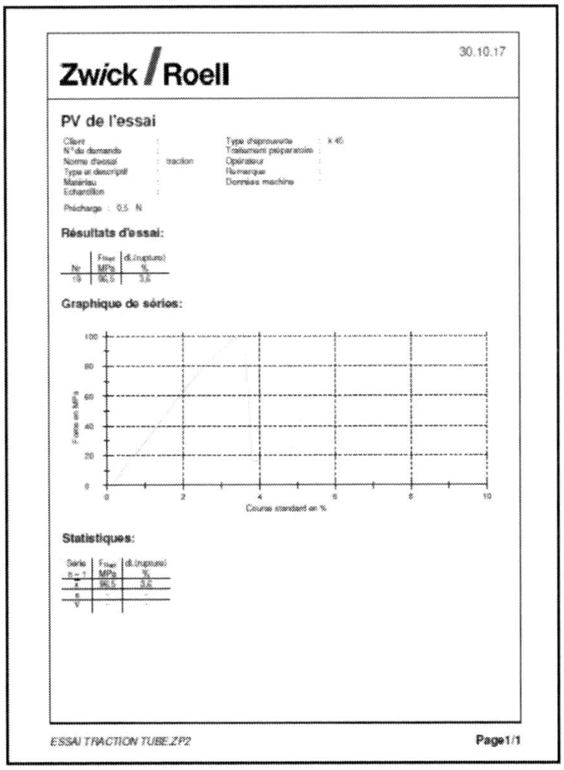

Fig.6-D. Stress-elongation diagram "EProof No. 4"

2.4 Comments

We can summarize by these remarques :
Demonstration of an intrinsic elastic behavior for this chosen direction of cutting specimens;

Identical reactivities of the virgin material to tensile stresses in similar conditions of experience during the traction test similar to the elastic behavior of the biomaterial (Figures 6);
Elastic behavior, characterized by the reversibility of strains during the suppression of stresses, is manifested only for constraints less than a limit value, noted "Re" which is called the elasticity limit, as shown in the figure above. However, beyond this limit value, the rupture occurs for all test specimens;

In the case of tensile tests on fibres, the ultimate constraint presents a stable shape as shown in the results part. This stems from the fact that the fracture is related to the orientation of pre-existing fibres (Figures 6).
This Type of behavior clearly shows that the biomaterial has a linear elastic behavior and reversible one. Its rupture occurs while it is still in a purely elastic diet.
At the end of the different results of elastic behaviors recorded by the traction diagram, we can see for the direction of the fibers involved (45 °) elastic performances such as:

- A limitation of rupture with very striking traction.
- A relatively small displacement.
- A significant module of longitudinal elasticity.

Also, this biomaterial does not undergo any permanent plastic deformation or necking before it ruptured: It is a rigid and fragile material (Figure 6).
As well this biomaterial can play, through all these optimum indicators, an important role in the bio-loading of other polymer materials and contribute, subsequently, in a respectful manner to the environment by the valorisation of organic waste, to a promotion of the Bio-diversity 14].

3. Conclusion

The development of more environmentally friendly materials has become a major concern for the industrial world and this responds to a growing society demand. Currently, the research axes on these materials relate to the materials Bio-sourced, biodegradable materials and recycled materials.
In this intention, we were able to contribute to the elastic characterization of a virgin material of animal origins by cutting of the test specimens depending on the orientation of its internal fibers and then testing its tensile strength on a traction bench under similar experience conditions. It is noted that in these tests the material underwent several extension tests and the most significant experimental values were selected.
The reactivity of the biomaterial at tensile stresses generated a similar elastic intrinsic behavior, knowing that four registered behaviors are likely to be present by their related characteristics i.e., maximum resistance to rupture, relative displacement and Young's modulus.
This study has for objective the development of typological classification of the virgin biomaterial according to its mechanical characterization. It shows that biomaterial has characteristics mechanical qualities according to the fibers orientation sense of the material: Anisotropic behavior is judged very favorable to the development of the bio-loads of animal origin for technological innovation of certain polymeric materials [6].

In terms of sustainable development, two main approaches will be developed in the research axis: The development of recycling animal's organ in end of life through improved identification performances and the development of bio-loaded materials of renewable resources from organic biomass.

Scientific barriers

The development of bio based, and biodegradable materials requires the development of specific tests to evaluate the biodegradation of the formulated materials. The end of life by composting must be maintained regardless of the physical properties of the materials and the surface treatments made. The development of materials from secondary first resources requires the obtaining of increasingly important purity in order to aim at high value-added applications.

References

1. A. KOUADRI, S. CRETIN, (2007): "Inseparable evolutions".
2. E. BRETAGNE, J. BREARD, v. Manasiya and V. Verney, (2012): Eco-Materials: The materials Go Green ", **No. 100**, 367-368.
3. D. Lice, (1991): (Shell chemistry), "economics-Ecology of Plastics Recycling", symposium of the French Group of studies and Applications of Polymers, Douai, 38-43.
4. K. HALADA, (1992) : Bull. Jpn. Inst. Metals **No. 31**, 505-512.
5. L. AUGIER, (2007): "Study of the elaboration of PVC/wood composite materials from carpentry waste: formulation, characterization, durability and recyclability". Vol. 2507, 11-12.
6. M. JAMMOUKH, K. Mansouri and B. Sadeghi, (2016): Experimental study of the elastic behaviour of a resource bio-load Renewable CSR-2016, 1, 61-66.
7. There. AGARI, Mr. SHIMADA. AND A. UEDA, (1997): Polymer, Vol. 38, **N ° 11**, 2649-2655.
8. Jc. Angola, Y. FUJITA, T. SAKAI, T. INOUE, (1988): "Compatibilizer-aided toughening in polymer blends consisting of brittle particles dispersed in a ductile polymer matrix", Journal of Polymer Science, 26 (4) 807-816.
9. F.S. BATES, (1991): "Polymer-polymer phase behaviour", Science, 251 (4996), 898-905.
10. C.A. BERNARDO, a.m. CUNHA AND M.J. OLIVEIRA, 1996: The Department of Polymer Engineering, Universidade Do Minho), "The Recycling of thermoplastics: Prediction of The Properties of mixtures of Virgin and reprocessed Polyolefins", 4800 Guimaraes, Portugal, Polymer Engineering and Science, Vol. 36, **No. 4.**
11. A.S. NEFF (FIBL), B. HUMI and R. STREIFF, (2015): The importance of horns in cows. **Vol. 6**, 1691
12. K. HALADA, ,(2003): "Progress of Ecomaterials Towards a sustainable society, Current Opinion in Solid State and Materials Science, **Vol. 7**, 209-216-368.
13. E. BRETAGNE, J. BREARD, v. Manasiya and V. Verney, (2012): Eco-Materials: The materials Go Green ", **No. 100**, 367-368.
14. ALLRED, (1996): Recycling process to scrap composites and Prepregs, Int. Sampe SyPm. Exhib, **41**, 21-31.

Climate Change Impact on Rainfall: How will Threaten Wheat Yield?

*K.*Tafoughalti[1*], *E.M.* El Faleh[1], *Y.* Moujahid[2], *F.* Ouargaga[1]

[1] Geosciences, Patrimony useful Substances lab, Department of Geology, Faculty of Sciences, UMI - Meknes, Morocco
[2] Regional Center for Agricultural Research - URGDRNESR, Meknes, Morocco

ABSTRACT. Climate change has a significant impact on the environmental condition of the agricultural region. Meknes has an agrarian economy and wheat production is of paramount importance. As most arable area are under rainfed system, Meknes is one of the sensitive regions to rainfall variability and consequently to climate change. Therefore, the use of changes in rainfall is vital for detecting the influence of climate system on agricultural productivity. This article identifies rainfall temporal variability and its impact on wheat yields. We used monthly rainfall records for three decades and wheat yields records of fifteen years. Rainfall variability is assessed utilizing the precipitation concentration index and the variation coefficient. The association between wheat yields and cumulative rainfall amounts of different scales was calculated based on a regression model.
The analysis shown moderate seasonal and irregular annual rainfall distribution. Yields fluctuated from 210 to 4500 Kg/ha with 52% of coefficient of variation. The correlation results shows that wheat yields are strongly correlated with rainfall of the period January to March. This investigation concluded that climate change is altering wheat yield and it is crucial to adept the necessary adaptation to challenge the risk.
Keywords: Climate change, wheat yield, rainfall variability, precipitation concentration index.

1. Introduction

Climate change is considered to be a world-wide phenomenon. The effect that it will have on agriculture is the biggest concerns. Morocco is especially notable for its vulnerability to climate change [1], climate is characterized by long periods of drought and a strong interannual variability in rainfall amount and distribution [2-4]. In the past few decades Moroccan agriculture have been suffering from severe water resources deficits [5-7].

* Corresponding author: khalid.tafoughalt@gmail.com ;

Various studies has reported varied effect of climate variations on agricultural production [1, 8-10]. In Morocco, cereals represents the main crop and are of great importance to national food security; the main part of the diet comes directly or indirectly from cereals [11]. Compared to other countries in the Mediterranean Basin, Production levels are generally poor [12]; the most due to direct exposure to nature. The heaviest bills are those of cereals, which represent almost 36% of imports in 2014 [13].

In earlier studies, Jarlan [14] has mentioned that the development of cereals is related to local environmental conditions, among which water availability are the most decisive. This discovery is consistent with previous works [15-17] and indicate the need for local specific studies.
Rainfall furnish the water which contributes to the transportation of the nutrients for plants development. In consideration of this essential role, deficiency of water supply, especially during critical developmental stages has negative effects on efficient plants growth, resulting enormous yield reductions.

Meknes is among the provinces that contribute most to national wheat production. Considering that 97% of total cultivated area is non-irrigated, this makes the agricultural system in the region vulnerable to dry spells and uneven distribution of the rainfall. What might straightly influencing the livelihood of the inhabitants.
Our ongoing study analyses rainfall variability at annual, seasonal and monthly scales and evaluate the correlation between rainfall and the yields of wheat, then it could be used in developing risk reducing strategies.

2. Material and methods

2.1. Study area

The domain of study was Meknes province north-west of Morocco (Figure 1). It is situated between longitude 33°88 N and latitudes 5°53 S. It has an area of 1692 km2. Characterized by a semi continental climate of Mediterranean type, with cool winters and rainy and hot summers and dry. The temperatures means range from 3 °C in cool season and 34 °C in hot season. The average annual rainfall ranges from 450 to 600 mm.

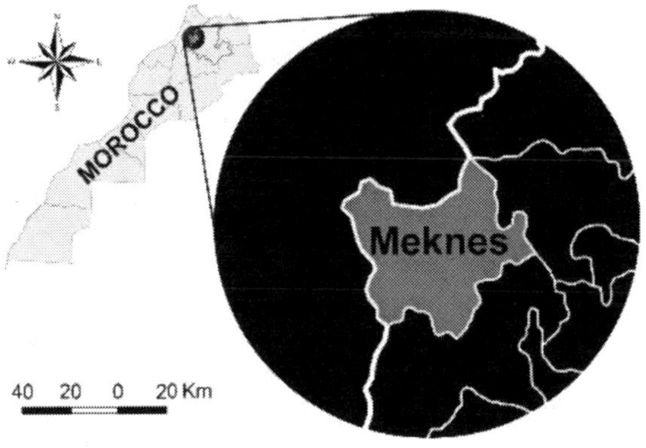

Fig.1. Location of Meknes in Morocco

2.2. Data and analysis

A rainfall data recorded over three decades (from 1985 to 2016) and crop yield records of soft wheat and hard wheat from 1995 to 2015 were provided by DPA; Provincial Department of Agriculture in Meknes. The mean and Coefficient of Variation (CV), Standard Deviation (SD) of monthly, seasonal and annual rainfall are computed by using

Excel software. The Precipitation Concentration Index (PCI, [18, 19]), was calculated on an annual scale, seasonal scale for winter (DJF), spring (MAM), autumn (SON), and on supra-seasonal scale for wet season (ONDJFM) by means of the following equation:

$$PCI = [CV * Pi]^2 + 1/n$$

Where Pi is the monthly precipitation in month i, CV is variance.

Each month is represented by its first letter e.g. ONDJFMAM = October November December January February March April May.

According to Oliver's classification:

A PCI values of less than 10 represent a uniform precipitation distribution; PCI values from 11 to 15 denote a moderate precipitation concentration; values from 16 to 20 denote irregular distribution and values above 20 indicates a strong irregularity (i.e., high precipitation concentration).

To evaluate the relationship between rainfall and wheat yields we used the correlation analysis. The calculation of the correlation coefficient is performed using Equation (1), in which rainfall represents the independent variable (X) and wheat yield represents the dependent variable (Y).

$$r = \frac{\Sigma (x - \bar{x})(y - \bar{y})}{\sqrt{\Sigma (x - \bar{x})^2 (y - \bar{y})^2}} \quad (1)$$

Where ẍ and ȳ are the mean of the matrix X and the matrix Y respectively.

3. Result and discussion

3.1. Rainfall pattern

The trend of the total annual rainfall (mm) from 1984-2015 in Meknes province is as shown in the Figure 2. The annual rainfall volume was variable from year to year and ranges from 284 mm to 876 mm. In seasons 2008-2009, 2009-2010 and 2012-2013 the annual rainfall amounts surpass 800 mm and fall below 300 mm in 1994-1995.

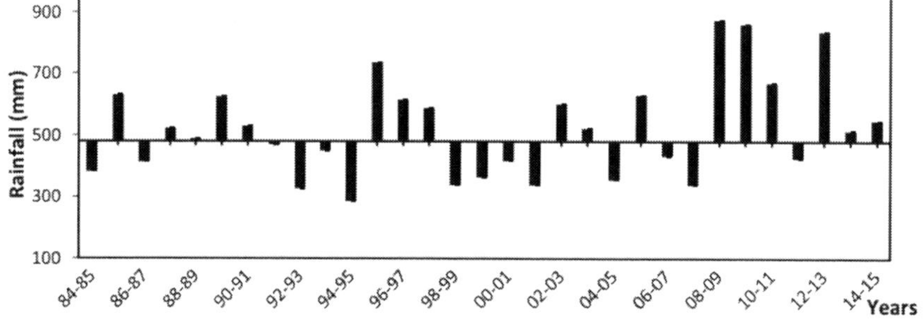

Fig.2. Annual rainfall compared with the median

From Table 1 one can learn that growing season account 93% of total annual rainfall where more than 46% occurs in winter-season and approximately 20% in the ending of autumn and 20% at the beginning of spring.

Table 1: Statistics for rainfall of multiple time scales from 1984 to 2016

	Annual	Growing Season	Supra-Season	Autumn	Winter	Spring
Mean	520	486	405	145	221	142
Coefficient of Variation	31	30	38	50	51	42
PCI (%)	17	15	12	13	11	13
Total %	100	93	76	28	41	28

3.2. Wheat yield trends

Historical records of soft and hard wheat yield from 1995 to 2015 as represented in Figure 3, show that, in general, the yield is low and variable in time. The coefficient of variation was very high, showed 52% for soft wheat and 53% for hard wheat. The highest observed yield was approximately 4500 kg/ha for both hard and soft wheat, registered in season 2014-2015, while the lowest yields were 210 kg/ha for hard wheat and 240 kg/ha for soft wheat obtained in season 1999-2000, it was a very dry year. Such variation has often exposed poor farmers to migration to cities.

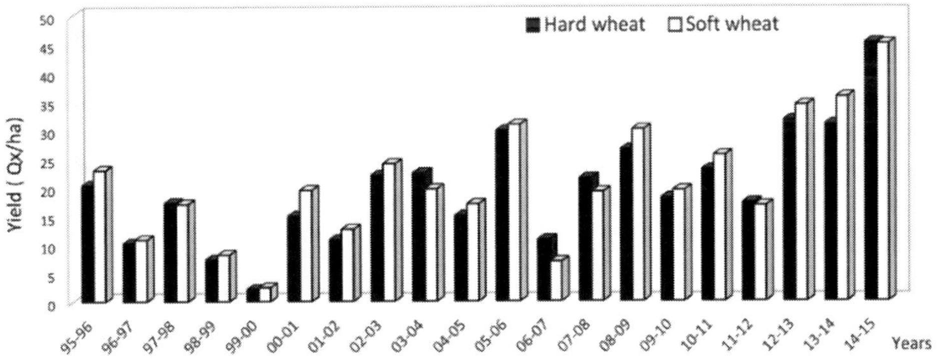

Fig.3. Variation of yield of hard wheat and soft wheat of Meknes from 1995 to 2015

3.3. Variability of rainfall and distribution

The rainfall variability resulted by coefficients of variation of seasonal and annual rainfall have shown that there existed a moderate variability, except for Autumn and Winter season where rainfall variability is high. Standard Deviation shows 72mm for Autumn and 112mm for winter, such values can lead to a success or failure of the crops. Additionally, on yearly basis, the computed PCI annual value of 17% indicates that, rainfall distribution over months is erratic. Generally dry years have recorded a high values (PCI ranges from 15 to 25). Nonetheless, the PCI value for season indicates moderate rainfall concentration (PCI<13).

Monthly average rainfall in Meknes is a clear seasonal cycle. Generally most of the precipitation occurs between November and January. This situation often fills the soil profile of the coarse-textured soil. A preview of the monthly rainfall trend over the period of two decades shows that monthly rainfall amount tend to decline in February after reaching its peak in December and January.

In respect to distribution, Figure 4 show that rainfall in the month of February and March are very erratic; the variation coefficient showed respectively 78% and 87%. This might present dangers of water stress and dry spells which has a proven effect on yield since it coincide with production phase.

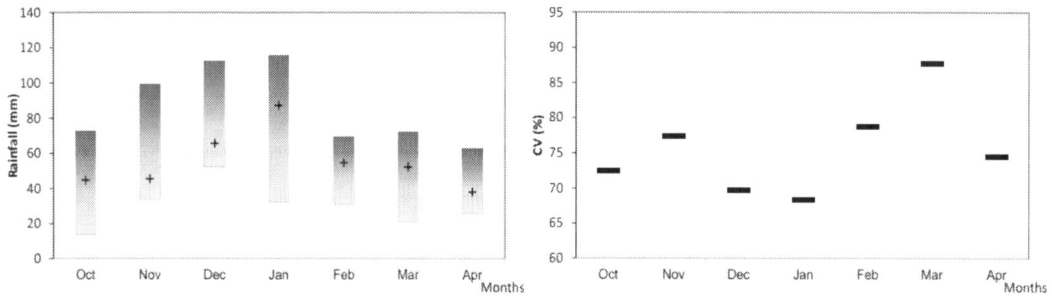

Fig.4. Box plot and coefficient of variation of monthly rainfall (Oct to Apr) from 1995 to 2016.

3.4. Relationship between rainfall and wheat yield

The Table 2 shows the relationship between rainfall of different period and wheat yield. Whereby all results, for seasons and inter-seasons are positive and significant.
For spring the correlation coefficient value was low. It showed 0.18; unfortunately, because the most rainfall of this season (March to May) coincides with wheat maturation. While a moderate amount of rainfall benefits wheat grain development, however, waterlogged soils during grain fill contribute to high disease pressure and can reduce wheat yields and test weight.

In contrast, a strong correlation were illustrated between wheat yield and rainfall at the other intervals; the correlation coefficients range from 0.52 to 0.76. Thus revealing that the beginning and the middle of growth period is the most sensitive stages of growth to rainfall. Owing to the fact that the water supply from rainfall during this period will affect the root yield, the survival and fertilization of ears and consequently of the grains. The highest value of correlation was 0.73 obtained between wheat yield and rainfall of January to March (JFM).

These results reveal on the one hand the importance of rainfall volume of the period January to March, on the other hand the damaging effect of mid-spring rains. Both factors independently influencing annual yield of soft and hard wheat.

Table 2: Coefficients of determination showing the relationship between wheat yield and rainfall at different time scales.

Crop/Time	Annual	Grow season	Supra-season	Autumn (SON)	Winter (DJF)	Spring (MAM)	*Inter-season*			
							ONDJ	JFMA	JFM	FMA
Wheat	0,66	0,72	0,76	0,57	0,52	0,18	0,62	0,64	0,73	0,53

4. Conclusion

The study quantifies the impact of climate change on wheat production by assessing the rainfall variability and its relationship with wheat yields in Meknes province of Morocco. It revealed moderate seasonal and irregular annual rainfall concentration. When the estimated correlation between wheat yield and rainfall at different scale showed a significant effect of the precipitation. The yield variability showed a higher relationship with precipitation at

small period (January to march). Seen the degree of rainfall variation of month in this period, wheat cropping would be more risky.
Having these results, without considering changes in practices of water-conservation, irrigation schemes, varieties and other, wheat yield might be in big trouble.

5. References

1. Barron, J., Rockström, J., Gichuki, F., & Hatibu, N. (2003). Dry spell analysis and maize yields for two semi-arid locations in east Africa. Agricultural and forest meteorology, 117(1), pp 23-37.
2. Born, K., Fink, A. H., & Paeth, H. (2008). Dry and wet periods in the northwestern Maghreb for present day and future climate conditions. Meteorologische Zeitschrift, 17(5), pp 533-551.
3. Chbouki, N., Stockton, C.W., Myers, D.E., 1995. Spatio-temporal patterns of drought in Morocco. Int. J. Climatol. 15 (2), pp 187–205.
4. FAO, 2006. The State of Food and Agriculture 2006: Food Aid for Food Security? FAO, Rome.
5. FAOStat, 2016. FAOSTAT database collections. Food and Agriculture Organization of the United Nations. http://faostat.fao.org.
6. Jarlan, L., Abaoui, J., Duchemin, B., Ouldbba, A., Tourre, Y. M., Khabba, S., ... & Chehbouni, G. (2014). Linkages between common wheat yields and climate in Morocco (1982–2008). International journal of biometeorology, 58(7), pp1489-1502.
7. Jouve, P., & Daoudi, A. (1984). Effets de la date de semis sur l'elaboration du rendement du blé tendre et de l'orge en zones semi-arides et arides (cas du Maroc atlantique). L'agronomie tropicale, 39(3), pp 216-228.
8. Lobell, D. B., & Field, C. B. (2007). Global scale climate–crop yield relationships and the impacts of recent warming. Environmental research letters, 2(1), pp 014002.
9. Laux, P., Kunstmann, H., & Bárdossy, A. (2008). Predicting the regional onset of the rainy season in West Africa. International Journal of Climatology, 28(3), pp 329-342.
10. MAPM, (ed.). 2015. L'agriculture marocaine en chiffres 2014, Statistique du ministère de l'agriculture et de la pêche maritimes, Rabat.
11. Karrou, M., & Oweis, T. (2014). Assessment of the severity and impact of drought spells on rainfed cereals in Morocco. African Journal of Agricultural Research, 9(49), pp 3519-3530.
12. Oliver, J. E. (1980). Monthly precipitation distribution: a comparative index. The Professional Geographer, 32(3), pp 300-309.
13. Pala, M., Oweis, T., Benli, B., De Pauw, E., El Mourid, M., Karrou, M., ... & Zencirci, N. (2011). Assessment of wheat yield gap in the Mediterranean: case studies from Morocco, Syria and Turkey. International Center for Agricultural Research in the Dry Areas (ICARDA), Aleppo, Syria. iv, pp 963-21.
14. Requier-Desjardins, M. (2010). Impacts des changements climatiques sur l'agriculture au Maroc et en Tunisie et priorités d'adaptation. Centre International de Hautes Études Agronomiques Méditerranéennes, Montpellier.
15. Sebbar, A., Badri, W., Fougrach, H., Hsaine, M., & Saloui, A. (2011). Étude de la variabilité du régime pluviométrique au Maroc septentrional (1935-2004). Science et changements planétaires/Sécheresse, 22(3), pp 139-148.

16. SEBBAR, A., HSAINE, M., FOUGRACH, H., & BADRI, W. (2012). Étude des variations climatiques de la région centre du Maroc. Les climats régionaux: observation et modélisation, pp 709-714.
17. Sowers, J., Vengosh, A., & Weinthal, E. (2011). Climate change, water resources, and the politics of adaptation in the Middle East and North Africa. Climatic Change, 104(3), pp 599-627.
18. Watts, D. G., & El-Mourid, M. (1988). Rainfall patterns and probabilities in the semi-arid cereal production region of Morocco.
19. Zeggaf, T. A., Nsarellah, N., Hadrabach, D., & Baidani, A. (2002). Caractérisation de la variabilitéspatiale du climat marocain en relation avec la productivité des céréales à petites graines. Rev. HTE, (124), pp 84-92.
20. Stocker, T. (ed.). (2014). Climate change 2013: the physical science basis: Working Group I contribution to the Fifth assessment report of the Intergovernmental Panel on Climate Change. Cambridge University Press.

Analysis of environmental impacts of renewable energy on the Moroccan electricity sector: A System Dynamics approach

M. CHENTOUF [a,*], *M*. ALLOUCH [a]

[a] Laboratoire des Matériaux et Valorisation des Ressources (LMVR), Faculté des Sciences et Techniques de Tanger, Ancienne Route de l'Aéroport, Km 10, Ziaten. BP 416 - Tangier, Morocco.

Abstract. Producing electricity at an affordable price while taking into account environmental concerns has become a major challenge in Morocco. Moreover, the technical and financial issues related to renewable electricity plants are still hindering their efficient integration in the country. In fact, the energy sector (both electricity and heat) accounted for more than half of all Greenhouse Gases (GHG) emissions in the kingdom due to the major reliance on fossil fuels for answering the growing local demand. The key strategies to alleviate this critical situation include the integration of more renewable energies in the total energy mix and the enhancement of energy efficiency measures in different sectors. This paper strives to (1) evaluate the potential of carbon dioxide mitigation in Moroccan electricity sector following the actual and projected strategies and (2) highlight the policy schemes to be taken in order to achieve the ambitious carbon dioxide mitigation targets in the mid-term. A system dynamics model was built in order to simulate different scenarios of carbon dioxide mitigation policies up to 2030. The results shows that the achievement of renewable energies projects by 2030 could save 228.143 MtCO2 between 2020 and 2030 and an additional 18.127 MtCO2 could be avoided in the same period by enhancing energy efficiency measures.

Keywords: Morocco, Renewable Energy, Energy Efficiency, GHG emissions, System Dynamics.

* Corresponding author: mohammedchentouf2@gmail.com

1. Introduction

Global warming and climate change are mainly attributed to the presence of a large amount of carbon dioxide (CO_2) in the atmosphere [1]. In fact, this heat-trapping gas is the major contributor to climate change among all human-influenced climate drivers [2]. This can be associated with its abundance in the atmosphere compared to other heat-trapping gases [3] and its large accumulation [4] due to its long life (20% of CO_2 in the atmosphere can still exist up to 800 years from the time of the emission [1]). Almost 76.7% of Greenhouse Gas (GHG) emissions are CO_2-based and fossil fuels on their own contribute to more than 73% of these emissions (of which 43% from coal, 36% from oil and 20% from natural gas [5]). According to the IEA [6], the power sector (electricity and heat) uses 41% of global fossil fuels followed by the transportation sector 22% and the industrial sector 20%. The remaining share is consumed in other sectors like agriculture and fishery.

The large share of fossil fuels in the power industry will most likely contribute to more adverse effects on the Earth's climate if no efficient measures were taken to limit its impacts. Hence, the further deployment of Renewable Energy (RE) technologies and Combined Heat and Power (CHP) plants seems to be an ideal solution for decarbonising this sector [7]. Moreover, the enhancement of energy measures is capable of avoiding the use of a considerable amount of these hydrocarbons and thus avoiding the resulted CO_2 emissions.

The aim of this work is to (1) evaluate the potential of carbon dioxide mitigation in Moroccan electricity sector following the current and projected strategies, and to (2) the policy schemes to be taken in order to achieve the carbon dioxide mitigation targets in the mid-term.

The layout of this paper is as follows: Section 2 provides a background analysis of the energy sector in Morocco. Materials and methods for conducting this research work are provided in section 3, followed by a detailed discussion of the results in section 4. Finally, a conclusion of this work is given in section 5.

2. Background analysis

2.1 *The energy situation in Morocco*

Morocco is a north-African country (only 14 km from Europe) with an area of 710,850 km2 and a coastline of 3500 km (shared between the Atlantic Ocean and the Mediterranean Sea). According to the population census of 2014 [8], Morocco's population is estimated at 33.8 million (an increase by 13.42% compared to 2014) with a density of 70.92 inhabitants/km2. The kingdom of Morocco is considered as the largest fossil fuel importer in North-Africa [9] due to financial and economic barriers for developing local energy production facilities (e.g. oil shale) and the cheaper cost of energy imports compared to the total cost of exploring and investing in the local energy potential.

Electricity consumption per capita is the fastest growing sector among different energy sectors in Morocco (an increase by almost 100% between 1985 (0.08 toe/capita) and 2005 (0.16 toe/capita)) and an additional 0.07 toe/capita has been recorded in 2015.This can be attributed to the implementation of the national programme for global rural electrification (PERG) in 1995, which enabled the connection of 12 million people to the national grid reaching an electrification rate of more than 99% in 2015 compared to only 18% in 1995 [11]. As a result of economic expansion and population growth, electricity demand is expected to continue its growth to reach 52 TWh in 2020 (compared to 24 TWh in 2008) and 95 TWh in 2030 [12].

Table 1 shows some key indicators of the energy sector in Morocco over the period 2000 – 2014 [13]. Total Primary Energy Supply (TPES) has increased by almost 72.23% during these 15 years. Consequently, electricity consumption has doubled during the same period reaching 30.93 TWh in 2014 compared to 14.11 TWh in 2000 (an increase by 119%) and resulting in an increase of carbon dioxide emissions by 0.55 tCO2/capita up to 2014.

Although carbon intensity is still fluctuating between 2.62 tCO2/toe and 2.80 tCO2/toe, the annual emissions of carbon dioxide have increased by 79.8% between 2000 and 2014. This carbon footprint reflects the magnitude of intensive CO2 emitting sectors (mainly electricity and transportation) in Morocco, and the high reliance on fossil fuels for ensuring local energy needs. Almost two thirds of carbon dioxide emissions are generated from electricity and transportation activities. The industrial sector generates 15% of these emissions followed by the commercial sector (including agriculture and fishery) with 12% of carbon emissions.

Years / Indicators	2000	2001	2002	2003	2004	2005	2006	2007	2008	2009	2010	2011	2012	2013	2014
TPES (Mtoe)	11.02	11.68	11.87	11.82	14.02	14.84	15.10	15.57	16.17	16.28	17.08	18.40	18.66	18.74	18.98
Electricity Consumption (TWh)	14.11	15.29	15.79	17.32	18.44	19.32	21.20	22.35	23.13	23.81	25.10	27.07	29.15	29.72	30.93
CO_2 emissions (Mt of CO_2)	29.54	32.19	33.33	33.06	36.29	38.89	40.18	40.96	42.75	42.72	45.95	50.26	51.75	50.47	53.11
CO_2/TPES (tCO_2/toe)	2.68	2.76	2.81	2.80	2.59	2.62	2.66	2.63	2.64	2.62	2.69	2.73	2.77	2.69	2.80
CO_2/population (tCO_2/capita)	1.02	1.10	1.13	1.11	1.21	1.28	1.31	1.32	1.36	1.35	1.43	1.55	1.57	1.51	1.57

Table 1. Main indicators of the energy sector in Morocco between 2000 and 2014

(Based on IEA data, www.iea.org/statistics, Licence: www.iea.org/t&c; as modified by M.Chentouf and M.Allouch.)

According to the IEA, 72.60% of CO2 emissions come from oil followed by 22.70% from coal and only 4.80% from natural gas. In fact, oil products accounted for 62% of TPES in 2014, while coal accounted for 21.26% and natural gas had a share of 5.3%. It must be highlighted that the country imports over 95% of fossil fuels from abroad ranging from 100% of coal and 99.99% of oil to 85% of natural gas.

2.2 The national response to climate change

The country has been striving during the last few decades to limit the negative impacts of climate change on its vital economic sectors (e.g., agriculture, fishery and forestry). In fact, Morocco is increasingly vulnerable to climate change with an average warming estimated at 1°C and rainfall variability between 3% and 30% [14]. Periods of droughts, floods and heat waves are expected to have adverse impacts in the upcoming years on water resources, agriculture and fishery products.

Several studies have been conducted in order to measure the impacts of climate change in Morocco including the work of Tramblay et al., [15] regarding the effects of climate change on extreme precipitation, the work of Simmonneaux et al., [16] on land use and climate change effects on soil erosion, and the work of Seif-Ennasr et al., [17] concerning climate change and adaptive water management measures in Morocco. The Moroccan endeavors to limit the impact of this global phenomenon dated back to 1992 when the country signed the United Nations Framework Convention on Climate Change (UNFCCC). In 1994, a national inventory on GHGs was achieved [18] and a National Committee for Climate Change (NCCC) was founded after the kingdom hosted the Conference of the Parties COP7 in 2001 in Marrakech. The third national communication submitted by the NCCC in 2015 includes 49 measures for mitigating GHG emissions to reach an average mitigation potential of 81.9 million tCO2 by 2040 [19].

Within this context, the National Energy Strategy (NES) adapted in 2009 stands as a key project among others for combating climate change in Morocco. The aim of the NES is to diversify the energy mix by introducing more renewable energies and supporting foreign investments in the sector [20]. Furthermore, the enhancement of energy efficiency measures is regarded as a promising tool for achieving the national mitigation targets of GHG emissions. Renewable energies will account of 42% of total installed capacity for electricity production (6 GW shared evenly between solar, wind and hydropower) by 2020 compared to 33% in 2015. An additional 10% of installed capacity is expected by 2030. Actually, Morocco is blessed with favorable conditions for developing renewable energy projects. The exploitable potential of wind energy is estimated at 25,000 MW with an Average Wind Speed (AWS) ranging from 7.5 m/s to 11 m/s at 40 m [11]. The achievement of the Moroccan Integrated Programme of Wind Energy (MIPWE) will result in avoiding the emission of 5.6 MtCO2/year and generating an annual 6,600 GWh of electricity. Solar energy is also regarded as a vital source for ensuring the growing energy demand in the country. In fact, Morocco has an abundance of solar resources in several sites with more than 3000 hours of sunshine per year and 5 kWh/m2/year in terms of received energy. The average Direct Normal Irradiation (DNI) is estimated at 2300 kWh/ m2/year being 30% higher than the best sites in Europe [21]. The technical potential of solar energy in Morocco is estimated at 10,829 MW and the fulfillment of the Moroccan Solar Plan will result in avoiding the emission of 3.7 MtCO2/year and generating an annual 4,500 GWh of electricity. Morocco has been using hydropower for generating electricity since 1929 and the potential of hydroelectric power is estimated at 3,800 MW. The NES defined the target of adding 580 MW of hydroelectric capacity to reach 2 GW of total installed capacity by 2020. This new capacity will be dedicated mainly to hydroelectric projects in Waste Water Treatment Plants (WWTPs) and micro-hydropower stations. The potential of biomass in Morocco is estimated at 950 MW (issued mainly from agriculture, forestry and municipal wastes) and targets of ensuring 200 MW by 2020 and 400 MW by 2030 were defined for developing the potential of bioelectricity in Morocco. Geothermal energy is still not used for electricity production due to economic and technical feasibility, but a study [22] in north-eastern Morocco has estimated the potential of geothermal electricity at 5 MWe. Simultaneously, the country set the target of saving up to 12% of consumed energy by 2020 (compared to 2011), and between 25% and 30% by 2030. These targets were introduced under the National Programme for Energy Efficiency (NPEE) in 2011. This programme defined five principal sectors for enhancing of energy efficiency measures including households, industry, agriculture, transport and tertiary (hospitals, hotels, restaurants, schools and other public services).

3. Materials and methods

In order to evaluate the potential of carbon dioxide mitigation in the Moroccan electricity sector, we will build a System Dynamics (SD) model depicting the current structure of this sector. SD is a modelling and simulation approach for analyzing complex systems by studying the interactions of their elements along with their feedbacks and delays. The main objective is to understand the evolution of the system and its causes [23].

The reason behind this selection lies in:

- The high level of technical details in modelling the national electricity market;
- The flexibility of this market due to different parameters and conditions; and
- The wide and successful use of this approach for addressing similar issues worldwide [24 and 25].

The data for conducting this study were collected from different sources including: governmental electronic publications, international studies and reports, and journal articles [26-29].

4. Results and discussion

4.1. Model scenarios

The time span of our model is defined over the period 2015-2030. This choice enables the validation of the model by comparing its outcomes with previous governmental studies and institutional reports for the same period. Moreover, this period covers the overall term of the NES.

Three scenarios were selected for simulating the model including:

- **The Base scenario:** this scenario follows the current trends of the Moroccan electricity sector. It suggests an annual electricity demand growth rate of 7.5%. Additionally, the energy mix will keep being dominated by fossil fuels (66%) and a limited share of renewables (34%). The role of this scenario is to serve as a reference for comparisons between other scenarios and their benefits for the overall electricity system.
- **The Renewable Energy (RE) scenario:** this scenario reflects the targets of the NES. Total installed capacity of renewables is expected to reach 52% by 2030 and the annual electricity demand growth rate is fixed at 7%.
- **The Renewable Energy (RE) and Energy Efficiency (EE) scenario:** this scenario is considered as an ideal yet a promising pathway for the future of the Moroccan electricity sector. Within the RE+EE scenario, the total installed capacity of renewables is expected to reach 59% by 2030 and the annual electricity demand growth rate has decreased to 6.5% due to the enhancement of energy efficiency measures.

4.2. Electricity supply and demand

Figure 1 and Figure 2 show the development of renewable and fossil installed capacity between 2015 and 2030. The initial installed capacity of renewables in 2015 was 2,822 MW in the three pre-defined scenarios, while the initial installed capacity of fossil fuels in 2015 was 5,478 MW in the same scenarios. According to the base scenario, the renewable installed capacity will reach 6,927 MW by 2030 compared to 14,222 MW in the RE scenario and 15,622 MW in the RE+EE scenario. On the other hand, the fossil installed capacity amounts to 16,518 MW by 2030 in the base scenario compared to 10,528 MW in the RE scenario and 9,728 MW in the RE+EE scenario. This enormous increase in renewable installed capacity is attributed mainly to the exploitation of solar (5,416 MW in the RE scenario and 5,916 MW in the RE+EE scenario by 2030) and wind (5,330 MW in the RE scenario and 5,830 MW in the RE+EE scenario by 2030) energies, while hydropower (3,076 MW in the RE scenario and 3,176 MW in the RE+EE scenario by 2030) and biomass (400 MW in the RE scenario and 700 MW in the RE+EE scenario by

2030) contribution will be limited following their available potential in the country. Oil is expected to decrease to reach 1175 MW in the RE scenario and 1075 MW in the RE+EE scenario compared to 6275 MW in the base scenario. This decrease in mainly attributed to the increasing prices of oil imports compared with coal and natural gas. Coal installed capacity is expected to increase in the three scenarios reaching 7748 MW in the base scenario, 4948 MW in the RE scenario, and 4448 MW in the RE+EE scenario. Natural gas will experience the highest increase due to its lower carbon footprint and adaptability with CHP plants. The installed capacity of natural gas will reach 2495 MW in the base scenario, 6080 MW in the RE scenario, and 5580 MW in the RE+EE scenario.

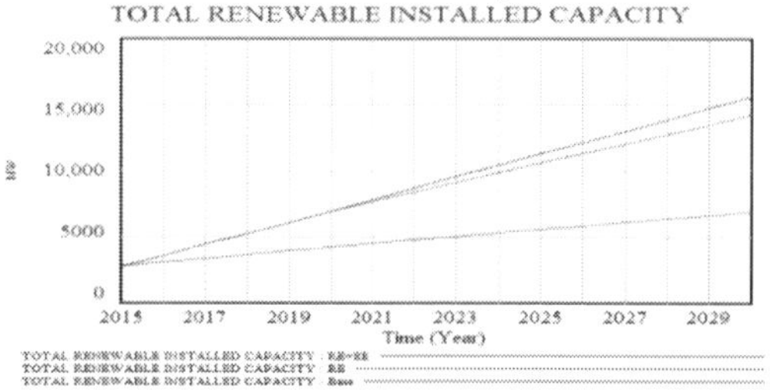

Fig.1. The total renewable installed capacity in Morocco up to 2030.

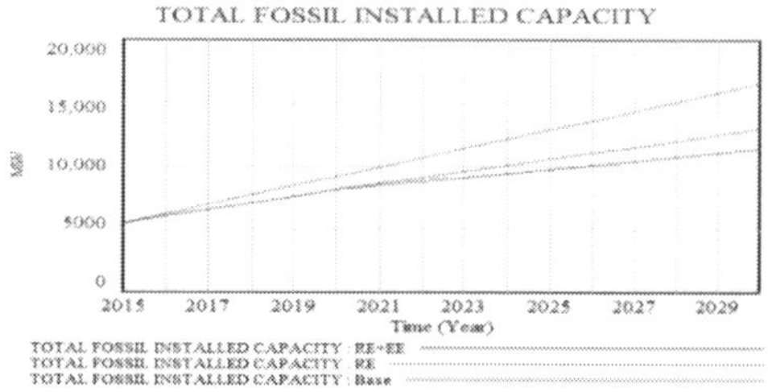

Fig.2. The total fossil installed capacity in Morocco up to 2030.

Electricity production from renewables and fossil fuels is following the same patterns as it is shown in Figure 3 and Figure 4. The initial renewable electricity generation in 2015 was 6095.92 GWh in the three scenarios, while the initial fossil electricity generation in 2015 was 22098.3 GWh in the same scenarios. It is noteworthy to mention that fossil fuels are still efficient in terms of energy output compared to renewables. However, wind and hydroelectric plants are witnessing progressive advances to close the gap between renewables and fossil technologies. According to the base scenario, renewable electricity generation will reach 14962.3 GWh by 2030 compared to 30719.5 GWh in the RE scenario and 33743.5 GWh in the RE+EE scenario. On the other hand, fossil generated electricity will amount to 66633.6 GWh by 2030 in the base scenario compared to 52252.4 GWh in the RE scenario and 45798 GWh in the RE+EE scenario.

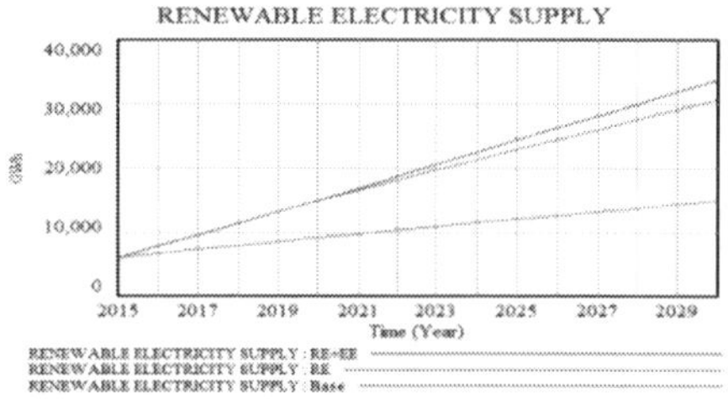

Fig.3. The renewable electricity supply in Morocco up to 2030.

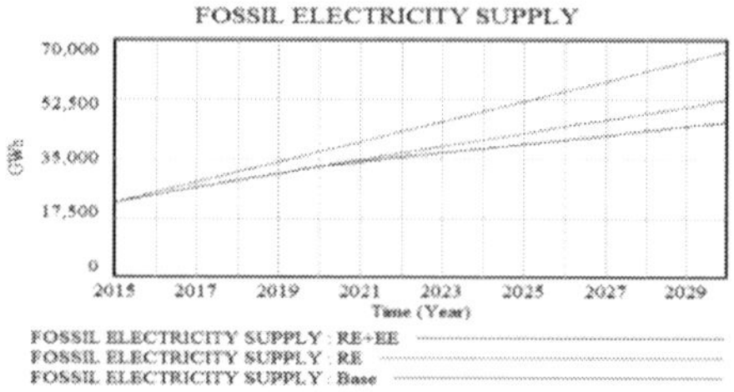

Fig.4. The fossil electricity supply in Morocco up to 2030.

Electricity demand will range between 97652.8 GWh in the RE+EE scenario, 104,760 GWh in the RE scenario and 112,349 GWh in the base scenario (see Figure 5). This suggests a gap of 30,754 GWh in the base scenario, 18755.1GWh in the RE scenario, and 18111.3 GWh in the RE+EE scenario. It must be highlighted that advances in renewable energy outcomes can close the supply/demand gap and even led the country to export the surplus of green energy using its connections with Spain and Algeria.

 https://doi.org/10.1051/e3sconf/20183703002

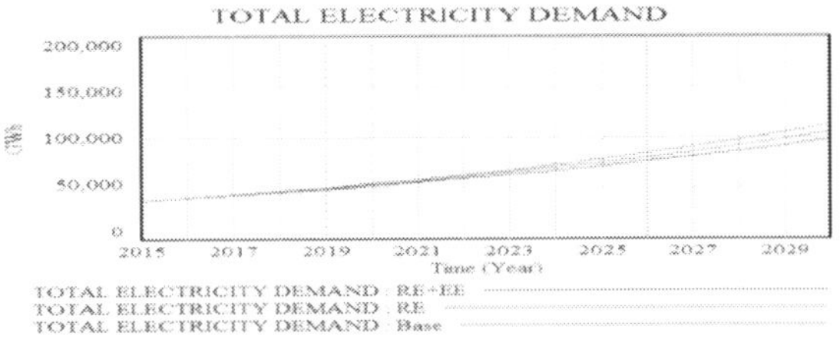

Fig.5. The total electricity demand in Morocco up to 2030.

4.3. CO2 emissions in the Moroccan electricity sector

The carbon dioxide emissions of the Moroccan electricity sector were calculated for the three scenarios by including direct and indirect CO2 emissions. The direct emissions enclose the emissions from the combustion/gasification process while the indirect emissions are linked to the fuel supply chain and the plant construction [37]. Depending on fuel type and technology, direct emissions could vary between 850 kg CO2/MWh in case of Circulating Fluidised Bed Combustion (CFBC) of coal to 60 kg CO2/MWh in case of Combined Cycle Gas Turbine with Carbon Capture and Storage (CCS). Accordingly, indirect emissions are ranging between 125 kg CO2/MWh in case of Integrated Gasification Combined Cycle (IGCC) or Pulverised Coal Combustion (PCC) to 70 kg CO2/MWh in case of Combined Cycle Gas Turbine with CCS. Regarding renewables, only biomass (6 kg CO2/MWh for solid biomass and 5 kg CO2/MWh for biogas) and solar (120 kg CO2/MWh for Concentrating Solar Power (CSP) in case of using natural gas for backup heat production) can emit direct CO2 emissions. Indirect emissions range from 6 kg CO2/MWh in case of hydroelectric plants (either small or large) to 45 kg CO2/MWh in case of photovoltaic plants.

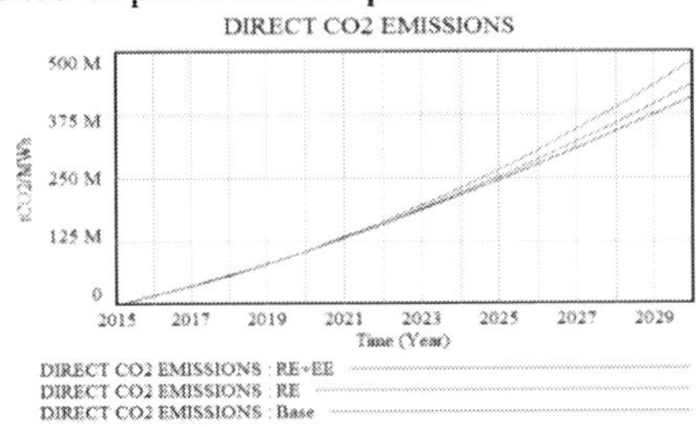

Fig.6. The cumulative direct CO_2 emissions of the Moroccan electricity sector up between 2015 and 2030.

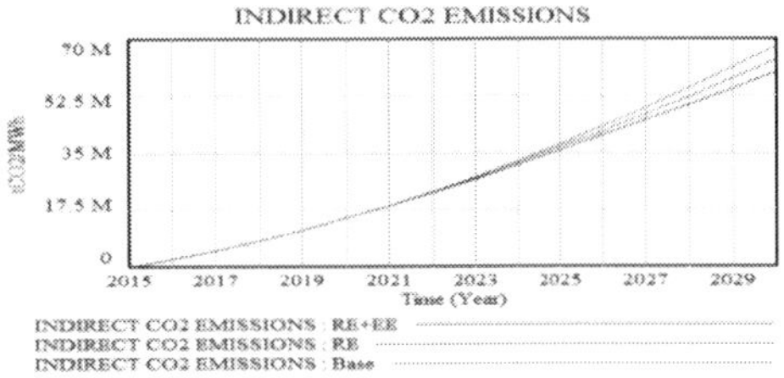

Fig.7. The cumulative indirect CO_2 emissions of the Moroccan electricity sector between 2015 and 2030.

According to Figure 6, the cumulative direct CO2 emissions of the Moroccan electricity sector over the period 2015 – 2030 are estimated at 480.37 MtCO2 in the base scenario compared to 433.75 MtCO2 in the RE scenario and 406.69 MtCO2 in the RE+EE scenario. This implies avoiding 10% (in case of the RE scenario) and 15% (in case of the RE+EE scenario) of direct carbon dioxide emissions of the Moroccan electricity sector if appropriate measures were taken to further deploy renewable energies and enhance energy efficiency measures.

The cumulative indirect CO2 emissions of the Moroccan electricity sector over the same period re illustrated in Figure 7. Since most of these emissions are emitted from fossil fuels, the same patterns of behavior can be seen in this Figure. The cumulative indirect emissions in the base scenario amounts to 67.81 MtCO2 compared to 63.65 MtCO2 in the RE scenario and 59.78 MtCO2 in the RE+EE scenario. Almost 6.13% (in case of the RE scenario and 11.67% (in case of the RE+EE scenario) of indirect carbon dioxide emissions could be avoided by achieving the pre-mentioned targets. The decrease in the share of avoided emissions reflects the indirect emissions of constructing and fueling some renewable energy plants. Overall, the total avoided CO2 emissions in the Moroccan electricity sector can amount to 9.28% (in case of the RE scenario) and 14.9% (in case of the RE+EE scenario).

It is noteworthy to mention that the results of this study seem to be in line with previous studies regarding the potential of carbon dioxide mitigation in the Moroccan energy sector.

4.4. Challenges and recommendations

It is obvious that the fulfillment of the mentioned plans will change the electricity sector completely in Morocco during the upcoming years and placing the country as the forefront of clean energy users worldwide. However, many challenges and barriers are still hindering the smooth deployment of these projects and must be addressed holistically by different stakeholders. The aim is to achieve the governmental targets and increase the attractiveness of this potential on a global scale. To this end, major challenges are categorized into five areas: political, technical, economic, social, and environmental barriers. The main obstacles of each category are provided below (based mainly on the guidelines of IEA [26]) along with key recommendations to alleviate their impacts on the implementation of the NES.

4.4.1. Political challenges and recommendations

This category encloses institutional and regulatory aspects of the electricity sector in Morocco. Major obstacles include the lack of an integrated approach for alternative energy deployment in the country. Moreover, the Moroccan electricity sector is still highly monopolized by ONEE although to a less extent than before. This situation is not suitable for increasing the share and the competitiveness of alternatives compared to conventional technologies. Finally, the lack of information and data is still hindering the access of different social segments and utilities to this green knowledge, which is urgently required for answering different issues of renewables in Morocco. To overcome these barriers, an inclusive approach - favoring partnership and cooperation - must be implemented in order to allow different stakeholders to take part in developing national policies for renewable and alternative energy. Furthermore, encouraging domestic and industrial IPPs to integrate the Moroccan electricity sector is capable of increasing the installed capacity (along with electricity generation) of renewables and security of supply (in contrast to monopolistic power generation). Finally, collecting and providing public access to relevant data and information on renewables will most likely accelerate the deployment of renewable and alternative energy in Morocco.

4.4.2. Technical challenges and recommendations

These obstacles relate mainly to the intermittent nature of renewables. This can affect energy security and widen the gap between supply and demand, which led to increasing energy imports. Moreover, the field of renewable energy is not mature yet in Morocco, and there is an urgent need for technology expertise and qualified personnel on a local scale. In addition to the importance of developing other renewable technologies (namely biomass and geothermal), a need for a flexible regulating energy is urgently required in order to balance supply and demand, and decrease energy imports. CHP stands as a suitable technology for Morocco due to its overall efficiency, which is high compared to separate heat and power generation. Moreover, CHP technologies are relatively cleaner in terms of CO2 emissions. It is noteworthy to mention the importance of increasing the installed capacity of renewables in different regions of Morocco in order to minimize the impact of their intermittency on security of supply. Simultaneously, energy storage stands as an important strategy for reducing the sector failures due to instability of renewables.

Additionally, promoting R&D projects can fill the gap in terms of local human potential in this area. These projects can be conducted through partnerships between Universities and power industries. In addition, technology transfer can play a major role within this scope, and Morocco could potentially benefit from successful experience of some leading nations in this domain.

4.4.3. Economic challenges and recommendations

These barriers are of great importance and they represent the main driving force behind the reluctant investment of capitals in the field of renewables. In fact, despite the advantages of renewables and their great potential in Morocco, the state is still facing complicated issues for attracting local and foreign investments in the area. This situation is due to the intermittent nature of renewables, their high construction and O&M (Operation and Maintenance) costs, their long span of time concerning return on investment, and the lack of financial aids from the Moroccan government to develop renewable projects. Morocco can use financial savings from cutting fuel subsidies to support local and foreign IPPs and R&D activities in the field of renewables. In addition, developing local businesses for installing and operating renewable projects can decrease their global costs and thus provide renewable electricity at an affordable price. Finally, different financial schemes must be implemented by the government in order to support the deployment of renewables. This include feed-in tariff for IPPs and taxes for conventional energy.

4.4.4. Social challenges and recommendations

The success of renewable energy deployment in Morocco relies among other reasons on social awareness and acceptance of these technologies in the energy mix. Unfortunately, the lack of educational programs on renewable energy is responsible for social lack of appreciation towards renewables and their enormous potential and opportunities. Moreover, the lack of best practices for energy efficiency is highly witnessed among different social layers in Morocco. Raising the understanding of renewables and their potential can be ensured through strong governmental policies including educational programs, informational campaigns, and engagement of different social segments in the process of sustainable development based on participative democracy.

4.4.5. Environmental challenges and recommendations

Although renewables present the ideal option for Morocco in order to achieve a sustainable development of the electricity sector, they also have some environmental impacts mainly in the area of hydropower and biomass. Moreover, the prospective reliance on nuclear energy can generate adverse effects on the ecosystem if safety measures were not taken. In fact, hydropower (particularly run-of-the-river technologies) is always linked to a massive disruption of the ecosystem resulting from the construction and the operation of hydropower plants near to some sensitive areas. Additionally, the intensive use of biomass (namely food crops) for generating electricity can lead to major impacts on the Moroccan agriculture and sylviculture sectors. Consequently, the environmental aspect of renewable energy has to be placed at the core of the NES in addition to the national plans for mitigating CO2 emissions. Major measures to reduce the ecological footprint of renewable plants (mainly hydropower and biomass) include afforestation, micro- and WWTP hydropower palnts, and R&D activities.

Conclusions

The present study shows that RE and EE measures are the main strategies to answer the growing electricity demand in Morocco and mitigate CO2 emissions. In fact, RE plants have a potential of ensuring up to 36% of total electricity generation by 2030 if the targets of the NES were achieved. Moreover, an additional 6% of total electricity generation can be gained from further investments and implementation of EE measures. Consequently, the cumulative potential of carbon dioxide mitigation over the period 2015 – 2030 could reach up to 50.78 MtCO2 in the case of RE scenario and 81.7 MtCO2 in the case of RE+EE scenario. This study can be conducted in other countries or regions having similar energy situation as Morocco. Moreover, the outcomes of this study can be enlarged by studying the mitigation potential of the Moroccan heat sector and also of other sectors like transportation, industry and construction.

Acknowledgments

- This research was supported by the National Center of Scientific and Technical Research (CNRST) in Morocco through the scholarship of excellence Number 22UAE2016.

References

1.IPCC, Climate change 2007 synthesis rerpot (2007).
2.Union of concerned scientists, Why does CO_2 gets most of the attention when there are so many other heat-trapping gases (greenhouse gases)? (available online) [accessed on 17 May 2017].
3. Vaughan, A.,Global carbon dioxide levels break 400ppm milestone, The Guardian, (May 2015).
4.Amos J., Carbon dioxide passes symbolic mark, BBC, (May 2013).
5.IPCC, Climate change 2014: Mitigation of climate change, chapter 7: energy systems, Cambridge University press, (2014)
6. IEA, Energy technology perspectives – part 3: fossil fuels and CCS, (2012).
7. IEA, Cogeneration and renewables: solutions for a low-carbon energy future, (2011).
8. HCP, RGPH 2014, (2015).
9. Altmanm K. A bright future for Morocco, (2012) 33-35.
10. HCP, Prospective Maroc 2030 – Energie 2030 : prospective énergétique du Maroc – Enjeux et défis, (2011).
11. ONEE, Le marché de l'énergie électrique au Maroc, (2015).
12. ONEE, Rapport d'activités, (2015).
13. IEA, statistics for Morocco (available online) [accessed on 12 April 2017].
14. Ministry Delegate of the Minister of Energy, Mines, Water and Environment, in charge of Environment, Moroccan Climate Change Policy (MCCP), (March 2014).
15. Tramblay Y., Badi W., Driouech F., El Adlouni S., Neppel L., Servat E., Climate change impacts on extreme precipitation in Morocco, Gloval and Planetary Change 82-83 (2012) 104-114.
16. Simonneaux V., Cheggour A., Deschamps C., Mouillot F., Cerdan O., Le Bissonnais Y., Land use and climate change effects on soil erosion in a semi-arid mountainous watershed (High Atlas, Morocco), Journal of Arid Environments 122 (2015) 64-75.
17. Seif-Ennasr M., Zaaboul R., Hirich A., Caroletti G., Bouchaou L., El Morjani Z., Beraaouz E., McDonnell R., Choukr-Allah R., Climate change and adaptive water management measures in Chtouka Ait Baha region (Morocco), Science of the Total Environment 573 (2016) 862-875

18. GIZ, Establishment of a national competence centre for climate change mitigation and adaptation in Morocco (4C Maroc) (available online) [accessed on 19 April 2017].
19. 4C Maroc, 3ème Communication Nationale du Maroc à la Convention Cadre des Nations Unies sur le Changement Climatique : Des mesures d'atténuation proposées à l'horizon 2040, (February 2016).
20. SIEM, Invest in renewable energies in Morocco (available online) [accessed on 25 April 2017].
21. Adriani B, Lignières P, Barges M, Bennis A, Mokhtari G, Lighting up the kingdom of Morocco: energy strategy and recent developments in power projects, (2013).
22. Zarhloule Y. Geothermal potentialities of Morocco, online (https://www.fsb.unizg.hr) [accessed 7 March 2017]
23. Forrester J.W., Industrial Dynamics, Pegasus Communications, (1961).
24. Saysel A., Hekimoglu M., Exploring the options for carbon dioxide mitigation in Turkish electric power industry: System dynamics approach, Energy policy 60 (2013) 675-686.
25. Robalino-Lopez A., Mena-Nieto A., Garcia-Ramos J.E.? System dynamics modeling for renewable energy and CO_2 emissions: A case study of Ecuador, Energy for Sustainable development 20 (2014) 11-20.
26. IEA, Morocco 2014 – energy policies beyond IEA countries, (2014).
27. The 1 Gigaton coalition, Renewable energy and energy efficiency in developing countries: contributions to reducing global emissions, (2016).
28. IEA, Clean energy technology assessment methodology pilot study: Morocco, (2016).
29. Commission of the European Communities, Energy sources, production costs and performance of technologies for power generation, heating and transport, (2008).

The climate changes in the sub-basin of the Oum Er rbia central and the impact on the surface waters.

Zhour Echakraoui [1],*, *Ahmed.* Boukdir[1], *Olaide.* Aderoju[2], *El Hassan* Ben – Saïd[1], *Abdelhamid. Zitouni*[1], *Rachid* El maslouhi[3] *and António* Guerner Dias[2]

[1]University Sultane Moulay Slimane, Faculty of Sciences and Technology, Laboratory of Industrial Engineering, Beni Mellal, Morocco
[2]University of Porto, Faculty of Sciences and Technology, Department of Geosciences, Environment and Spatial Planning
[3]OumEr Rbia Hydraulic Basin Agency, Morocco

Abstract. Observations and model results indicate that climate trends in North Africa show both drying and warming over the past few decades, according to the latest Intergovernmental Panel on Climate Change (IPCC) assessment. During the last decades, due to changes in climatic and environmental conditions, water resources available in Morocco are decreasing. They are, moreover, subject to extreme cyclical variations and to soaring water demands because of rapid population growth, improvement of living standards, industrial development and expansion of irrigated agriculture. The pressure on these water resources is accompanied by a growing and increasingly serious degradation of their quality. This is found at the level of average of air temperatures that are continuously growing and at the level of precipitation with an average potential of water in the area with a significant decrease in the last forty years.

The purpose of this work is to make a study on the impact of climate change on water resources that exist in the basin of the Oum Er Rbia Central, and to give justifiable results regarding the evolution of climate change over time. From the created database, we brought out diagrams, curves and maps of the evolution of climate change that show the results below:

The study of the evolution of rainfall recorded since 1934 and the breaks in time series highlighted two methodologically distinct periods: a wet period with high rainfall (1986 - 1971), a dry season and low rainfall (1970/71 to 2007).

Observed climatic trends, calculated over the period of 1935-2007 and reported in the study, indicate the following:

* Corresponding author: zhourgat@gmail.com

- On an annual basis, changes in precipitation were not significant and varied from one region to another. On the other hand, spring rainfall declined significantly in the northern part of Morocco at a rate of 0.5 mm / day per decade;

- That the area experienced an average annual rainfall reduction of 70 mm (20%) over this period compared to 1940-1980. The area is a hydraulic region that is already experiencing a water deficit. The sharp decline in water supplies since 1980-2007 (by 40% compared with 1940-1980) and the increase in demand and water degradation by different causes;

- The annual average flows measured at the the central Oum Er Rbia stations were reduced by considerable hydrological deficits ranging between 40.8 and 49.5%.

Global warming and rainfall regression are added to the intrinsic conditions of sub basins of the Oum Er Rbia Central (especially waterproofing of land and the lack of groundwater reservoir) to increase its vulnerability to water scarcity. This critical situation requires adapting good management methods of meteorological water as the only source of water in this basin.You should leave 8 mm of space above the abstract and 10 mm after the abstract. The heading Abstract should be typed in bold 9-point Arial. The body of the abstract should be typed in normal 9-point Times in a single paragraph, immediately following the heading. The text should be set to 1 line spacing. The abstract should be centred across the page, indented 17 mm from the left and right page margins and justified. It should not normally exceed 200 words.

1 Introduction

Climate change has become progressively more widespread since the Rio conference in 1992, when the world's countries recognized this danger. At this conference is launched the United Nations Framework Convention on Climate Change. These countries were committed to addressing this climate change through the mitigation of GHG emissions and by adapting development policies and programs to limit the negative impacts that could result from global warming.

Each of the last three decades has been successively warmer on the Earth's surface than any previous decade since 1850. The years 1983 to 2012 are probably the warmest 30-year period in the northern hemisphere in the last 1400 Years (average confidence). The linear trend in the global average of surface temperature data combining land and oceans indicates a warming of 0.85°C over the period of 1880-2012 (IPCC 2014).

Observations and model results indicate that climate trends in North Africa show both drying and warming over the past few decades. According to the latest Intergovernmental Panel on Climate Change (IPCC) assessment, the region's climate is projected to become even hotter and drier in the next century, which will expose additional millions of people to further water stress. It is projected that climate change would likely increase the pressure on the already over-exploited surface and groundwater resources. With the added complexity of population growth, economic growth, the potential for increasing frequency of heat waves, and declining snow-packs, Morocco's water scarcity problem is likely to amplify in the future.

The climate change is confirmed with strong impacts notably on the water cycle. This is particularly true in the Mediterranean and in Africa where this warming is important and where the impact on water resources accentuates and accelerates the supply-demand deficit that many countries in the region are experiencing, logically with a strong growth in demand for water, pressure on water resources Countries in recent decades.

In Morocco, the trend towards global warming has been confirmed, also with the intra-annual and interannual irregularity of precipitation that has been in place since the 1970s and more extreme events, with consequence in dysregulation of the seasonal cycle. To address this situation and its impact on the water sector in Morocco, commendable efforts have been made, with a reinforced dams policy, structured and organized water management planning and adaptive measures and actions, have been more reactive than preventive.

Today, 30 years after the first signs of the hydro-climatic problem experienced during the droughts of the 1980s, this instability / irregularity of the climate reached a threshold whose impact on the management of water resources no longer allows to be considered without adequate planning and anticipation. There is a clear need to integrate this climate change in a structural way and its possible effects on water management in the country, both in policy choices for water use and in the implementation of these policies.

2. Geographic location and presentation of the watershed

2.1 Physical characteristics the Oum Er Rbia area

The Oum Er Rbia Basin is one of the Kingdom's largest basins. It spreads over an area of 35 000 km2 with an extension of 550 km. The river Oum Er Rbia originates in the Middle Atlas at 1,800 m of altitude to 40 km of the city of Khénifra and to 26 km Of the town of M'Rirt and crosses the chain of the Middle Atlas, the plain of Tadla and the Coastal Meseta before sinking in the Atlantic Ocean to 16 km of the city of El Jadida. It has a length of about 600 km.

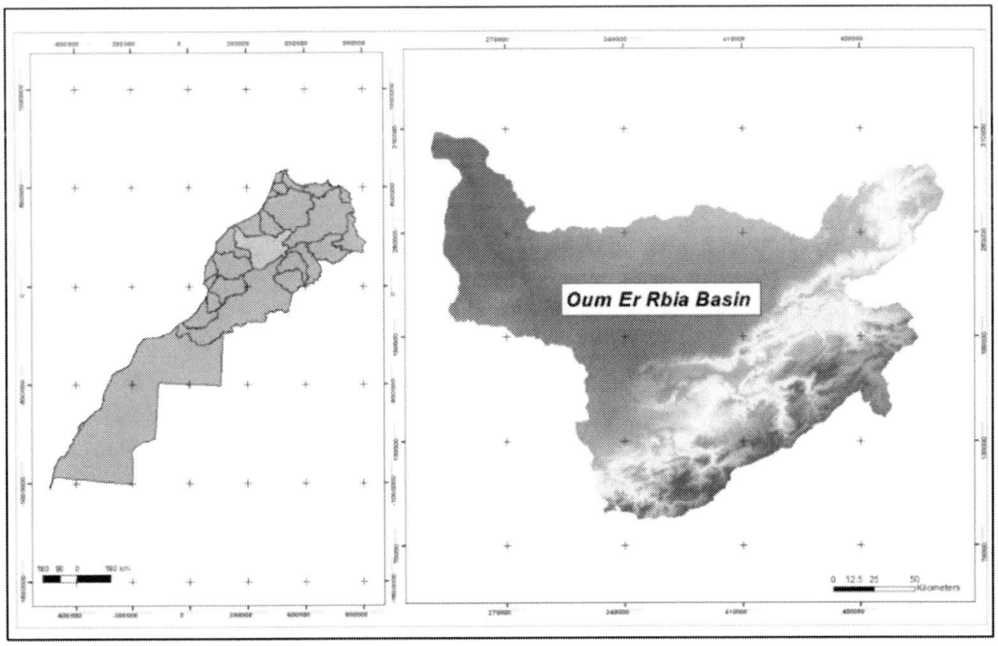

Fig. 1. Location of the Oum Er Rbia basin

The hydrographic network of the basin is formed by the Oued Oum Er Rbia and its tributaries: Tassaout, El Abid, Lakhdar, Derna, Melloul, Ouamana, Srou, Chbouka, Ouirine, Gzef, El Touim and Faragh. Thanks to rainfall varying from 300 mm to 1100 mm, the average supply at the mouth of the Oum Er Rbia river is significant and is estimated at 3 680 Mm3 / year

These contributions come mainly from Oum Er Rbia river and its main tributaries:

- Tassaout;
- Lakhdar;
- El Abid.

The contributions of numerous sources associated with those of the melting of the snow guarantee a very low water level for the Oum Er Rbia making it the most regular course of the Kingdom.

Average inputs in this zone are estimated at 3,730 Mm3 / year with a maximum of 8,300 Mm3 / year and a minimum of 1,300 Mm3 / year.

The area includes a mobilizable water potential of 4,000 Mm3 of which 3,365 Mm3 are already mobilized.

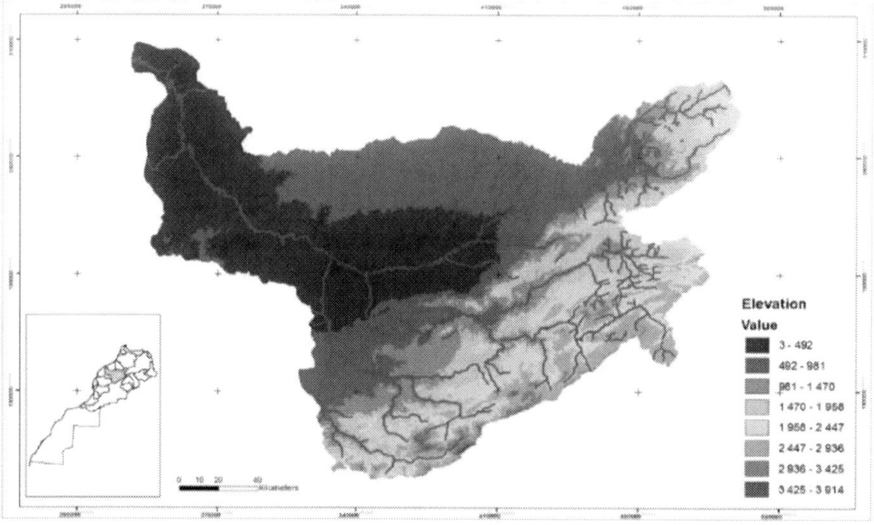

Fig. 2. River systems in Oum Er Rbia basin

The climate in the basin is arid to semi-arid, with decreasing precipitation from West to East (1100 to 250 mm/year). However, in the upstream portion of the basin, at the high altitude of the Atlas Mountains, the annual rainfall may reach 1000 mm, with a significant amount of snowfall at altitudes above 1500m.

The average rainfall in the basin is about 500 mm/year and varies between 1100 mm in the Middle Atlas and 300 mm in the area downstream of the River basin. The temperature varies between 10 and 50 °C. The lowest and highest temperatures are 3.5°C in January and 38°C in August.

Potential Evaporation can reach 1600 to 1800 mm/year. On the average, it is about 1600 mm per year in the coastal area and 2000 mm within the basin with a monthly maximum of 300 mm in July and August.

Most of the basin water resources are generated in the High Atlas Mountains, upstream of the basin. The hydrological network consists of the main river, the Oum Er Rbia (550 km), with three mains tributaries: the Tessaout River, the Lakhdar River, and El Abid River. The annual average flows of the river basin are estimated to 3,250 Mm3/year with a maximum of 8,300 Mm3 and a minimum of 1,300 Mm3.The title is set in bold 16-point Arial, justified. The first letter of the title should be capitalised with the rest in lower case. You should leave 22 mm of space above the title and 6 mm after the title.

2.2 Steps for extracting catchment basin of the Oum Er Rbia basin

The extraction of the basins in all the cases followed the procedure described by the algorithm of O'Callaghan and Mark (1984). The various steps in this procedure are all functions implemented in Arc / Info-Grid. These different functions were carried on Arc-View by the University of Texas team that developed the CRWRPrePro extensions. Detailed descriptions of these procedures are given in Maidment et al. (1997).
Briefly, this sequence of proceedings consists of:

1. Filling of the "wells" of the MNT (FILL): This function locates the closed depressions (either real or created by interpolation during the construction of the MNT) And fills them;
2. calculation of the Mana flow directions;
3. calculation of the rate of accumulation for each of the cells (number of upstream cells;
4. choice of an accumulation threshold for the individualization of the hydrographic network;
5. hydrographic networks from DTM is the individualization of watercourses in flat terrain.

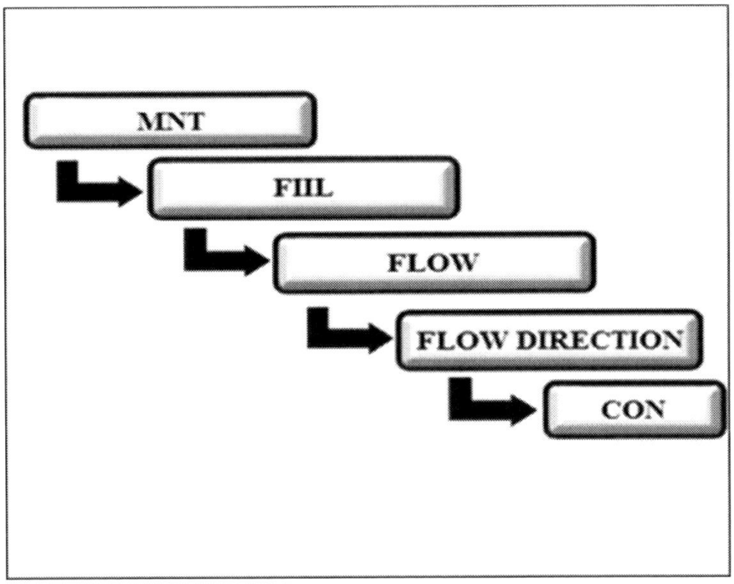

Fig. 3. Extraction steps tributary catchments

Much work has been devoted to this problem and the most commonly adopted solution is to digital terrain model using a pre-existing network. The use of a pre-existing network to allow the automatic extraction of the hydrographic network from a digital terrain model may seem inconsistent. But the modeling objective (and this is the case for many distributed hydrological models using physical bases) requires that a coherent hydrographic network be obtained with a mesh digital topographic basis. (Frédérique Seyler and all).

Table 1. The different sous basins

Abbreviation	Sous basins name
TS	(Oued Tessaout)
HO	(Haut Oum Er Rbia)
AB	(Oued El Abid)
CO	(Oum Er Rbia central)

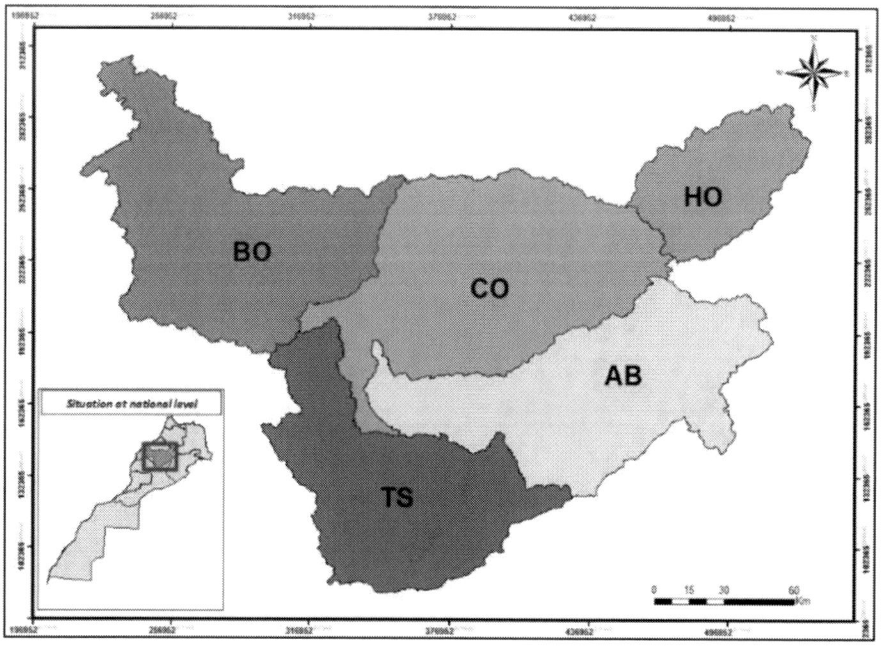

Fig. 4. locations of the 5 tributary catchments in Oum Er Rbia basin

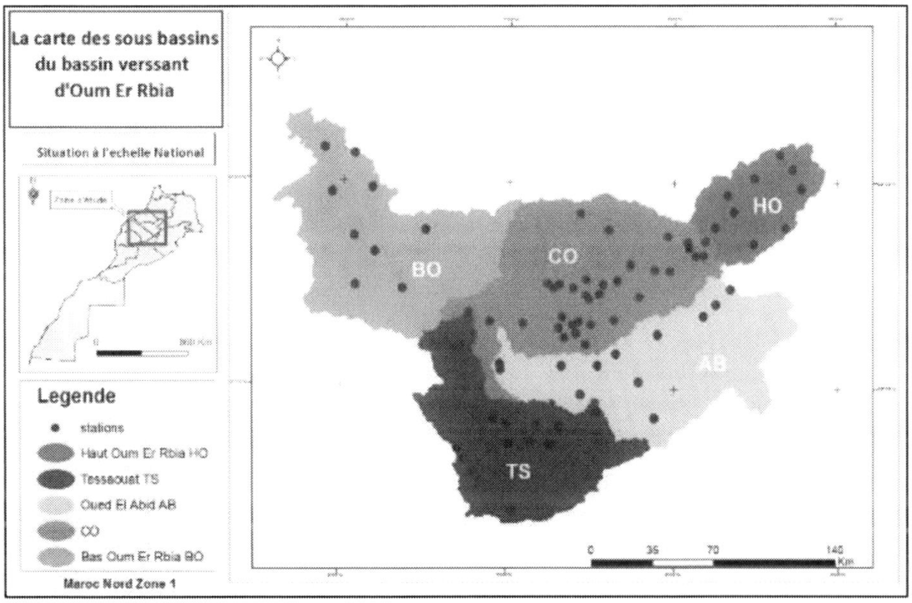

Fig. 5. hydrological and thermal stations in Oum Er Rbia basin

2.3 Geographic location of the watershed

So, to facilitate the work, we will choose the basin Oum Er Rbia central for assessment and identify the climate change from the rainfall series.

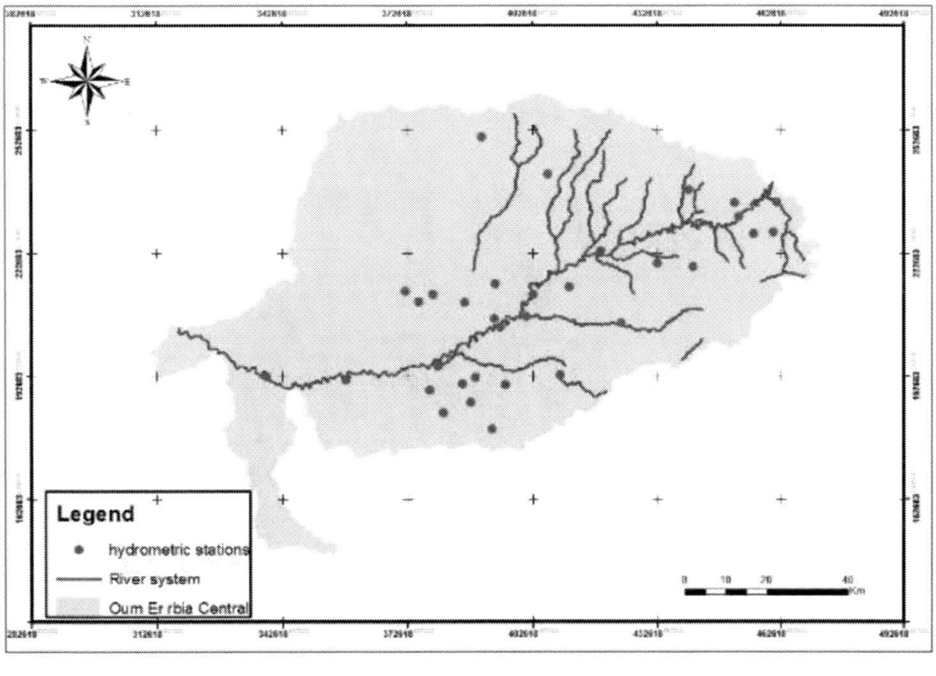

Fig. 6. Geographic location of the basin Oum Er Rbia central

3 Data and methods

3.1 Data

The rainfall data collected from the Oum Er Rbia basin agency were basically pre-checked before being supplied to users.

The task aspects studied for this rainfall analysis are presented as follows:

•Data collection, data analysis and review;

•Filling of data;

•The homogenization of rainfall data;

3.2 Methods

The analysis of rainfall measured at different meteorological stations is usually carried out in well-defined years (Morel, 1995).

The three criteria for assessing the quality of a rainfall network are:

- The distribution: the network must be spatially homogeneous so as not to favor a region.
- The density: stations must cover the basin from upstream to downstream. For these stations it's important to maintain the interpretations as correct as possible;
- Duration: the choice is made for those with the longest measurement periods (25 to 50 years).

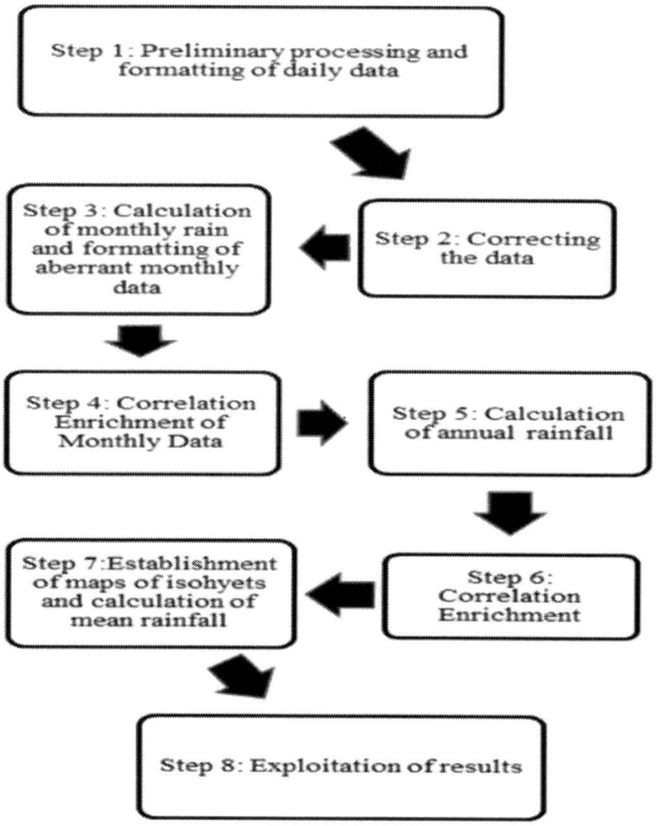

Fig. 7. The analysis of rainfall

 https://doi.org/10.1051/e3sconf/20183703003

4 Results

The results of the analysis of rainfall is presented in graphs below:

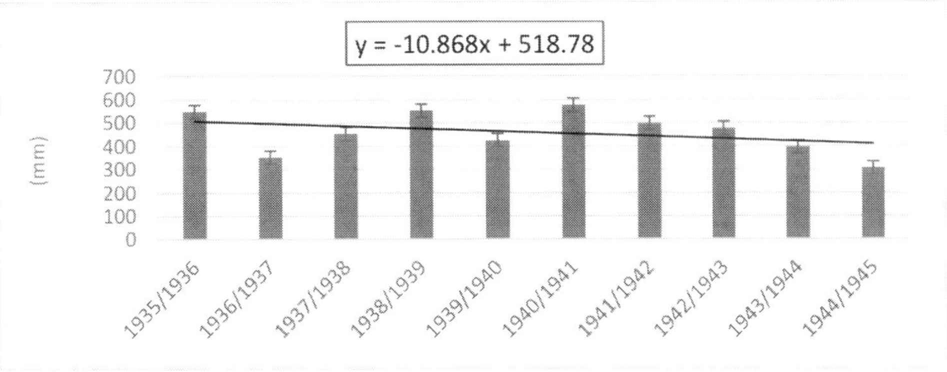

Fig. 8. Evolution of the interannual mean rainfall over the period 1935-1945 of the catchment basin CO

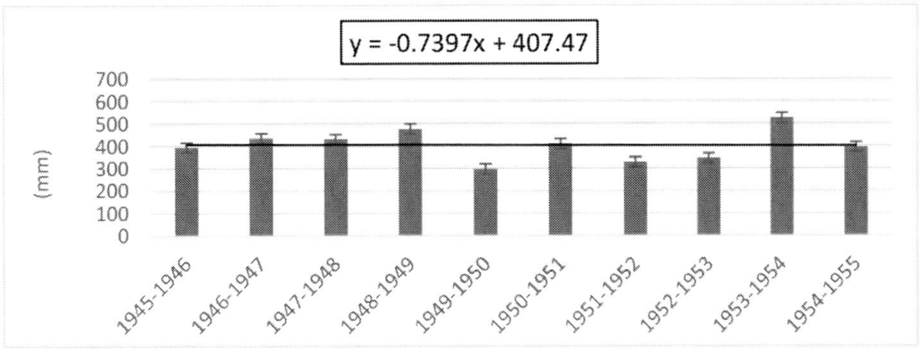

Fig. 9. Evolution of the interannual mean rainfall over the period 1945-1955 of the catchment basin

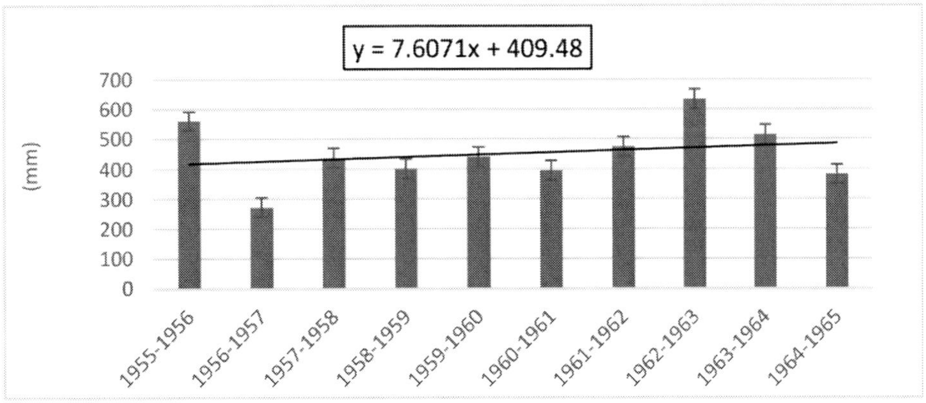

CO

Fig. 10. Evolution of the interannual mean rainfall over the period 1955-1965 of the catchment basin CO

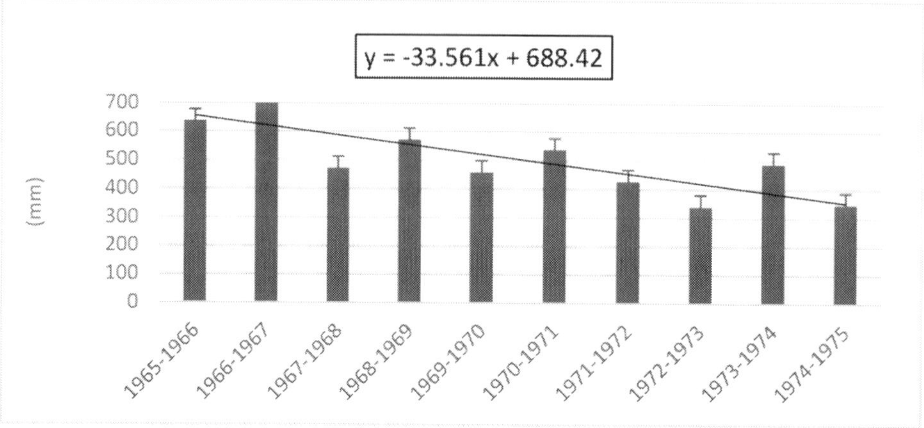

Fig. 11. Evolution of the interannual mean rainfall over the period 1965-1975 of the catchment basin CO

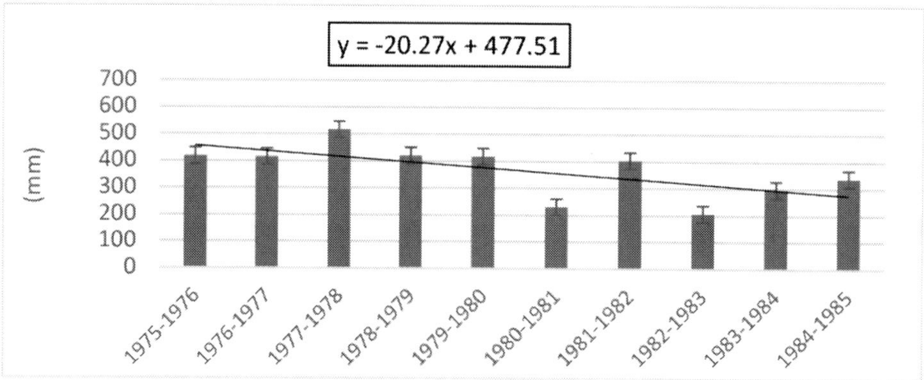

Fig. 12. Evolution of the interannual mean rainfall over the period 1975-1985 of the catchment basin CO

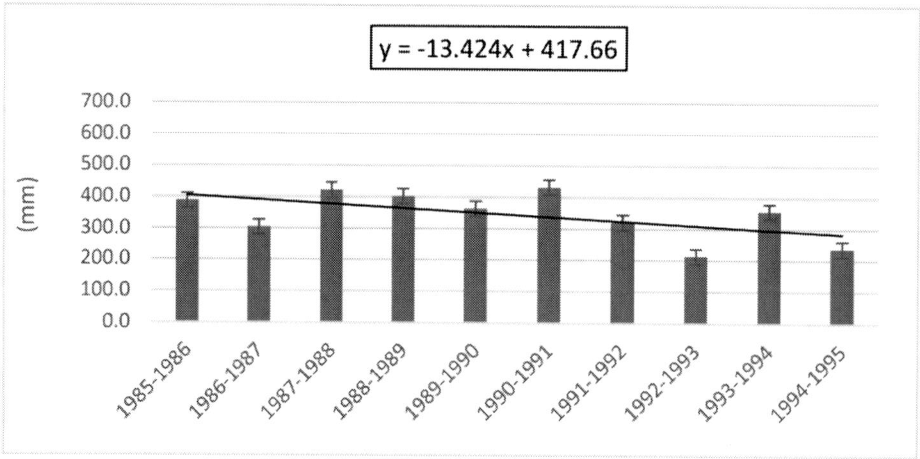

Fig. 11. Evolution of the interannual mean rainfall over the period 1985-1995 of the catchment basin CO

Fig.

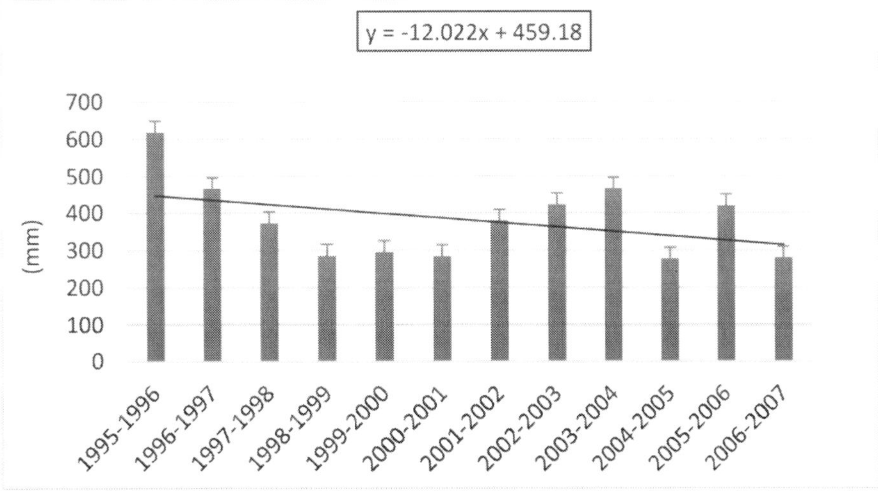

13. Evolution of the interannual mean rainfall over the period 1995-2007 of the catchment basin CO

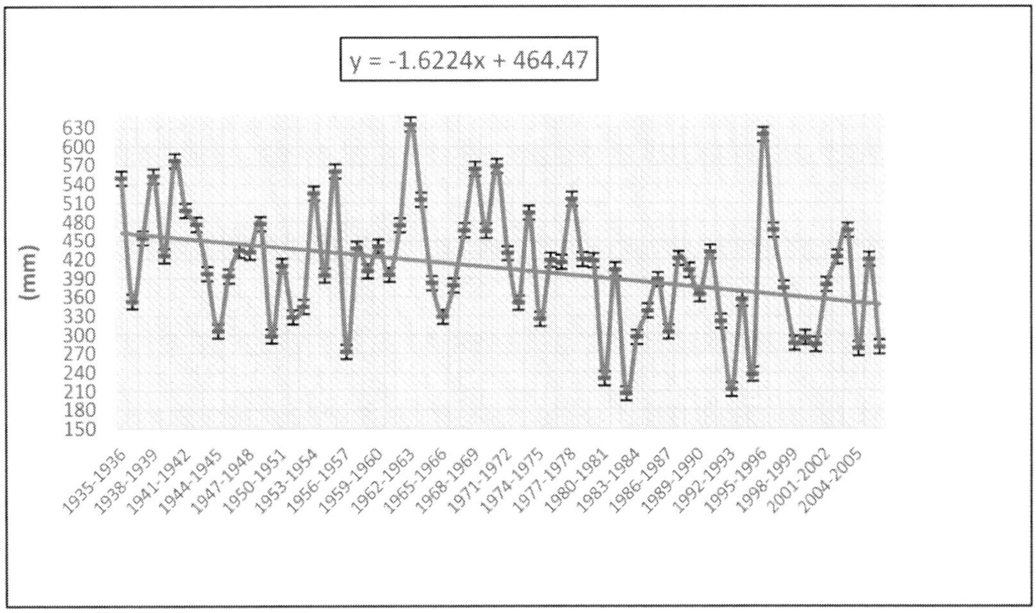

Fig .14. Evolution of the interannual mean rainfall over the period 1935-2006 of the watershed CO

The Rainfall regression and mitigation of drought weather will have a negative impact on runoff and water resources in the central Oum Er Rbia watershed.

Table 2. Station

Station	X	Y	Z
Dchar El Oued	4525501	231550	591
Mechra Eddahk	394975	204720	406
Taghzirt	423900	205600	565
Moulay Bouzekri	401310	207400	431
Ouled Sidi Driss	338950	192600	336
Taghzout	460950	235700	974

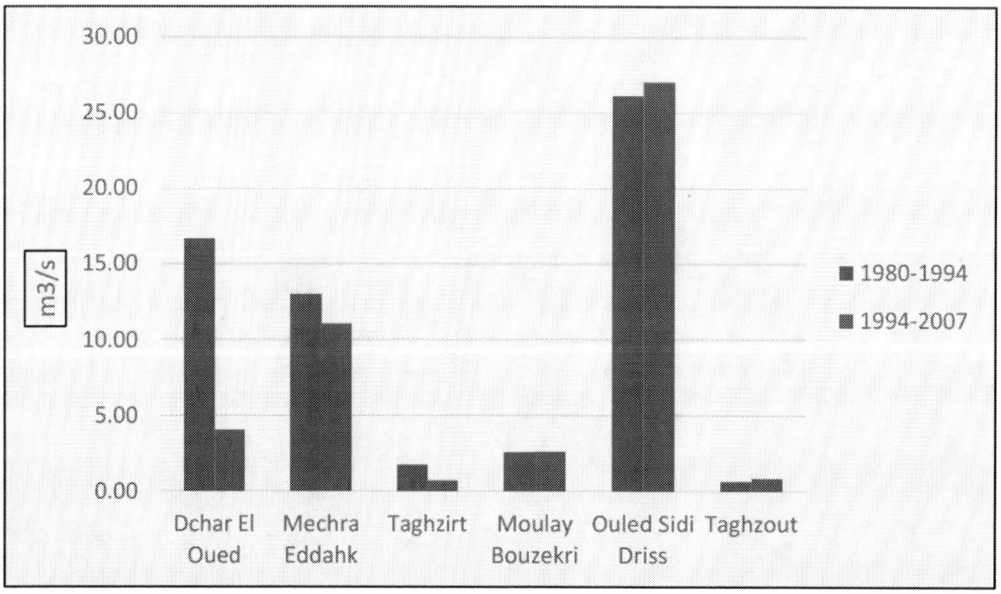

Fig. 15. Interannual average flows between two periods

5 Discussion

Observed climatic trends, calculated over the period 1935-2007 and reported in the study, indicate the following:

On an annual basis, changes in precipitation were not significant and varied from one region to another.

On the other hand, spring rainfall declined significantly in the northern part of Morocco at a rate of 0.5 mm / day per decade

That the area experienced an average annual rainfall reduction of 70 mm (20%) over this period compared to 1940-1980. The area is a hydraulic region that is already experiencing a water deficit. The sharp decline in water supplies since 1980-2007 (by 40% compared with 1940-1980) and the increase in demand and water degradation by different forms of situation.

The annual average flows measured at the central Oum Er Rbia stations were reduced by considerable hydrological deficits ranging between 40.8 and 49.5%.

Climate change forecasts for this region point to an increase in global warming by 2030, with increased losses due to evaporation and evapotranspiration. According to these forecasts, precipitation levels are expected to stagnate or decline slightly compared to the current situation.

The region's water scarcity and recurring droughts will also be part of the region's hydro-climatic landscape for the next 20 years.

The adaptation strategy of the region's water resources management in the face of climate change must consider this new hydro-climatic situation.

References

1. G. Wang, J.Zhang, DOI 10.1007/s11027-015-9664-x, (2015).
2. Y. Tramblay, W. Badi ,82-83 (2012) 104–114, (2012).
3. J. Schilling, P. Freier, 156 12– 26, Ecosystems and Environment (2012).
4. T. Ou, D. Chen, HW. Linderholm, Journal of Meteorology & Oceanography 65.51393-1399, (2013).
5. W. Liu YLi, J. Zhou, Journal of Acta Meteorologica Sinica (in Chinese), 72.2 277-290, (2014)
6. A. Voldoire, E. Sanchez-Gomez, Journal of Climate Dynamics, 40.9-102091-2121, (2013).
7. IPCC (2007) Climate change, (2007)
8. IPCC (2008) Climate Change and Water, (2008)
9. IPCC (2013) Climate change, (2013)
10. IPCC (2014) Climate change, (2014)

Radiative forcing of the desert aerosol at Ouarzazate (Morocco)

Abdelouahid TAHIRI[1*] and *Mohamed* DIOURI[1]

[1]Atmosphere Physics Team (LME), Faculty of Sciences, Mohammed First University, Oujda, 60000 Morocco

Abstract. The atmospheric aerosol contributes to the definition of the climate with direct effect, the diffusion and absorption of solar and terrestrial radiations, and indirect, the cloud formation process where aerosols behave as condensation nuclei and alter the optical properties. Satellites and ground-based networks (solar photometers) allow the terrestrial aerosol observation and the determination of impact. Desert aerosol considered among the main types of tropospheric aerosols whose optical property uncertainties are still quite important. The analysis concerns the optical parameters recorded in 2015 at Ouarzazate solar photometric station (AERONET/PHOTONS network, http://aeronet.gsfc.nasa.gov/) close to Saharan zone. The daily average aerosol optical depth τ_{aer} at 0.5μm, are relatively high in summer and less degree in spring (from 0.01 to 1.82). Daily average of the Angstrom coefficients α vary between 0.01 and 1.55. The daily average of aerosol radiative forcing at the surface range between -150 W/m^2 and -10 W/m^2 with peaks recorded in summer, characterized locally by large loads of desert aerosol in agreement with the advections of the Southeast of Morocco. Those recorded at the Top of the atmosphere show a variation from -74 W/m^2 to +24 W/m^2.

1 Introduction

The desert aerosol observed in Morocco is composed of mineral particles of Saharan origin of size between 0.1 and 10 microns. These particles are raised by the wind over arid and semi-arid regions, remain omnipresent in the atmosphere, and can move over several hundred kilometres. Aerosols in the atmosphere have different physical and chemical characteristics, the effects of which depend on the wavelengths of light. The evolution in time and space of aerosols has a significant influence on the accurate assessment of atmospheric transmission related to their presence. The determination of the physical properties of the tropospheric aerosol constitutes an important field of research enabling the improvement of climate models. Aerosols still represent one of the main sources of error in the forecast of the climate, by the significant uncertainty that they introduce in the evolution of the radiative forcing. This fact results from the great spatial and temporal variability of the tropospheric aerosol, its concentration and composition are quite variable, as it comes from different sources. This study presents the data concerning the optical properties of the Moroccan desert aerosol based on the measurements obtained by the AERONET /

[*]dyourim@gmail.com

PHOTONS network. Several studies have used solar photometers from the ground to characterize the aerosol in different regions of the globe [1], [2], [3], [4], [5] and [6]. Measurement campaigns to determine Saharan mineral dust microphysical characteristics such as the SAMUM measurement campaign carried out in May-June 2006 near Zagora (Tamgroute Village) and Ouarzazate [7], [8] and [9]. In the same, the AMMA campaign took place from mid-January to 2 February 2006 in the region (Niger, Benin and Nigeria) to characterize dynamics during the dry season in West Africa. We will analyze the results of the measurements made for the Moroccan site of Ouarzazate. These analyzes concern the aerosol optical thickness (AOT), the Angstrom coefficient (α), the volume distribution and the radiative forcing at the surface and at the top of the atmosphere (ARF).

2 Materials and methods

2.1. AERONET/PHOTONS network

The Goddard Space Flight Centre (NASA, USA) and the Atmospheric Optics Laboratory (CNRS, France) set up the AERONET / PHOTONS network. It collects measurements made on the ground by solar photometers to specify the optical and microphysical properties of the particles contained in the atmosphere. The data are available from more than one thousand one hundred stations across the globe. The network instruments checked and calibrated with an identical protocol. The data processed in the same way regardless of the date and place of acquisition, whether for the inversion itself [10] or for the rejection of the affected measurements in the presence of clouds [11]. The CIMEL CE-318 solar photometer (Figure 1) is an instrument allowing the restitution of the vertically optical integrated aerosols on the atmospheric column. Photometric measurements are possible only during the day when the sun is visible and in the absence of clouds. The automatic photometer, developed in France by the Laboratory of Atmospheric Optics (Lille), description made by [12]. The photometer is equipped with a robot with two axes allowing movements in the zenith and azimuth planes and can target any point of the celestial cost with an accuracy of 0.05° and a field of view of 1.2°. It has a filter wheel in eight channels (0.34; 0.38; 0.44; 0.50; 0.675; 0.87; 1.02 and 1.64 μm, Table 1). These spectral filters with a width of 0.01 μm correspond to atmospheric windows in which the absorption of solar radiation by the gaseous compounds is low. A geostationary satellite transmits data, every four hours, to NASA's Goddard Space Flight Centre. The photometer provides direct solar view measurements as well as scattering measurements according to the angular luminance distribution of the sky ($W/m^2.Sr.\mu m$) in the Almucantar (circle of the same solar elevation including the sun, retaining the zenith angle θ_v constant, Figure 2). Several morning and afternoon of Almucantar measurement sequences performed. For a sequence of measurements according to the directions Almucantar A_i over 2π (rad), on average 35s.

There are different levels of data processing available: level 1.0 (raw data), level 1.5 (the cloud mask applied while the final calibration not necessarily performed) and level 2.0 (the cloud mask applied as well as the final calibration). We use for our study level 2.0. The algorithm used in AERONET [10] and [13] characterizes aerosols by equating simulated luminance with a radiative transfer code to luminance measurements (direct sunlight luminance and sky luminance) obtained at 4 wavelengths. (0.440, 0.675, 0.870 and 1.020 μm). Radiative transfer is related to the refractive indices (and wavelength dependent) of aerosols and their distribution. The use of this radiative transfer code makes it possible to determine by inversion the desired optical properties with a minimum of initial hypotheses (spherical particles, homogeneous and lognormally distributed). Several studies with other radiative transfer codes involving nonlinear inversion methods [14] and [15] allowed the determination of the optical properties of the desert aerosol. The most recent [16] proposed for particles of non-spherical shapes.

 https://doi.org/10.1051/e3sconf/20183703004

2.2 Ouarzazate solar-photometer

The Ouarzazate site (30.92 °N, 6.91 °W, 1136 m) is associated with the meteorological station. Ouarzazate is a small town with almost no industrial activity. It is located southeast of the Atlas and north-west of the Sahara. Ouarzazate has a pre-Saharan climate characterized by low rainfall. In summer, the climate is hot and dry. The average maximum temperature in July is 40 ° C, the average minimum is 25 °C. In January, the minimum and maximum average temperatures are respectively 2 °C and 16 °C, in the middle of the night, the temperature can go down to -4°C. In the spring, clouds are rare and storms are exceptional. Desert winds (Sirocco and Chergui) play an important role in the climate of Ouarzazate (http://fr.wikipedia.org/wiki/Ouarzazate, AERONET).

Fig.1. Ouarzazate site

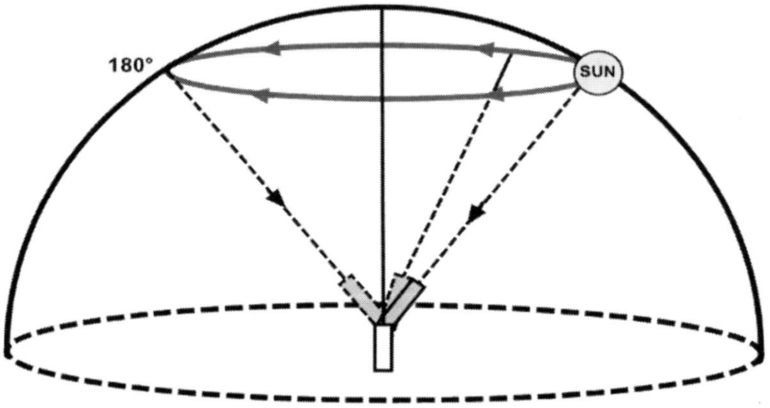

Fig. 2. ALMUCANTARAT position

Table 1. Spectral wavelengths (μm)

λ Standard (μm)	Level 2.0
0.340 (0.002μm)	Ray, NO_2, O_3
0.380 (0.004μm)	Ray, NO_2
0.440 (0.010μm)	Ray, NO_2
0.500 (0.010μm)	Ray, NO_2, O_3
0.675 (0.010μm)	Ray, O_3
0.870 (0.010μm)	Rayleigh
1.020 (0.010μm)	Ray, H_2O
1.640 (0.025μm)	Ray, H_2O, CO_2, CH_4

3 Results and analysis

3.1. Aerosol Optical Thickness (AOT) and Angstrom Coefficient (α)

The aerosol optical thickness (EOA), denoted τ_{aer} (λ) is dimensionless, it quantifies the extinction of incident radiation in a column of atmosphere, both by absorption and by diffusion. τ_{aer} is a key physical parameter for the aerosol characterization, it gives an indication of atmospheric aerosol content. The study of the annual evolution of the AOT allows analyzing the changes in the aerosol type and concentration between the different seasons. Aerosol optical thicknesses can reach very high values, for sandstorms, it is common practice to observe aerosol optical thicknesses at 0.5 μm exceeding 2, which can even exceed 4 above the continent African [17]. The direct measuring the aerosol optical thickness is the measurement of the attenuation by the atmosphere of direct solar radiation. The decrease in solar flux expressed by Bouguer law:

$$I(\lambda) = I_0(\lambda) e^{-m_{air}\tau(\lambda)} \qquad (1)$$

τ (λ) is the total optical thickness that accounts for the contribution to aerosol attenuation, Rayleigh scattering and atmospheric gases.

$$\tau(\lambda) = \tau_{aér}(\lambda) + \tau_{Ray}(\lambda) + \tau_{gaz}(\lambda) \qquad (2)$$

$$\tau_{aér}(\lambda) = \frac{1}{m_{air}} \ln\left(\frac{I_0(\lambda)}{I(\lambda)}\right) - \tau_{Ray}(\lambda) - \tau_{gaz}(\lambda) \qquad (3)$$

$$\tau_{Ray}(\lambda) = \frac{p}{p_0} 0.00877.\lambda^{-4.05} \qquad (4)$$

I (λ) : solar flux (W / m2);
I_0 (λ) : extraterrestrial solar flux (W / m2)
τ_{gaz} (λ) : optical thickness due to gas absorption (O_3, NO_2, CH_4 and H_2O) [19] and [20];
τ_{Ray} : Optical thickness due to Rayleigh scattering [21]
mair : air mass [22];
θ_v : zenithal angle (°);
P: atmospheric pressure; P_0 : standard atmospheric pressure at sea level.

The turbidity coefficient of Angström β, represents the optical thickness of the wavelength 1 μm. For a wavelength different from 1 μm, the AOT is related to this coefficient by:

$$\tau_{aér}(\lambda) = \beta\lambda^{-\alpha} \qquad (5)$$

β varies in the range 0 to 0.5, it is an index which represents the quantity of the aerosol present in the atmosphere in a vertical direction (atmospheric load), β can exceed the value 0.5 for a very charged atmosphere.

The Angstrom coefficient α depends on the aerosol particle size distribution. α close to 0 is associated with large loads of large particles introduced by thermal instability or by a sandstorm (the optical thicknesses vary slowly with the wavelength) whereas α around 4 indicate that the particle population is mainly nucleation and accumulation (very fine to fine particles, optical thicknesses vary more rapidly with wavelength). A population of large particles whose number distributed in one mode may have a slightly negative Angstrom coefficient. The upper limit of α corresponds to the spectral behavior of the attenuation optical thickness associated with particles much smaller than the wavelength, such as gaseous molecules of the atmosphere (Rayleigh scattering). The lower limit corresponds to the case where the particle size is much greater than the wavelength. The micro droplets of water contained in the clouds correspond to this condition; in this case, all the visible wavelengths diffuse in the same way. To quantify the variation of the Angstrom coefficient, the data of the AOTs use for two respective wavelengths. We define the Angstrom coefficient:

$$\alpha_{1-2} = -\frac{\ln\left(\frac{\tau_{aér}(\lambda_1)}{\tau_{aér}(\lambda_2)}\right)}{\ln\left(\frac{\lambda_1}{\lambda_2}\right)} \tag{6}$$

For our study, we put $\lambda_1 = 0.44\mu m$ and $\lambda_2 = 0.87\mu m$.

The monthly averages of the aerosol optical thicknesses at 0.5 μm (Figure 4) and the Angstrom coefficients (0.44-0.87 μm) observed in Ouarzazate for the year 2015. The peaks of AOT register in summer. The type of aerosol alternately changes in the Ouarzazate region due to the low influence associated with the Atlantic air masses that moderates the air dust loads of East and Saharan origin on summer. The minimum values recorded during the fall and winter. During the SAMUM experimental measurement campaign carried out in May-June 2006 in Ouarzazate, the daily average of the AOTs at 0.5 μm wavelength is 0.28 [23], in agreement with the AERONET data (Figure 4). For the Angstrom coefficients, it noted that the largest values of the optical thicknesses of the aerosol are associated with small values of the Angstrom coefficients α characteristic of the presence of desert dust in summer coming mainly from the Southeast.

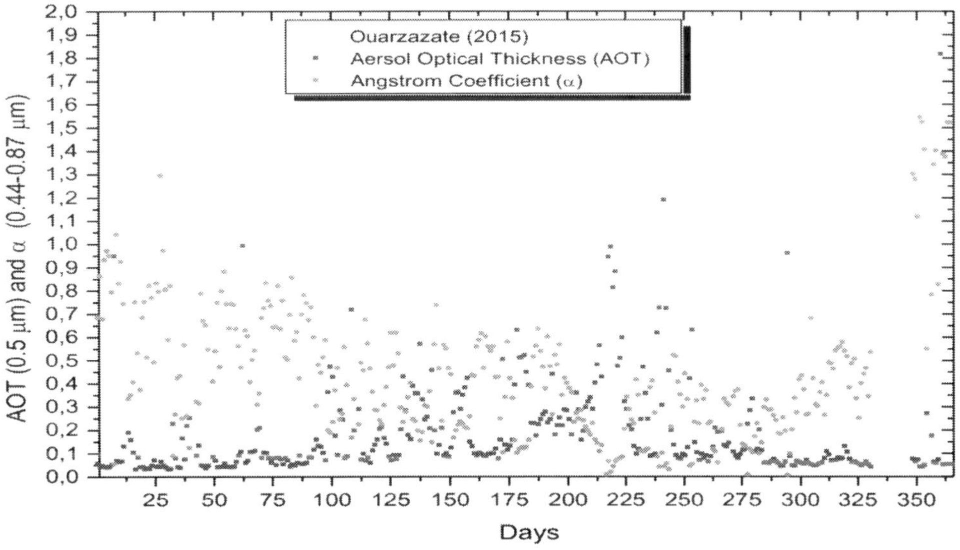

Fig. 3. Time series of AOT (0.5μm) and Angstrom Coefficient (0.44-0.87 μm) observed at Ouarzazate (2015)

3.2 Aerosol Particle Size Distribution

The monthly averages of volume particle distributions show two modes (Figure 4), the main is that of coarse particles. The fine particles are around 0.15 μm and the coarse close 2 μm. The annual volume concentration of fine particles varies between 0.010 and 0.021 $\mu m^3/\mu m^2$, whereas for large particles it varies between 0.086 and 0.126 $\mu m^3/\mu m^2$. The parameters of the distributions record high values in summer where one observes important advections of the masses of air coming mainly from Southeast of Morocco.

 https://doi.org/10.1051/e3sconf/20183703004

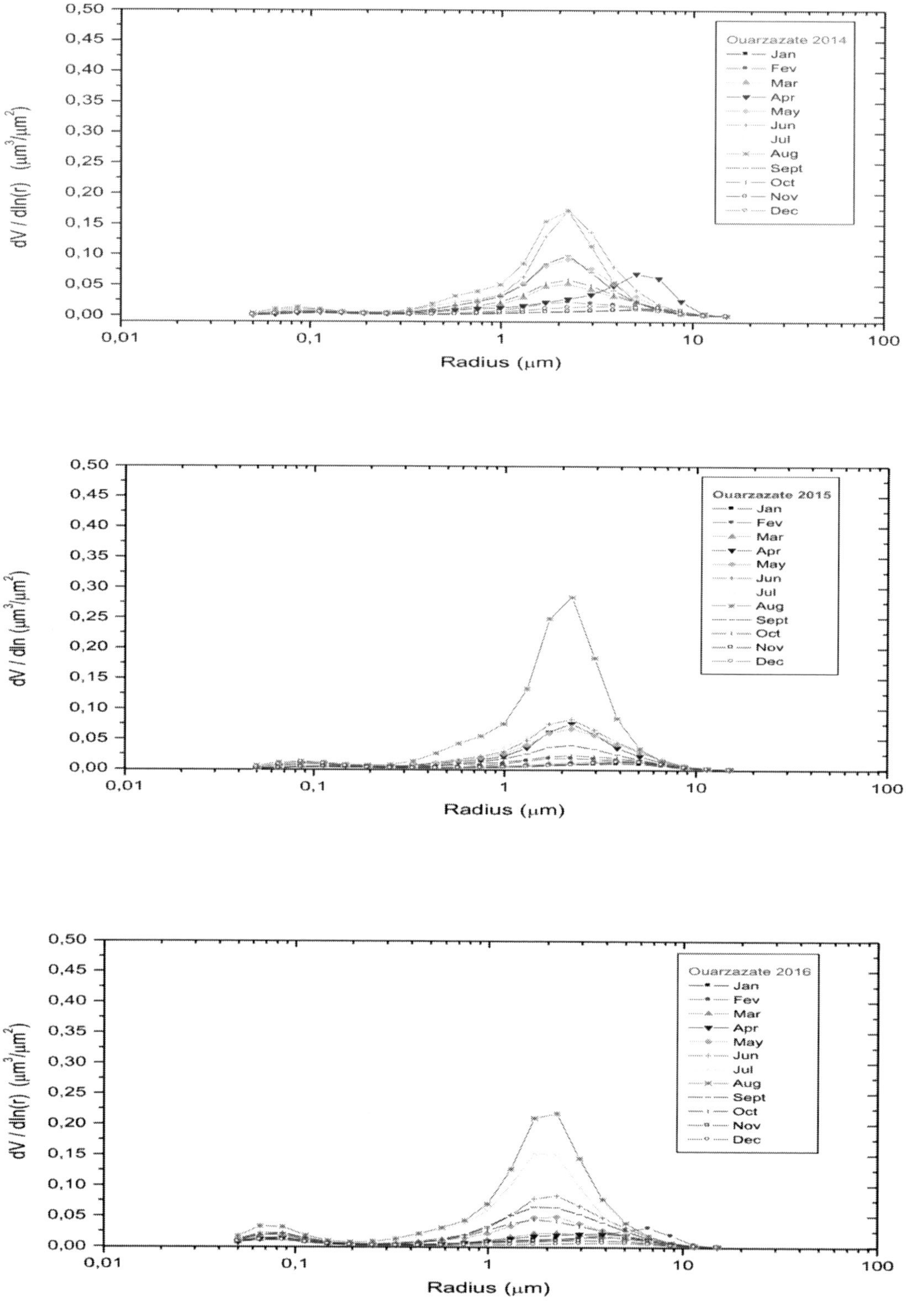

Fig. 4. Monthly averages of volume distributions (2014, 2015 and 2016)

3.3 Aerosol Radiative Forcing ARF

The aerosol radiative forcing (W / m^2) corresponds to any modification of the net radiative flux at the altitude z due to atmospheric aerosol. Positive radiative forcing tends to warm up the Earth's system, whereas a negative forcing tends to cool it. It quantified at the top of the atmosphere or at the ground surface according to the equation:

$$\Delta F_z = \left(F^{\downarrow} - F^{\uparrow}\right) - \left(F^{\downarrow 0} - F^{\uparrow 0}\right) \tag{7}$$

$F^{\uparrow}$ (visible + infrared) is the total ascending flow and $F^{\downarrow}$ (visible + infrared) is the descending one. The indices ($\uparrow^0$) and ($\downarrow^0$) correspond to the flows calculated in absence of aerosol, case of very clear sky. A clean atmosphere without aerosol characterized by a maximum of global irradiance, a minimum of diffuse irradiance and low optical thicknesses. The radiative forcing of aerosols depends on the intrinsic properties of the aerosol but also on the relative humidity of the air as well as the albedo of the soil surface.

The daily average radiative forcing values at the surface and at the top of the atmosphere (Figure 5) show a variation from -150 to -10 W / m^2 with significant variations observed in spring and summer. For these seasons, there are decreases in net flux to the Earth's surface. The values at the top of the atmosphere vary between -74 and +24 W / m^2, the negative values mean the increase of the flux at the top of the atmosphere due to the diffusion of the radiation towards the space by the aerosols thus contributing to the significant cooling of the Earth-Atmosphere system. While, the positive values indicate a greater amount of radiative energy reflected by the cirrus passengers.

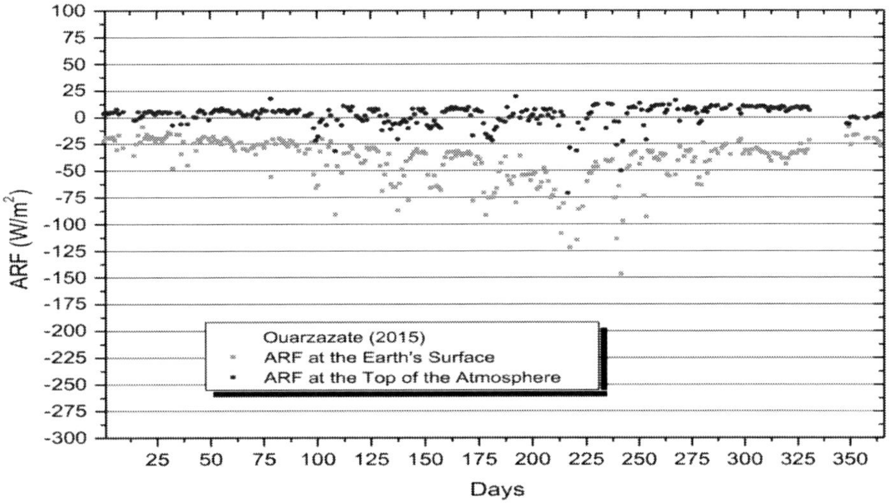

Fig. 5. Aerosol Radiative Forcing at the Earth's Surface and at the Top of the Atmosphere Observed at Ouarzazate in 2015.

4 Conclusion

Monthly averages of atmospheric aerosol optical thicknesses obtained at the Ouarzazate site confirm the importance of the influence of desert aerosol with relatively high values in summer and to a lesser degree in spring. The daily mean values of the α-Angstrom coefficients vary between 0.01 and 1.55. The low characteristic values of large particles and corresponding to the presence of desert dust appear in summer. The volume distributions of the aerosol mainly characterized by a mode at 2μm and a second mode of very small importance at 0.15μm. The trace of presence of the desert aerosol emitted regularly from the South and Southeast of Morocco confirmed at the aerosol optical parameters cyclically observed for the three years.

Monthly averages of aerosol radiative forcing at the surface and at the top of the atmosphere obtained at the Ouarzazate site for 2015 confirm the influence of the desert aerosol. The large ARF values correspond to the large values of the AOTs. The values at the top of the atmosphere vary between -74 and + 24 W/m^2 , the negative values correspond to the increase of the flux due to the diffusion of the radiation towards the space by the aerosols contributing significantly to cooling the Earth-Atmosphere system. Positive values indicate a greater amount of radiative energy reflected by the importance of the cirrus passengers. At the surface, ARF values vary between -150 W/m^2 and -10 W/m^2, are quite representative of the importance of forward scatter of the coarse particle modes characteristic of desert aerosols which cause a decrease in the net flux at the ground surface.

Aknowledgements

Authors want to thank Principal Investigator of AERONET Site (Ouarzazate): Emilio Cuevas-Agullo and M. Taoufik Zaidouni.

References

1. P.C.S. Devara, G. Pandithurai, P.E. Raj, S. Sharma. Investigations of aerosol optical depth variations using spectroradiometer at an urban station, Pune, India. J. Aerosol Sci., **27**, 621- 632(1996).
2. M. Diouri et D. Chabane, 2010. AERONET/PHOTONS (http://aeronet.gsfc.nasa.gov/).
3. L. El Amraoui and Diouri. M. Characterisation of different air mass influences in terms of aerosol optical parameters.J.Aerosol Sci.Vol **32** Supp 1 pp S643-S644 (2001).
4. I. El Aouadi. Estimation de la quantité de vapeur d'eau précipitable et du forçage radiatif de l'aérosol atmosphérique à partir de la télédétection passive à Oujda (Maroc). Thèse de Doctorat,Université Mohamed Premier, Faculté des sciences, Oujda, Maroc (2005).
5. A. Tahiri. Etude de l'aérosol désertique avec des données actuelles, DESA. Université Mohamed Premier, Faculté des sciences, Oujda, Maroc (2006).
6. W. von Hoyningen-Huene et Posse. P. Nonsphericity of aerosol particles and their contribution to radiative forcing. J. Quant. Spectrosc. Radiat. Transfer, **57**, 651-668 (1997).
7. K. Kandler. Et al. Size distribution, mass concentration, chemical and mineralogical composition and derived optical parameters of the boundary layer aerosol at Tinfou, Morocco, during SAMUM 2006. Tellus **61B**, 32-50 (2009).
8. A. Petzold et al. Saharan dust absorption and refractive index from aircraft-based observation during SAMUM 2006. Tellus **61B**, 118-130. DOI: 10.1111/j.1600-0889.2008.00383.x (2009).
9. A. Tahiri et Diouri. M. Premier colloque francophone sur les Polluants Organiques Générés par l'Agriculture et les Transports POGAT 2011. Epaisseur optique de l'aérosol désertique observé à Ouarzazate et Zagora. 24 - 27 Octobre 2011, Agadir, Maroc.

10. O. Dubovik and King M.D. A flexible inversion algorithm for retrieval of aerosol optical properties from sun and sky radiance measurements. Journal of Geophysical Research, **105**, 20673-20696 (2000).
11. A. Smirnov., Holben. B. N., Eck T.F., Dubovik. O & Slutsker. I. Cloud screening and Quality control algorithms for the AERONET data base. Remote Sensing of Environment, **73**, 73337-73349 (2000).
12. B. N. Holben et al. AERONET- A federated instrument network and data archive for aerosol characterization. Remote Sensing of Environment, **66**, 1-16 (1998).
13. O. Dubovik., A. Smirnov., B. N. Holben., M. D. King., Y. J. Kaufman., T. F. Eck and I. Slutsker. Accuracy assessment of aerosol optical properties retrieval from AERONET sun and sky radiance measurements. Journal of Geophysical Research, **105**, 9791-9806 (2000).
14. M. Diouri and Sanda. S.I. Deduction of particle size distribution from aerosol optical depth.CLEOPATRE-I code. J. Aerosol Sci, **28**, S459 (1997).
15. W. von Hoyningene-Huene and Wendisch. M. Possibility of refractive index determination of atmospheric aerosol particles by ground based solar extinction and scattering measurements. Atmos. Environ, **28**, 785-795(1994).
16. M.I. Mishchenko., Travis, L.D and Lacis, A.A. Multiple Scattering of Light by Particles. Radiative Transfer and Coherent Backscattering, Cambridge Univ Press, New York (2006).
17. A. Slingo et al. Observations of the impact of a major Saharan dust storm on the atmospheric radiation balance, Geophysical Research Letters, **33**, l24817, doi :10.1029/2006GL027869 (2006).
18. W. v. Hoyningen-Huene. Atmospheric aerosol –A special topic in environmental physics . IUP, Universitat Bremen, 16/04/2003.
19. J.P.Burrows., Dehn, A., Deters, B., Himmelmann, S., Richter, A., Voigt, S and Orphal, J., Atmospheric Remote-Sensing Reference Data from GOME: Part 1. Temperature-Dependent Absorption Cross-sections of NO_2 in the 231-794 nm Range, JQSRT, **60**, 1025-1031 (1998).
20. J.P.Burrows., Richter, A., Dehn, A., Deters, B., Himmelmann, S., Voigt, S. and Orphal J., Atmospheric remote -sensing-reference data from GOME: 2. Temperature -dependent absorption cross sections of O_3 in the 231-794 nm range, JQSRT, **61**, 509-517, 1999.
21. E.G. Dutton., Reddy. P., Ryan.S., Deluisi.J.J. Features and effects of aerosol optical depth observed at Mauna Loa, Hawaii: 1982-1992, J. Geophys. Res., **99**, 8295-8306, (1994).
22. F. Kasten and Young. A.T. Revised optical air mass tables and approximation formula. Applied Optics, **28**, 4735-4738 (1989).
23. C. Toledano et al. Spectral aerosol optical depth characterization of desert dust during SAMUM 2006.Tellus **61B**, 216-228.Doi: 10.1111/ j.1600-0889. 2008. 00382.x (2009).

Hydrological and hydroclimatic regimes in the Ouergha watershed

Karim Msatef [1(1)], *Lahcen* Benaabidate [(1)], *Aziz* Bouignane [(2)]

[1]Laboratoire de Géo-ressources et Environnement, Faculté des Sciences et Techniques de Fès, Université Sidi Mohamed Ben Abdellah, Route d'Imouzzer, BP 2202, Fès, Maroc

[2]Agence du bassin hydraulique, Fès, Maroc

Abstract: This work consists in studying the hydrological and hydroclimatic regime of the Ouergha watershed and frequency analysis of extreme flows and extreme rainfall for peak estimation and return periods, in order to prevention and forecasting against risks (flood...). Hydrological regime analysis showed a regime of the rain type, characterized by rainfed abundance with very high winter flows, so strong floods. The annual module and the different coefficients show hydroclimatic fluctuations in relation to a semi-humid climate. The water balance has highlighted the importance of the volumes of water conveyed upstream than downstream, thus confirming the morphometric parameters of watershed and the lithological nature. Frequency study of flows and extreme rainfall showed that these flows governed by dissymmetrical laws based on methods Gumbel, GEV, Gamma and Log Pearson III.

1 Introduction

The watershed Ouergha is one of the main rivers of Morocco, it has severe hydrological characteristics that can cause risks. In order to prevent and forecast against these risks, an understanding of the hydrological and hydroclimatic functioning of the entire watershed is necessary. The study was started by a morphological characterization of the studied area, followed by a geological and climatological synthesis, and finally an analysis of the hydrological behavior. The Sebou watershed is located in the north-west of Morocco has 30% of the national surface water reserve and 50% of these reserves are provided by the Ouergha sub-watershed. The Ouergha Wadi extends over most of the Rif Mountain and forms a passage between the Rif and the Middle Atlas over an area of 6190 Km², this watershed represents 16.5% of the Sebou watershed of which it is part, this geographical situation favors the penetration of the freshness and humidity of the air

* Karim Msatef : karim.msatef@gmail.com; mobile: +212649375893

masses to allow it to be the most rainy region in Morocco especially at the level from its climax Jbel Outka.

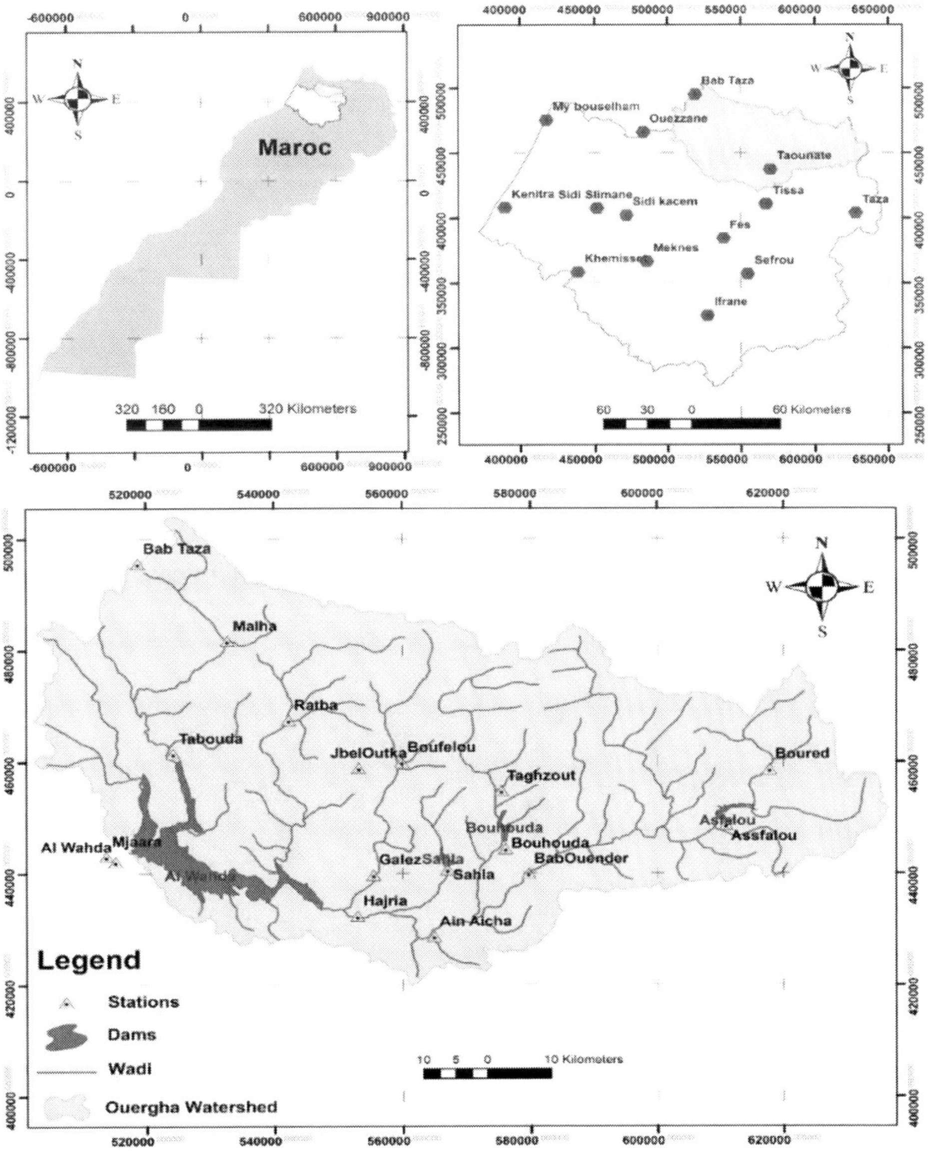

Fig. 1. Map of the geographical location of the Ouergha watershed

2 Materials and methods

2.1 Data

The data used must respect two important criteria: first, the length of the chronicles (cover the greatest possible period of time) and secondly, the quality of the data (the least possible missing data).

 https://doi.org/10.1051/e3sconf/20183704001

The hydroclimatic study is based on rainfall data at seven meteorological stations (Mjaara, Ain aicha, Bab Ouender, Bab taza, Boured, Jbel Oudka et Tabouda) which are represented in the following table provided by the management of the Sebou Hydraulic Basin Agency (Tab.1) and temperatures at the station (hajria) during the period (1981 to 2011).

Table 1: Coordinates of rainfall stations in the Ouergha watershed during the period (1981 to 2007)

Station	N° Station	X(Km)	Y(Km)	Z(m)
Mjaara	5128	513,6	443,2	92
Ain aicha	236	564,7	428,8	230
Bab ouander	1568	579,5	440,1	312
Bab taza	1586	518,5	495	902
Boured	2338	617	458	815
Jbel oudka	4626	553	459	1115
Tabouda	7283	524,25	461,6	180

Analysis of hydrological behavior based on flow and precipitation data at the stations (Ain Aicha, Galaz and Tabouda) during the period (1981 to 2011).

The figure below shows the different stations of the Wadi Ouergha watershed (Fig.2):

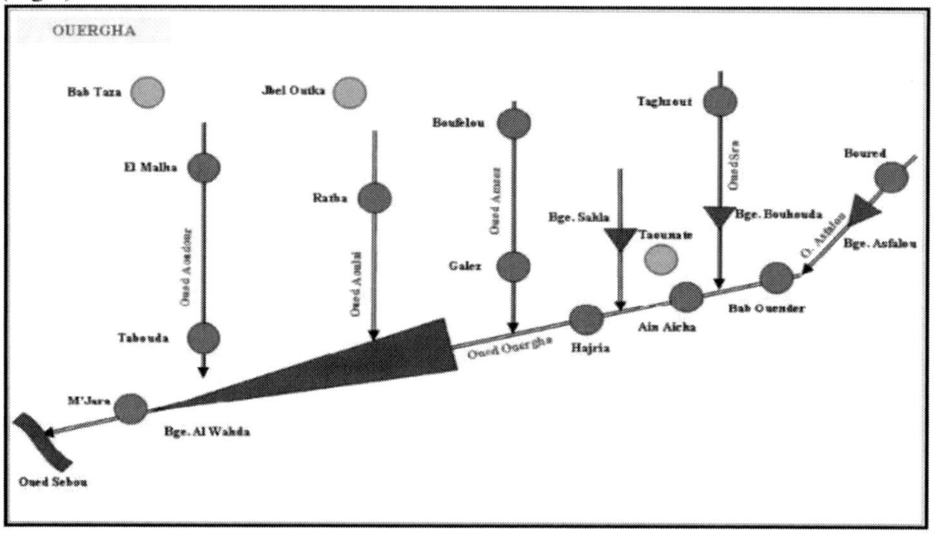

Fig. 2. Ouergha wadi watershed stations

2.2 Methods

The study was divided by four chapter:

- The first chapter presents general data on the basin and determines the physical characteristics of the watershed: it's about determining geometric characteristics (area, morphology, shape, topography, relief) and its hydrographic network. The purpose is to determine the role of these characteristics in the basin's hydrologic response (flow regime).

- The second chapter presents the geological context which is a determining element in the understanding of superficial and underground flow mechanisms in this basin. Substrate geology influences not only groundwater flow but also surface runoff. The geological study of a watershed in the context of a hydrological project is mainly intended to determine the permeability of the substratum and the influence of geology on the differentiation of the hydrogeological context.
- The third chapter is interested to study of the climatic characteristics whose study is necessary to the understanding of the mechanisms of water supply and circulation of the superficial waters. It intervenes in the establishment of the equation of the hydrological balance essentially by the precipitations, secondarily by the temperatures and thirdly by evapotranspiration.
- The fourth chapter will focus on hydrological characterization, to determine the flow regime in the watershed, as well as a frequency analysis to determine the period of flood return and flow adjustment by statistical laws.

3 Results and discussion

3.1 The physical characteristics of the watershed

3.1.1 Relief

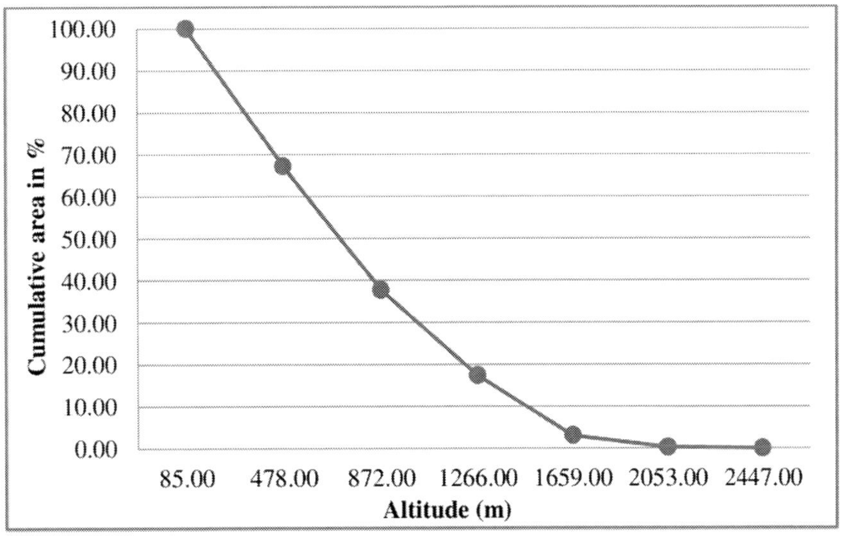

Fig. 3. Hypsometric curve of the Ouergha watershed

The analysis of the hypsometric curve confirms that there is an old basin in which the area is small compared to the altitude change. The differences in altitudes are very marked, which will make it easy to circulate the waters inside the basin and consequently larger earth losses.

3.1.2 Study of the slope

The index of the global slope (Ig) makes it possible to determine the importance of the relief on the basin. This index characterizes the relief of the basin. its formula is:

$$I_g = \frac{DU}{L_{eq}} = \frac{H_{5\%} - H_{95\%}}{L}$$

Ig : Overall slope index in m / km
Leq: Length of the equivalent rectangle (Km)
H5% : Altitude corresponding to 5% of the total area of the watershed
H95% : Altitude corresponding to 95% of the total area of the watershed

Table 2: Classification of relief according to (Ig) given by ORSTOM

Relief	Value of Ig
very low relief	Ig<0,002
low relief	0,002<Ig<0,005
rather weak relief	0,005<Ig<0,01
moderate relief	0,01<Ig<0,02
strong enough relief	0,02<Ig<0,05
strong relief	0,05<Ig<0,5
very strong relief	Ig>0,5

The overall slope index given by the formula is **7.3** or **0.07%.** According to the classification of relief given by ORSTOM (Tab.2), the value of the overall slope index of the Ouergha catchment shows that **the relief is strong.**

3.1.3 Morphological characteristics

The determination of the various physical parameters of the Ouergha watershed enabled us to define the topographic characteristics. The topographic watershed has a perimeter of 462Km, an area of 6190Km2 and a watercourse of 255Km long. The analysis of the distribution of the altitudes shows that the average altitude of this watershed is 776m. The watershed of Ouergha shows that the relief is strong, we have summarized the different physical parameters studied in the table (Tab.3):

Table 3: Morphological characteristics of the Ouergha watershed

Parameters	Values
Area	6190 Km^2
Perimeter	462 Km
Gravelius compacite index	1,64
Length of the main watercourse	255 Km
Length of the equivalent rectangle	199,35 Km
Width of the equivalent rectangle	31,04 Km
Maximum altitude	2744 m
Minimum altitude	85 m
Average altitude	1266 m
Average slope	0,06%

3.2 Climatic characteristics

3.2.1 Emberger's climagram

The rainfall quotient of Emberger (Q2) corresponds to a synthetic expression of the Mediterranean climate taking into account the average annual rainfall and temperature (Sauvage, 1960). These last taken into consideration of a part, the average of the minimums of the coldest month «m» and secondly, the average of the maximums of the month the lime «M». These two extreme thermal values make it possible to evaluate the average temperature (M + m) / 2 and the average extreme thermal amplitude (M-m).

$$Q_2 = \frac{P}{\frac{(M+m)(M-m)}{2}} \times 1000$$

Q2 : The rainfall quotient of EMBERGER
P : Average annual precipitation (P=741mm)
M : Average of maximum of the hottest month (M=303.25°Kelvin)
m : Average of the minima of the coldest month (m=284.15°Kelvin)
Q2 =132

Emberger's climate is subdivided into zones corresponding to various Mediterranean bioclimatic floors (Fig.4).

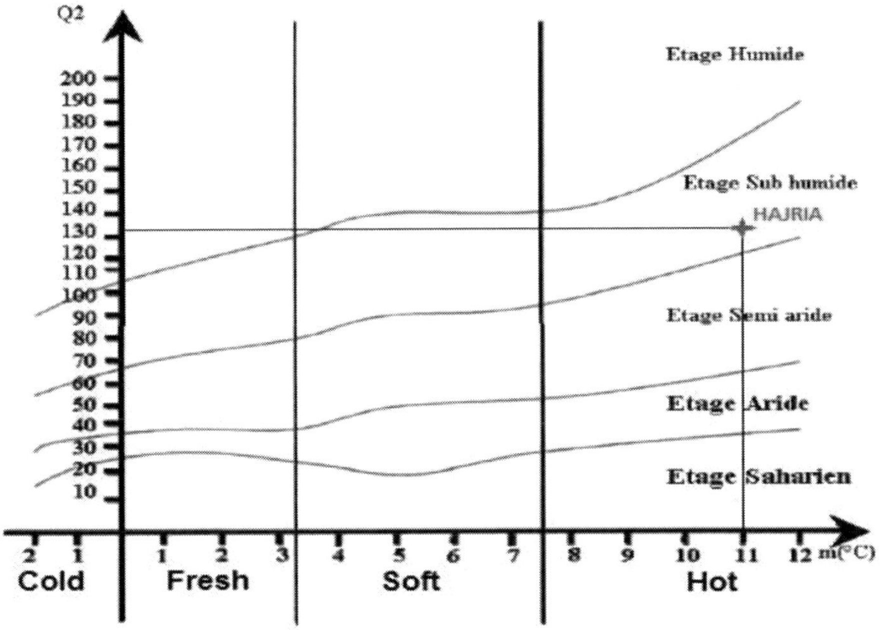

Fig. 4. Emberger's climagram in the watershed Ouergha (Hajria)

The result of this climagram shows that the watershed of Ouergha, at the Hajria station, is in a **semi-humid** climatic stage.

3.2.2 Water balance

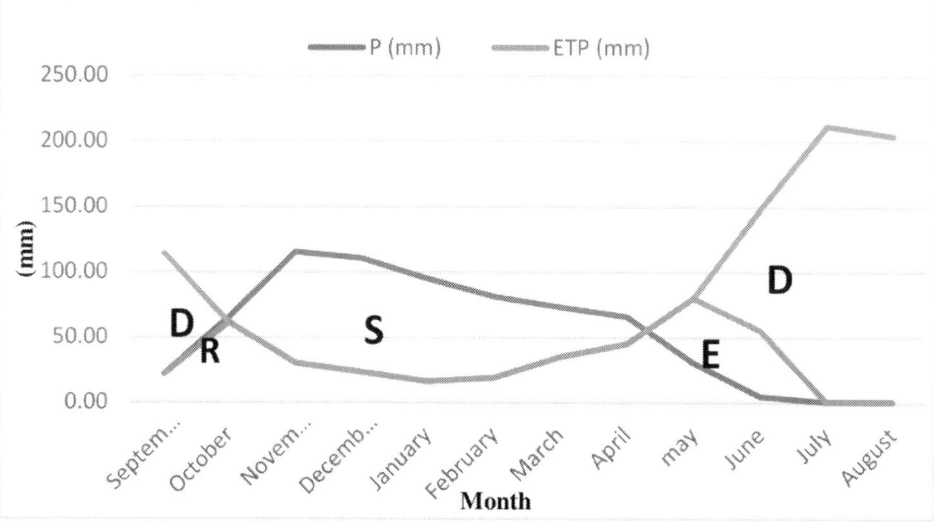

Fig. 5. Water balance of the Ouergha watershed at Hajria, estimated from precipitation, ETP and ETR from the Thornthwaite method

The future of the water brought by the precipitation during the year can be described by the water balance. The latter makes it possible to reveal the nature of each period of the year according to the entries and exits. According to the water balance of the Ouergha watershed (Fig.5), from the average balance sheet of the station Hajria, we can distinguish the periods:

- Phase D : considered a deficit phase and extends from June to the month of September.
- Phase R : it is a phase of reconstitution of soil water reserves, used during the previous months. This reconstitution is favored by an inversely proportional evolution of precipitation and evapotranspiration during the month of October.
- Phase S : It start from November to April.This is a phase where a surplus of water appears because of precipitation that far exceeds evapotranspiration. This surplus will be streamed and ensures the supply of groundwater.
- Phase E : It start from April until June. In this phase, evapotranspiration be done at the expense of easily usable reserves which are gradually depleted and disappear in the month of July. According to the water balance of the Ouergha watershed, four periods can be distinguished: a deficit period from June to September. The second period during the month of October corresponds to the period of return of the soil water reserves. A surplus is observed in the third period starting from November to April. The last the period starts from the month of April until the month of June it is considered as a phase of exhaustion of the reserves easily usable.

3.3 Hydrological characteristics

3.3.1 Relationship between flows and precipitation

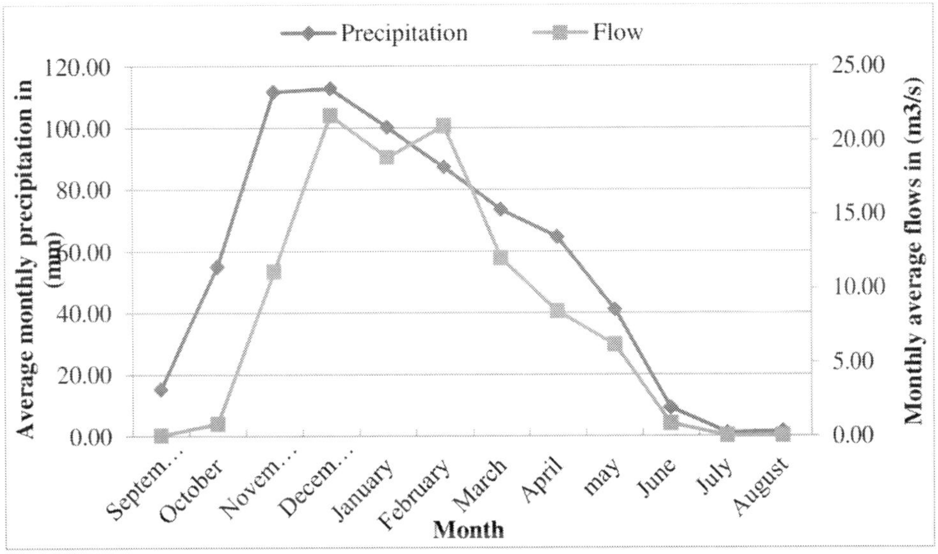

Fig .6. Evolution of precipitation and flow at the Galaz station

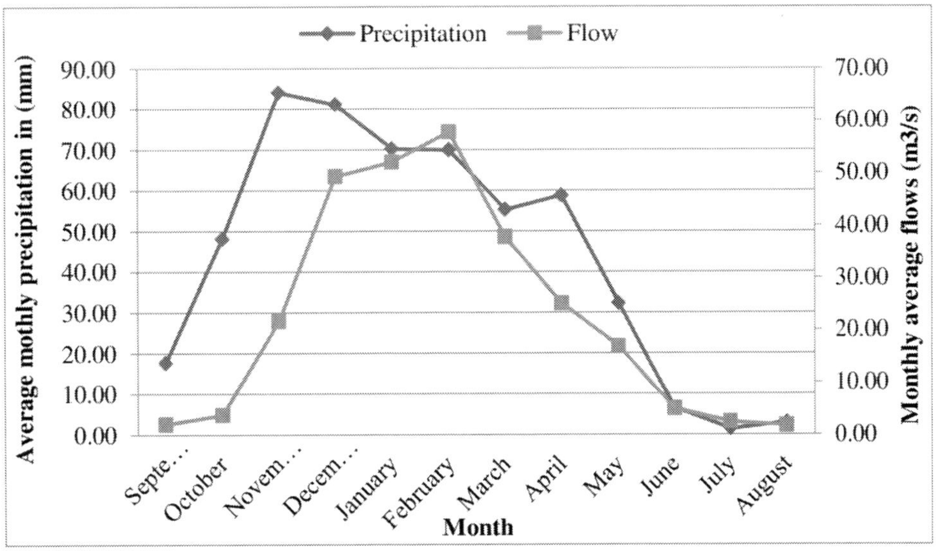

Fig. 7. Evolution of precipitation and flow at the Ain Aicha station

The analysis of these figures shows that there is a succession of floods and showers. For the Galaz station, there is a maximum of precipitation and flow in December, contrariwise Ain Aicha station has a maximum of precipitation in November followed by a maximum of flow with a delay of two months in February. This shift can also be explained by the lithological nature of the Ouergha watershed (semi-permeable to permeable). Thus, the waters that precipitate there runoff immediately after their arrival on the ground. The delay due to the lapse of time separating the triggering of the rain and its circulation after infiltration is very low and thus denotes the impermeability of the medium.

However, the monthly mean flow curve shows for both stations that runoff responds only to heavy rainfall. The low intakes of the months of March, April,

May, June and July can be explained by intense evapotranspiration and may also be by irrigation.

3.3.2 Flood power

Floods are characterized by their suddenness and violence. The power of a flood can be determined by the coefficient of Mayer and Coutagne «I» whose formula is:

$$I = \frac{Q}{\sqrt{S}}$$

Q: The maximum flow of flood (m3/s)

S: Watershed area (Km2)

The values of the Mayer and Coutagne «I» index (Fig.8, Fig.9) at the Ain Aicha station are much larger than that of the floods at the Galaz station, which confirms a spatiotemporal variation of the rainfall and the brutal and violent floods especially upstream of the basin.

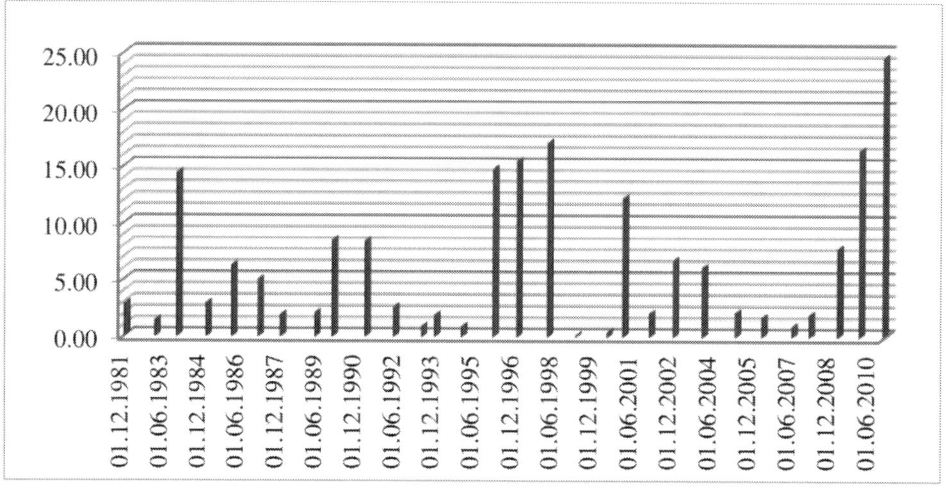

Fig. 8. Evolution of Coutagne and Mayer indice in the Ain Aicha station

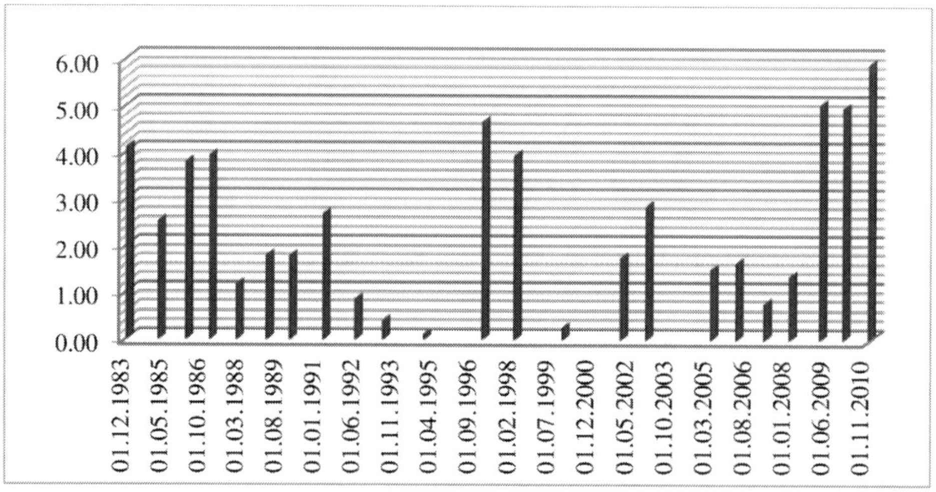

Fig. 9. Evolution of Coutagne and Mayer indice in the Galaz station

3.3.3 Quantiles estimated for different return periods

The return period T of an event is the average duration that separates two floods greater than or equal to this event. Conversely, a return period flood T is a flood where the average is equal to or exceeded at all time units return periods T. In general, the unit of time is the year. However, a return period flood T = 100 year, does not mean that it occurs once every 100 years, but this flood has a 1% chance of arriving each year for 100 years (Zian, 2011).

The results of the estimate quantiles for different return periods T, at the two stations, are shown respectively in (Fig.10, Fig.11).

The analysis of the fugures shows that the estimated quantiles are very close for the low return periods (T≤20), but beyond that, they deviate according to the law of adjustment in the two stations. For large periods of return (T≥2000years), the quantiles estimated by the law of GEV are very important at Ain Aicha level and for the other laws, the evolution is globally constant. For the Galaz station, the quantiles estimated by Gumbel and Gamma law are larger, but those estimated by the law of GEV and Log-pearson III are lower.

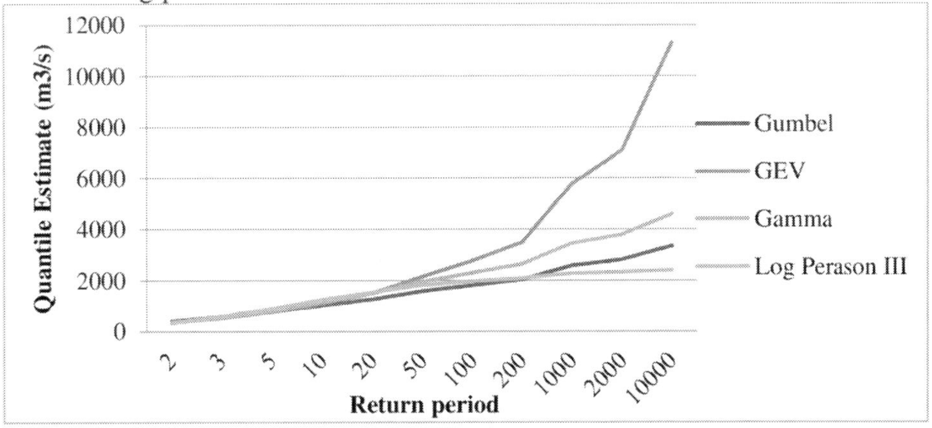

Fig. 10. Return times corresponding to the annual maximum flows recorded at the Ain Aicha station

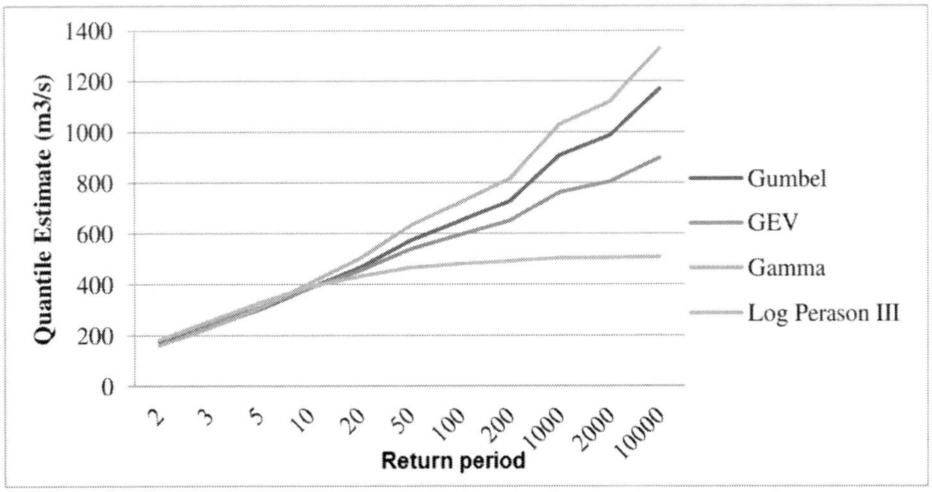

Fig. 11. Return times corresponding to the annual maximum flows recorded at the Galaz station

4 Conclusion

The catchment area of Ouergha is spread over an area of 6190 Km^2 and a perimeter of 462 Km. The watershed of Ouergha is elongated, the hypsometry of which is characterized by decreasing altitudes from North North-East to South South-West (barrage Al Wahda). According to the hypsometry curve, the average altitudes occupy the majority of the basin area. The relief is strong according to the index of the overall slope.

This watershed has three alignments oriented substantially E-W, located in the furrows, synclines Miocene and in the middle and upper valley. After the Mjaara station, the Ouergha takes a N-S direction along a fault line.

Rain has an annual average of 741mm throughout the basin with an average annual temperature of 18.7 ° C. The combination of precipitation and temperature, by different methods, has shown that the basin is in a semi-humid climate. The average value of the actual evapotranspiration is 523.6 mm / year.

The hydrological regime can be considered as a rain regime, characterized by the abundance of rain with very strong winter flows, thus of strong floods.

The estimated quantiles are very close for the low return periods ($T \leq 20$) for the two satations, On the other hand the large return periods ($T \geq 2000$ans), the quantiles estimated by the law of GEV are very important (11000m3/s) at the level of Ain Aicha and for the satation Galaz the law of Gumbel and Gamma are more important (1300m3/s).

References

1. ABHS-Fès : Source of data (rainfall, flow and temperature) (1981 à 2011).
2. Benaabidate L., Zian A., EL wartiti M. & EL rhaouti S., 2006 : Caractérisation climatique et hydrologique du bassin versant du R'dom. 18ème Colloque des Bassins Sédimentaires Marocains, Fès, 10-12, Novembre, 2005.
3. Coutagne A., 1954: Quelques considérations sur le pouvoir évaporant de l'atmosphère, le déficit d'écoulement effectif et le déficit d'écoulement maximum. La houille blanche, pp:360-369.
4. Sauvage C., 1960 : Le quotient pluviométrique d'Emberger, son utilisation et la présentation géographique de ses variations au Maroc. Ann .Serv . phys. globe météo r. I.S.C.
5. Thornthwaite 1955: "The water balance", Climatology, 8:1-104, 1955.
6. Zian A., 2011: Hydrologie, Qualité des Eaux de Surfaces et des sédiments et géochimie des eaux thermales du bassin versant de l'oued R'dom. Thèse doct.3éme cycle, FST FES, 276p.

Modelling soil salinity in Oued El Abid watershed, Morocco

El Mouatassime Sabri[1,*], *Ahmed* Boukdir[2], *Ismail* Karaoui[3], *Abdelkrim* Arioua[4], *Rachid* Messlouhi[5], *Abdelkhalek* El Amrani Idrissi[6]

[1]Sultan MoulaySlimane University, Department of Geology, Beni Mellal, Morocco
[2]Sultan MoulaySlimane University, Department of Geology, Beni Mellal, Morocco
[3]Sultan MoulaySlimane University, Department of Geology, Beni Mellal, Morocco
[4]Sultan MoulaySlimane University, Department of Geology, Beni Mellal, Morocco
[5]Agence de bassinhydraulique Oum Er Rbia|, Beni Mellal, Morocco
[6]Agence de bassinhydraulique Oum Er Rbia, Beni Mellal, Morocco

Abstract. Soil salinisation is a phenomenon considered to be a real threat to natural resources in semi-arid climates. The phenomenon is controlled by soil (texture, depth, slope etc.), anthropogenic factors (drainage system, irrigation, crops types, etc.), and climate factors. This study was conducted in the watershed of Oued El Abid in the region of Beni Mellal-Khenifra, aimed at localising saline soil using remote sensing and a regression model. The spectral indices were extracted from Landsat imagery (30 m resolution). A linear correlation of electrical conductivity, which was calculated based on soil samples (ECs), and the values extracted based on spectral bands showed a high accuracy with an R^2 (Root square) of 0.80. This study proposes a new spectral salinity index using Landsat bands B1 and B4. This hydro-chemical and statistical study, based on a yearlong survey, showed a moderate amount of salinity, which threatens dam water quality. The results present an improved ability to use remote sensing and regression model integration to detect soil salinity with high accuracy and low cost, and permit intervention at an early stage of salinisation.

Introduction

Soil salinity is a worldwide environmental problem [1] especially in arid and semi-arid climates [2]. Generally, it is calculated based on the electrical conductivity (EC) of the soil. In the last decade, many studies have aimed to identify the soil salinity using remote sensing based on satellite images [3]. In Morocco, as in many developing countries[4], the combination of poverty and population growth in fragile environments results the degradation of non-renewable resources. Salinity threatens about 500,000 hectares of land there, where most of it located upstream dams of water and belong to irrigated zones which treats natural resources as well as human livestock [5].

The phenomenon is controlled by several factors, including soil characteristics, climate factors, and anthropogenic factors, which make the control of the phenomenon difficult [6]. The sodium-rich soils have a pH of more than 8.2 with a preponderance of sodium carbonate and bicarbonate. This results in sodium saturation of the clay fraction, and leads

Corresponding author: e.sabri@usms.ma.

to a compact and impermeable soil, which prevents aeration and penetration of water necessary for normal plant growth[7,8].

The watershed of Oued El Abid has a moderate rate of soil erosion, which is related to soil salinity [9]. Erosion, in turn, influences the spatial distribution of salinity, for example causing high salinity in the deposit area farther down the watershed, caused by inundation event. Furthermore, the accumulated water in the Bin El Ouidane dam, which is the main sources of irrigation the region, threatens agricultural lands downstream [10]. After irrigation, the water added to the soil is used by the crop or evaporates directly from the soil. The salt, however, is left behind in the soil and, if not removed, by (calcium sulfate di-hydrate ($CaSO4*2H2O$)), accumulates. This negatively impacts the limited agricultural land, reduces production, and often leads to land being abandoned, leaving a bare surface with occasional shrubs [11].

This study covers the watershed of Oued El Abid with an area of 7686 km^2 in the upper part of the immense OumEr-Rbia watershed (50,000 km^2) in the region of TadlaAzilal between the High Atlas and the plain of Tadla. The main water course is the Oued El Abid, a standout amongst the most essential water assets of Morocco, which aliment with the Oued Assif Ahensal the Bin El Ouidane dam, utilized for water system and hydropower (Fig. 1).

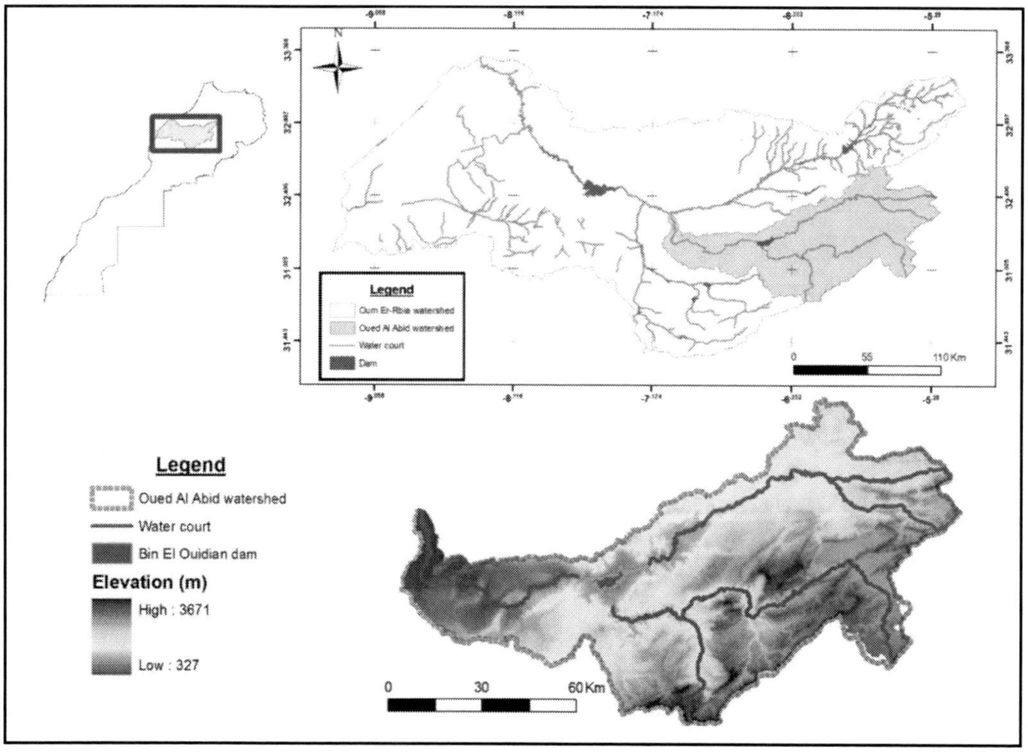

Fig. 1. Study area

The overall aim of this study was to develop effective combined spectral-based statistical regression models using Landsat images to introduce an approach to map salinity hazard, the application of which should help track down the occurrence of salt in the subsoil at an early stage.

Methodology

To localize the affected regions in the watershed, an approach using an integration of remote sensing and regression model was used (Fig. 2). The usefulness of remote sensing has been improved by many studies, in particular those based on visible spectral bands [12].

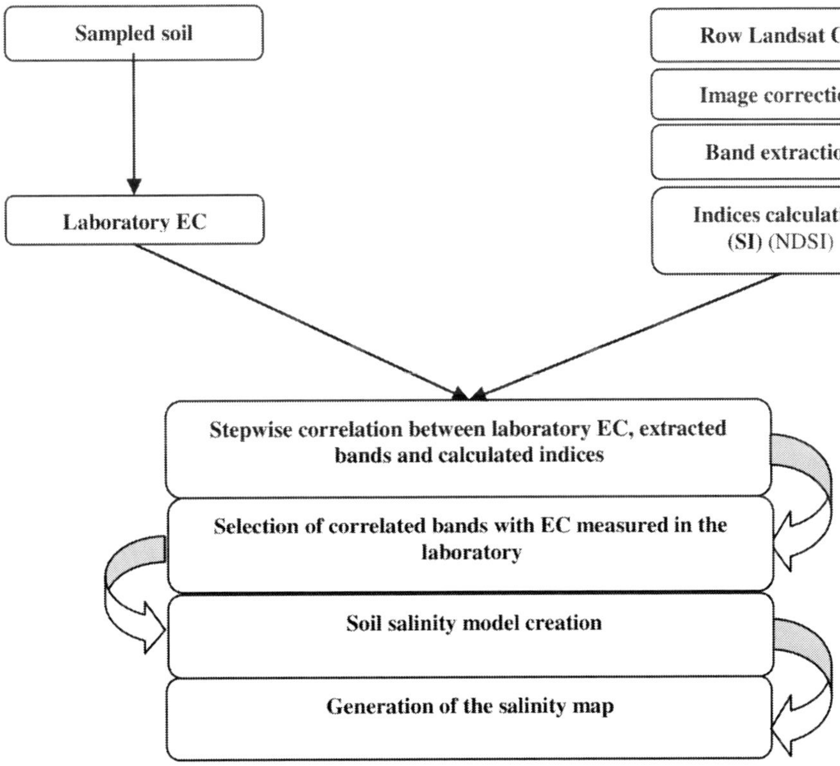

Fig. 2. Methodology

The EC was calculated in the laboratory based on saturated paste extracts from soil samples (Tab. 1) collected in the field [12].

Table 1. Bands data of soil samples.

Point	X	Y	Z	EC	B1	B2	B3	B4	B5	B6	B7	NDSI	SI	Panchro
units	m	m	m	µS/cm	-	-	-	-	-	-	-	-	-	-
P 1	345637	184470	263	533	0.18	0.18	0.2	0.26	0.35	0.41	0.37	-0.13	0.19	0.24
P 2	344284	189718	266	578	0.12	0.1	0.09	0.05	0.52	0.15	0.07	0.27	0.1	0.07
P 3	346298	173082	285	589	0.17	0.16	0.18	0.24	0.32	0.37	0.34	-0.15	0.18	0.22
P 4	368941	159905	560	640	0.12	0.1	0.07	0.07	0.18	0.13	0.09	0.05	0.09	0.07
P 5	390459	170007	670	731	0.11	0.08	0.05	0.04	0.08	0.05	0.03	0.14	0.08	0.06
P 6	399180	167559	845	752	0.14	0.13	0.13	0.15	0.29	0.31	0.25	-0.07	0.14	0.11
P 7	411017	171767	915	805	0.12	0.1	0.09	0.12	0.16	0.2	0.18	-0.14	0.1	0.1
P 8	413438	177963	1433	845	0.16	0.16	0.19	0.26	0.36	0.42	0.36	-0.15	0.17	0.22

P 9	426061	179763	909	851	0.12	0.1	0.07	0.08	0.1	0.08	0.07	-0.05	0.09	0.07
P 10	419074	182648	1416	892	0.1	0.08	0.06	0.05	0.06	0.06	0.05	0.07	0.08	0.04
P 11	428600	181453	1098	933	0.1	0.08	0.05	0.03	0.07	0.03	0.02	0.2	0.07	0.04
P 12	440598	189712	1114	946	0.15	0.13	0.14	0.16	0.2	0.24	0.23	-0.06	0.14	0.13
P 13	448196	191857	1404	983	0.11	0.09	0.09	0.12	0.2	0.26	0.23	-0.15	0.1	0.08
P 14	460365	194792	1555	993	0.13	0.11	0.12	0.13	0.28	0.3	0.22	-0.06	0.12	0.11
P 15	473463	193341	1530	1000	0.15	0.14	0.14	0.16	0.24	0.26	0.23	-0.06	0.14	0.16
P 16	477042	203032	1562	1017	0.12	0.11	0.11	0.15	0.21	0.29	0.25	-0.15	0.12	0.13
P 17	474299	200025	1384	1104	0.11	0.09	0.07	0.07	0.12	0.1	0.08	0.02	0.09	0.09
P 18	473671	207147	1535	1109	0.12	0.1	0.09	0.1	0.19	0.16	0.13	-0.1	0.1	0.08
P 19	463220	207420	1454	1143	0.13	0.11	0.11	0.13	0.21	0.23	0.19	-0.09	0.12	0.11
P 20	410229	164435	1211	1148	0.13	0.11	0.12	0.17	0.23	0.28	0.25	-0.17	0.12	0.12
P 21	423739	159237	1137	1155	0.13	0.12	0.12	0.13	0.28	0.23	0.17	-0.04	0.12	0.13
P 22	428461	155976	1151	1167	0.13	0.12	0.11	0.11	0.17	0.18	0.14	-0.02	0.12	0.1
P 23	434055	147979	1367	1183	0.12	0.11	0.11	0.13	0.23	0.25	0.21	-0.07	0.12	0.1
P 24	435443	153220	1734	1184	0.12	0.11	0.11	0.14	0.32	0.28	0.21	-0.11	0.12	0.13
P 25	438287	164126	1294	1215	0.15	0.15	0.2	0.29	0.41	0.51	0.43	-0.19	0.17	0.22
P 26	448972	163708	1486	1216	0.16	0.16	0.18	0.23	0.32	0.37	0.32	-0.12	0.17	0.18
P 27	418806	185929	1707	1267	0.17	0.17	0.19	0.23	0.28	0.31	0.26	-0.09	0.18	0.21
P 28	434579	187102	1409	1287	0.12	0.12	0.12	0.17	0.26	0.34	0.28	-0.19	0.12	0.13
P 29	440235	196160	1600	1338	0.12	0.11	0.11	0.15	0.19	0.29	0.24	-0.13	0.12	0.09
P 30	444117	208698	1052	1383	0.14	0.13	0.14	0.18	0.34	0.37	0.28	-0.12	0.14	0.1
P 31	452653	207356	1322	1508	0.14	0.13	0.15	0.25	0.37	0.49	0.4	-0.25	0.15	0.19
P 32	467816	199214	1420	1563	0.11	0.09	0.08	0.08	0.16	0.15	0.1	0	0.1	0.06
P 33	476825	183970	1752	1580	0.12	0.1	0.08	0.07	0.08	0.08	0.07	0.05	0.09	0.06
P 34	469864	182667	2118	1639	0.13	0.11	0.12	0.14	0.21	0.25	0.21	-0.1	0.12	0.13
P 35	462673	179683	2079	1643	0.12	0.11	0.13	0.17	0.24	0.35	0.31	-0.15	0.12	0.14
P 36	454659	176501	2154	1756	0.14	0.13	0.14	0.16	0.19	0.26	0.23	-0.07	0.14	0.14
P 37	448580	174642	2362	1846	0.12	0.11	0.11	0.13	0.16	0.2	0.18	-0.07	0.12	0.12
P 38	438156	185927	1434	2805	0.12	0.11	0.1	0.1	0.16	0.16	0.13	-0.04	0.11	0.09
P 39	396756	167243	1222	2917	0.13	0.12	0.11	0.12	0.2	0.22	0.18	-0.05	0.12	0.11

P 40	389967	157998	1367	3080	0.13	0.11	0.12	0.14	0.23	0.28	0.23	-0.09	0.12	0.11
P 41	394236	150337	1765	3300	0.1	0.07	0.04	0.03	0.06	0.03	0.01	0.24	0.06	0.04
P 42	410169	131201	2223	5240	0.09	0.07	0.05	0.04	0.04	0.04	0.03	0.09	0.07	0.04
P 43	430213	137871	2025	5440	0.13	0.13	0.15	0.19	0.29	0.33	0.3	-0.13	0.14	0.15

To obtain accurate analytical results, each of the 43 soil samples was reported based on the mean of five subsamples taken within a 1600 m^2 area. The mean sample was attributed to the central sample. Each subsample consists of about 1 kg of top soil taken at 30 cm depth. The samples' ECs ranged from 533 to 5440µs/cm, with a mean salinity value equal to 1454µs/cm.

Remote sensing was used to extract indices based on Landsat 8 Oli images from the same date as the field sampling. Those images are captured by the Landsat sensor launched on February 11, 2013, which consists of two science instruments, the Operational Land Imager (OLI) and the Thermal Infrared Sensor (TIRS). These two sensors composed from 11 bands (Tab.2) provide seasonal coverage of the global landmass at a spatial resolution of 30 meters (visible, NIR, SWIR), 100 meters (thermal), and 15 meters (panchromatic) [13].

Table 2. Landsat 8 Band specifications, Bands 1 to 9 are Oli spectral bands while bands 10-11 are TIRS spectral bands.

Spectral Band	Wavelength (um)	Resolution(m)
Band 1- Coastal/ Aerosol	0.433- 0.453	30
Band 2- Blue	0.450- 0.515	30
Band 3- Green	0.525- 0.600	30
Band 4- Red	0.630- 0.680	30
Band 5- Near Infrared (NIR)	0.845- 0.885	30
Band 6- Short Wavelength Infrared (SWIR)	1.560- 1.600	30
Band 7- Short Wavelength Infrared (SWIR)	2.100- 2.300	30
Band 8- Panchromatic	0.500- 0.680	15
Band 9- Cirrus	1.360- 1.390	30
Band 10-Long Wavelength Infrared	10.30- 11.30	100
Band 11- Long Wavelength Infrared	11.50- 12.50	100

The images were used after a radiometric- atmospheric treatments and classification to obtain high-accuracy results using ENVI, the radiometric correction consists of conversion from the sensor measurement to a physical quantity, when the sensor is measuring a radiance from the top of the atmosphere, the radiance is normalized into a reflectance values. Where the Dark Object Subtraction (DOS) has used for atmospheric correction to calibrate the luminance measured in the image. The map layout and indices extraction were done in ArcGIS software.
The validation of the obtained values from Landsat images by comparison to the soil sample analysis was done using a regression model in JMP13 [15].

Results

The results show a high correlation between the Landsat Oli image bands (B1 and B4) and EC with an acceptable R^2 (0.80) and root-mean-square error (RMSE = 488.56) (Tab. 3).

Table 3. Results of regression model

Squares Deviations Sum (SDS)	freedom Degree of error (FDE)	RMSE	R square	R square adjusted	Cp	p	AICc	BIC
9547841.4	40	488.5653	0.8024	0.7926	15.23675	3	660.4382	666.4304

Curent estimation

Lock	Grasped	Coefficient	Estimation	Freedom Degrees of the numerator	Squares Sum	"Report F"	"Prob.>F"
✓	✓	Constant	-4794.0544	1	0	0	1
		B5	0	1	672122.5	2.953	0.09363
		NDSI	0	1	1061142	4.876	0.03317
		SI	0	1	105193.7	0.434	0.51368
		B6	0	1	2184699	11.572	0.00156
		B7	0	1	2815671	16.311	0.00024
		Panch	0	1	382144.8	1.626	0.2098
	✓	B1	48583.2346	1	9463436	39.646	1.81E-07
		B2	0	1	171716.3	0.714	0.40319
		B3	0	1	22891.78	0.094	0.76112
	✓	B4	281.300919	1	3748.665	0.016	0.9009

Based on these results, a model (Eq. 1) was integrated with Landsat imagery to produce a soil salinity map (Fig. 3).

$$EC = 48583.23*(Band\ 1) + 281.30*(Band\ 4) - 4794.05 \qquad (1)$$

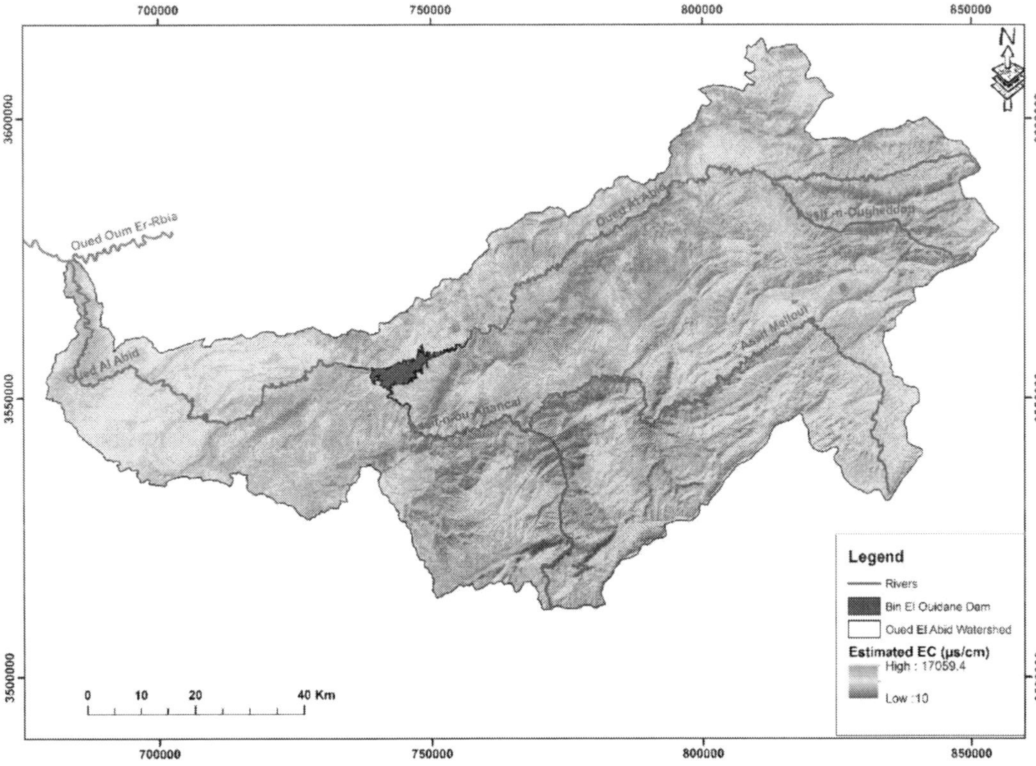

Fig. 3. Soil salinity map

A survey was done to compare the measurements obtained from the model and the salinity measured in field in order to validate the model results, as shown in the table below (Tab. 4).

Table 4. Measurement validation in µs/cm

Sample	units	P1	P2	P3	P4	P5	P6	P7
X	m	348075.9	375184.9	438346.0	414870.2	444773.9	470485.5	481943.9
Y	m	170743.4	159564.5	183878.7	150062.4	183040.3	203441.9	176053.4
EC M	µS/cm	2800.67	2025.27	733	413.67	1249.02	1583.88	1280.54
EC E	µS/cm	2820.86	2017.19	721.84	414.42	1256.95	1590.21	1286.31

EC M: Electrical conductivity measured
EC E: Electrical conductivity estimated

Discussion

The main contribution of this study is the use of remote sensing to localize area affected by salinisation. Our results prove the capability of the methodology to identify affected areas and show a high correlation between laboratory analyses and the model presented

The acceptable R^2 demonstrates the usefulness of Landsat satellite imagery even at medium resolutions—high resolution is not necessary to get an accurate result.
The map shows moderate and distributed salinity upstream of Bin El Ouidane dam, where soil erosion occurs at high rates [9] and increases downstream of the dam, which explains the agricultural lands affected by saline irrigation water [16, 17, 18].

The survey (Tab. 4) shows similarity of result of measured and model estimation with a standard deviation of 10 µs/cm.

Indeed, the obtained results has its limits by lack of historical datasets and unpublished studies concern the study area which could be helpful to validate accurately the model, adding some limits related to survey area as the topography which restrict our mobility and limits the spatial distribution of soil samples, the things which could affect the accuracy of the results and validity of the model.

Conclusion

This study shows the ability of remote sensing to map soil salinity by coupling satellite image bands to a regression model. With the simplicity of the model and accuracy of the result, remote sensing becomes a reliable source to map soil salinity and could provide important support to decision makers and planners to simulate scenarios for the evolution of salinity and plan interventions against it. This study also demonstrates that even a relatively low satellite image resolution of 30 m can yield a highly accurate map of soil salinity, which may be especially useful in large study areas where high image resolution is an obstacle.

References:

1. A. Allbed, L. Kumar, Adv. Remote Sens. **2**, 373-385 (2013)
2. K. Nagaz, F. E. Mokh, M. M. Masmoudi, N. B. Mechlia, JAEID. 111, 207-223 (2017)
3. G.I. Metternicht, J.A. Zinck, Remote Sens. Environ. **85**, 1–20 (2003)
4. P. Blaikie, The political economy of soil erosion in developing countries. Third World Planning Rev. **7**, 356 (2016)
5. K.F. Benbrahim, M. Ismaili, S.F. Benbrahim, A. Tribak, Sci. Chang. Planétaires Sécheresse. **15**, 307–320 (2004)
6. J. Wu, P. Li, H. Qian, Y. Fang, Environ. Earth Sci. **71**, 3465–3475 (2014)
7. A. Khalil, L. Hanich, A. Bannari, L. Zouhri, O. Pourret, R. Hakkou, J. Geochem. Explor. **125**, 117–129(2013)
8. A.Bannari, G. Kadhem, A. El-Battay, N.A. Hameid, M. Rouai, Adv. Remote Sens. **5**, 315 (2016)
9. E.M. Sabri, A. Boukdir, M. Rachid, A.E.L.Mahboul, V.M. Romaric, A. Zitouni, Z. Echakraoui, Agrofor Int. J. **2**, 99–108 (2017)
10. H. Ouatiki, A. Boudhar, Y. Tramblay, L. Jarlan, T. Benabdelouhab, L. Hanich, M.R. El Meslouhi, A. Chehbouni, Climate. **5**, 1(2017)
11. M.N. Qadir, A.D. Karajeh, B. F George, IWMI. **5**, 29 (2015)
12. R.E. Ontiveros, L. Diakite-Diakite, M.E. Álvarez-Sánchez, P.M. Coras-Merino, Tecnol. Cienc. Agua. **4**, 127–140 (2013)
13. S.Nawar, H. Buddenbaum, J. Hill, J. Kozak, Remote Sens. **6**, 10813–10834 (2014)

14. T. Dube, O. Mutanga, ISPRS J. Photogramm. Remote Sens. **101**, 36–46 (2015)

15. R. Carver, SAS Institute. Second edition (2014)

16. M. Barbouchi, R. Abdelfattah, K. Chokmani, N. Ben aissa, R. Lhissou, A. El Harti, IEEE J. Sel. Top. Appl. Earth Obs. Remote Sens. **8**, 3823–3832 (2015)

17. A. El Harti, R. Lhissou, K. Chokmani, J. Ouzemou, M. Hassouna, E.M. Bachaoui, A. El Ghmari, Int. J. Appl. Earth Obs. Geoinformation. **50**, 64–73(2016)

18. I. Karaoui, A. Arioua, A. El Amrani Idrissi, W. Nouaim, D. Elhamdouni, K.A. Ouhamchich, M. Hssaisoune, R. Hnini, J. Geosci. Environ. Prot. **5**, 121–132(2017)

Quantification of soil loss in various lithological areas of the western Middle Atlas Central: application to the Ras-Elma, Tamelalet and Sebab watershed (Tigrigra watershed, Morocco)

Hassan Achiban[1,*] *Ali* Taous[1], *Ismail* El-Khantoury[1], *Mohamed* El Mderssa[2], *Ali* Amechrouq[3]

[1] Laboratory for Geo-Environmental Analysis, Planning - DD, Sidi Mohammed Ben Abdellah University - Fez.
[2] Laboratory of Soil Microbiology and Environment, Moulay Ismail University, Meknes.
[3] Laboratory of Molecular Chemistry and Natural Substance, Moulay Ismail University, Meknes

Abstract. The present study proposes to evaluate the extent of erosion according to the lithology in three sub-watersheds (Ras Elma, Sebab and Tamelalet) belonging to the Tigrigra basin and evolving in humid climatic context. The methodology adopts the revised universal soil loss equation (RUSLE). The results obtained make it possible to establish erosion class maps via GIS. A clear spatial difference in soil loss is observed, between the three sub-basins and in proportion to the lithology: on average 42.15 t/ha/year on Paleozoic schistose soils, against 17.06 t/ha/year on carbonate substrates Mesozoic and 8.46 t/ha/year on quaternary basalts. Correlations between soil loss and RUSLE factors are established. Soil infiltration regimes on different substrates are studied.

Introduction

Water erosion is the main factor in the degradation of the edaphic resource. It has become a major environmental concern, and its consequences are well felt in soil loss, ecosystems and siltation of dam reservoirs. Several models for predicting potential erosion, such as long-term models and empirical models at the scale of a watershed or plot have been developed in recent decades. The Revised Universal Soil Loss Equation (RUSLE) has become the most widely used method because it can be applied in many situations, even in topographically complex landscapes [1].

By adapting the above method, the present study focuses on the evaluation of potential soil losses in three sub-catchments belonging to the Tigrigra basin in the Azrou region (Fig. 1). In fact, the studied sub-basins (Ras Elma, Sebab, Tamelalet) evolve under relatively similar climatic-ecological conditions, but their lithology is variable, which influences the process of soil loss.

* Corresponding address: hassan.achiban@usmba.ac.ma

It is recalled that the watershed of the wadi Tigrigra is located astride the limestone plateau of the central Middle Atlas and the central Hercynian schist massif. The limestone plateau is composed of dolomite (lower Lias), limestone and marly limestone (Middle Lias). The ensemble is based on the saline red clays and doleritic basalt of the Triassic [2], which appear along structural and unconformity on the mainly schistose substratum, with some quartzite and sandstone beds [3]. The limestone plateau is pierced by Plio-Quaternary volcanoes [4] having emitted basalts which cover large areas in the Tigrigra Valley.

The forest cover of the Tigrigra watershed is very diversified, with a dominance of Atlas cedar (Cedrus atlantica) which occupies 49% of the forest, the holm oak (Quercus rotundifolia) covers 35% of the forest. Quercus canariensis and secondary species occupy 11.01% of the forest area of the watershed. In the end, reforestation with Cedrus atlantica, Pinus halepensis, Cupressus arizonica and Populus alba and nigra occupies 4.99% of the total area of the forest.

The climate of the study area is Mediterranean with a wet and cold winter, which often records heavy snowfall, while the summer is dry and hot. The seasonal distribution of rainfall is: W.S.A.S (Winter, Spring, Autumn, Summer) where the maximum is recorded during winter and spring.

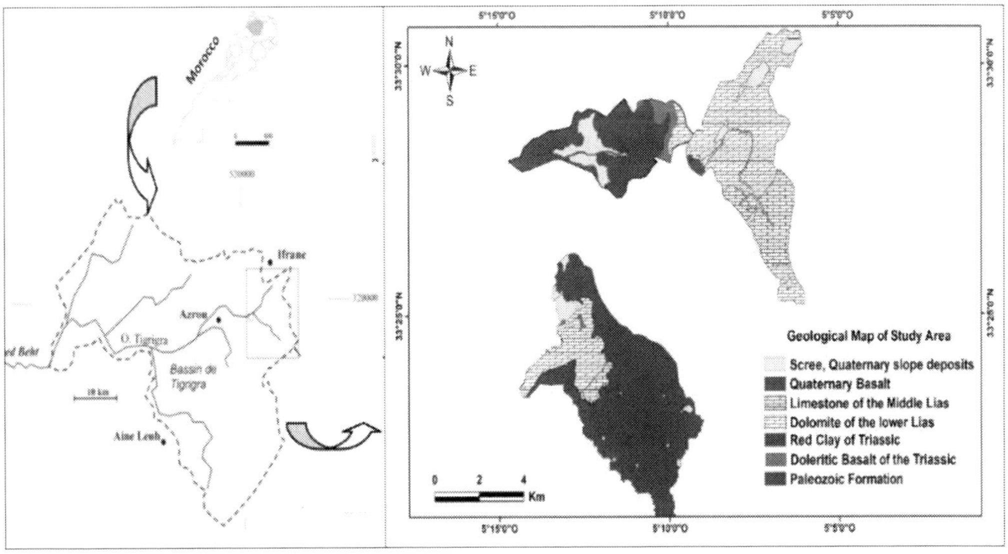

Fig. 1. Geological Map of Study Area

1. Materials and methods

1.1. Sources and Materials

The methodology adopted for the evaluation of water erosion in the sub-watersheds studied is founded on the spatialization and cross-referencing of altimetric, topographic, climatic, forest and geological data. It was based on a digital elevation model (DEM) of 20 m resolution, topographic maps at 1: 50,000 (Azrou and Ifrane sheets), the forest inventory map of the Azrou forest and Rainfall data over 30 years from six weather stations in the study area. The lithological characterization of the three watersheds was based on the Azrou 1: 50,000 and Elhajeb 1: 100,000 geological maps.

The particle size analysis of the developed soils on the various lithological facies as well as their contents in organic matter was determined experimentally in the laboratory. The permeability tests of the different soils were carried out by means of a disk infiltrometer.

1.2. Methodological approach

The erosion quantification is relying on the application of the USLE soil loss equation established by [5] and modified by [6]. This model makes it possible to quantify the sheet erosion, through the implication of the different parameters according to the relation:

A = R. K. LS. C. P (A, is the annual rate of soil loss expressed in t/ha/yr (1)

R is the rainfall erosivity factor, calculated for each station by the relation of [7]:

$$Ln\ R = 1.74\ log\ \Sigma(Pi^2 / P) + 1.29 \qquad (2)$$

with Pi = the average of the monthly precipitations and P = Average of the annual precipitations). The results obtained were interpolated by means of GIS on the three basins.

K is the erodibility of soils. Not only does it take into account the particle size and the percentage of organic matter, but also the soil structure and its permeability class. Its calculation was hinged on the granulometric analyses outcomes, in addition to the organic matter of 51 samples of the first horizons of soils results. The sampling sites were chosen based on the lithological variation and the nature of the land cover. The formula used to calculate erodibility is that of [8]:

$$100\ K = 2.1*10\text{-}4*M1.14*(12\text{-}a) + 3.25*(b\text{-}2) + 2.5*(c\text{-}3) \qquad (3)$$

with M is the product of the percentage of silt between 0.1 and 0.002 mm by the percentage of all that is not clay; a is the percentage of organic matter; b is the soil structure code used in the classification; c expresses the class of permeability of the profile.

LS is a dimensionless factor which represents the inclination (S in%) and the length of slope (L in m), it was calculated by the formula of [8]:

$$LS = (L/22,13)^{m}*(0.065 + 0.045*S + 0.065*S2) \qquad (4)$$

where m is a parameter such that m = 0.5 if the slope is> 5%, m = 0.4 if the slope is 3.5 to 4.5 %, m = 0.3 if the slope is 1 to 3% and m = 0.2 if the slope is < 1%. Topographic data were obtained from GIS treatments of the 20 m resolution digital terrain model covering the study area.

C is a dimensionless factor that represents the effect of vegetation cover; its determination is grounded on the density and height of the strata. It has been mapped from the classification of strata on the forest inventory map of the study area; based on the Wischmeier & Smith tables which give pre-established values according to the nature of the plant occupation of the soil.

P is also a dimensionless factor that takes into account cultural techniques and anti-erosion practices. For the study area, these practices are almost absent and the factor P has taken the value 1.

2. Results and discussion

The application of the Wischmeier & Smith formula [8] through the cross-mapping of the five main factors involved in soil water erosion yielded soil loss maps expressed in t/ha/yr. at all points of the three basins of Ras Elma, Tamelalet and Sebab (Fig. 2). For the purposes of map readability and to keep more detail, the different values have been grouped into five classes that have consistent values of the USLE factors. The classification of erosion takes into account the soil tolerance threshold which is defined as the balanced earth loss with soil formation through rock alteration and which is assumed to have no negative impact on soil productivity. This tolerance varies from 1 to 12 t/ha/year depending on the climate, type of rock and soil thickness [6]. The threshold considered in this study is equal to 7 t/ha/yr, value used in other erosion studies in Prerif Mountain [9], [10] and in the Beht Basin [11]. Above 20 t/ha/year, the loss is high and soils are much degraded [9]. However, the small losses below the threshold of tolerance of Wischmeier (<7 t/ha/year) at

 https://doi.org/10.1051/e3sconf/20183704003

the level of the three basins Sebab, Ras Elma and Tamelalet concern respectively 76,2%, 51,5% and 49,8%, while the strong erosion exceeding 20 t/ha/year affects only 9.8% and 12.1% respectively for the basins of Sebab and Ras Elma which are dominated by carbonate formations and basaltic formations, compared to 36.3% for the sub-basin of Tamelalet which drains the mainly schistose terrains of the western Paleozoic Meseta (Central Morocco).

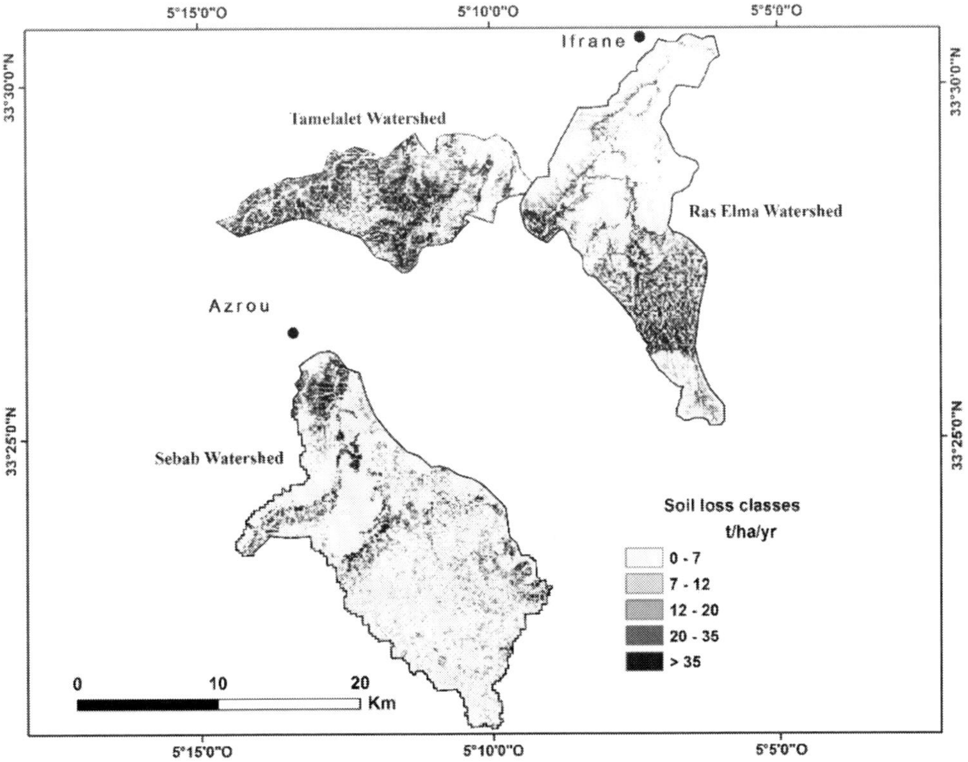

Fig. 2. Map of the water erosion of the basins of Sebab, Ras Elma and Tamelalet

The distribution of soil loss classes according to the lithological nature of the Sebab sub-basin shows that 52% of the formations of the Upper Ordovician undergo strong and very strong erosions whereas 80% of Quaternary basalts, which cover more than 74% of the basin, are subject to erosion below the tolerance threshold of Wischmeier (Tab. 1). In the predominantly carbonate-dominated Ras Elma watershed, 40% of the limestones of the Middle Lias undergo strong erosion compared to only 7% in the Lower Lias dolomite (Tab. 2).

Table 1. Distribution of soil loss classes according to the lithological nature (Sebab Watershed)

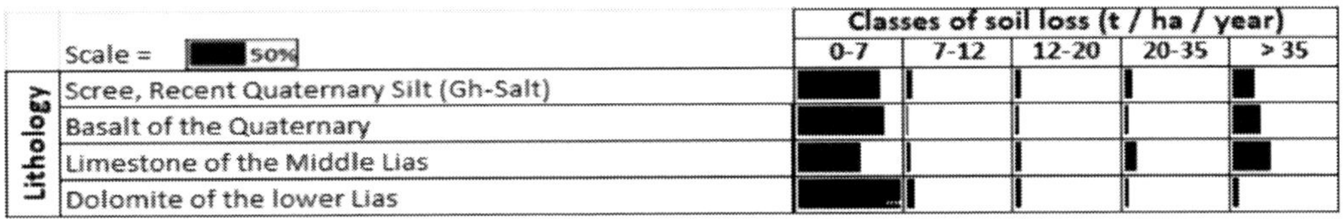

Table 2. Distribution of soil loss classes according to the lithological nature (Ras Elma Watershed)

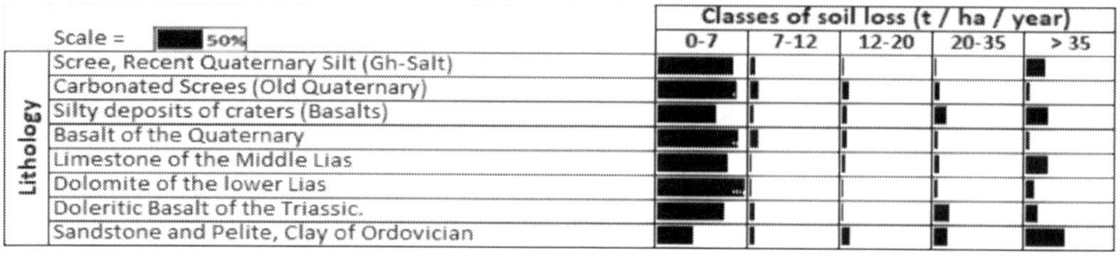

In the Tamelalet sub-basin, Silurian, Middle Devonian and Upper Viséan shales are the most affected by erosion, with 67%, 64% and 55% respectively of their areas. The Secondary formations are the least affected by the strong erosion: 95% of Triassic basalts show an erosion of less than 7 t/ha/year, as do 75% of Lower Lias dolomites and 54% of Quaternary basalts (Tab. 3).

Table 3. Distribution of soil loss classes according to the lithological nature (Tamelalet Watershed)

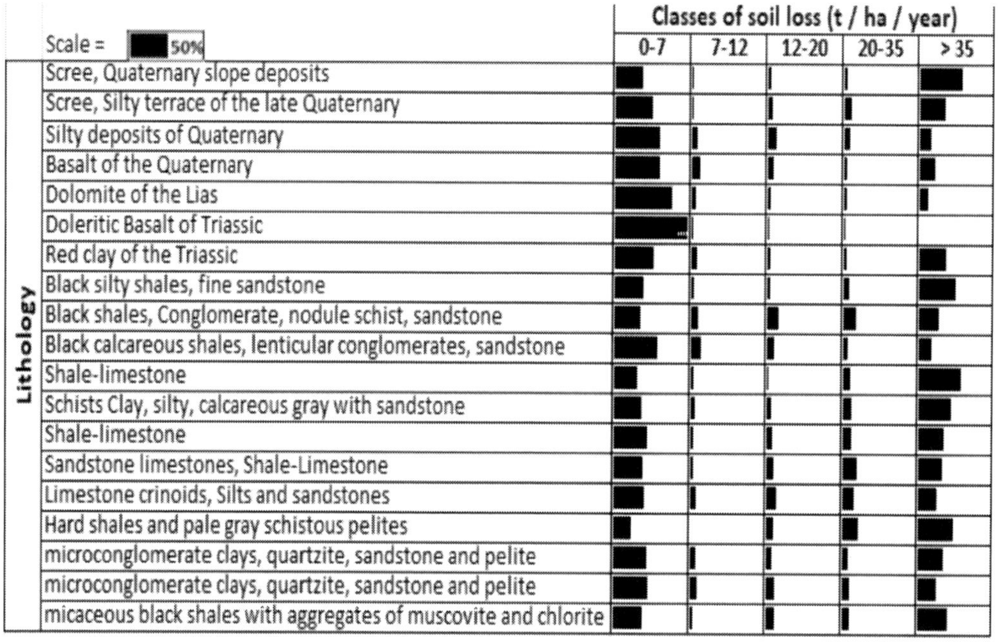

When considering the average losses according to the lithology, the results obtained by the empirical models show that the Paleozoic metamorphic formations are the most affected by the strong erosion, with an average of 47.3 t/ha/year; they are followed by the Triassic clays (36.6 t/ha/year), then by the limestones of the middle Lias (26 t/ha/year); while Quaternary basalts and Lower Lias dolomite are the least influenced, with mean values of 10.3 t/ha/year and 6.4 t/ha/year, respectively (Fig. 3).

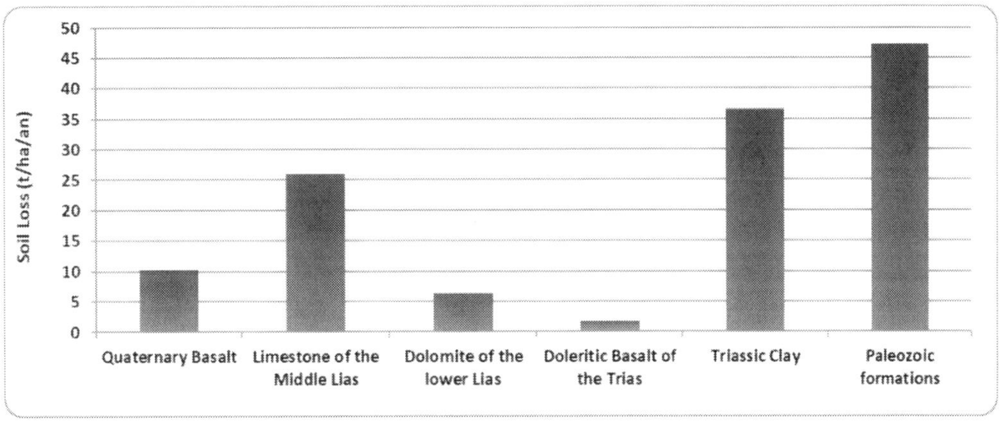

Fig. 3. Mean loss in t / ha / yr by lithologic formation

The distribution of soil losses in the three sub-basins is unequal for the same lithological formations because of the spatial variability of the various factors considered. To demonstrate the influence of each factor, the mean loss results were correlated with the values for each factor of the Wischmeier equation, using the Pearson (R) correlation test (Fig. 4). The results obtained show that all the factors act in the process of soil loss in the three sub-basins, but with a stronger influence of the K and LS factors. Thus, at the scale of the Tamelalet sub-basin, the correlation coefficient between the soil loss and these two factors is respectively 0.51 and 0.48, against 0.17 and 0.36 respectively when correlated with climatic aggressiveness (R) and vegetation cover (C).

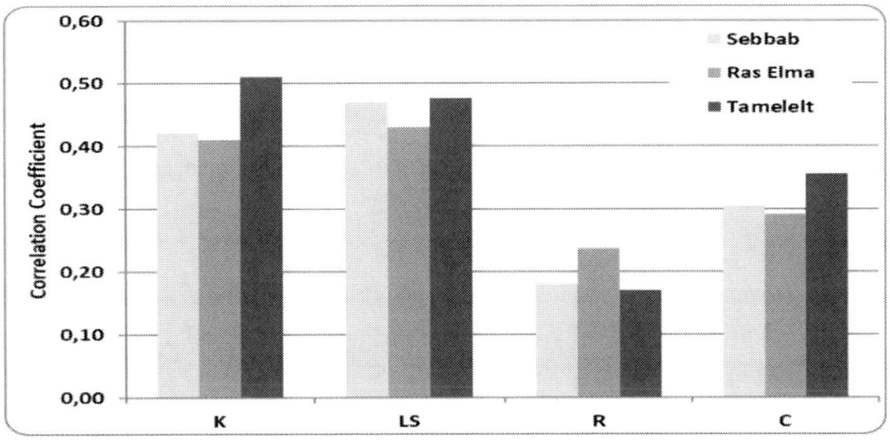

Fig. 4. Correlation between Soil Loss and RUSLE Factors

Various soil infiltration tests were performed using a disk infiltrometer at each lithological facies of the study area. The results obtained are shown in the figure 5. These results show that the doleritic basalt of the Triassic has a high infiltration capacity exceeding on average $26*10^{-2}$ m /s, followed by Lower Lias dolomites ($19.4*10^{-2}$ mm/s) and Quaternary basalts ($14.78*10^{-2}$ mm/s). In contrast, Paleozoic schists have the lowest values ($2.59*10^{-3}$ mm/s), followed by Triassic red clays ($1.8*10^{-2}$ mm/s) and limestone of Middle Lias ($2.96*10^{-2}$ mm/s). Thus, the low infiltration of certain formations, associated with the topographic factor, generates more runoff which causes the detachment and the displacement of particles of the surface of the ground.

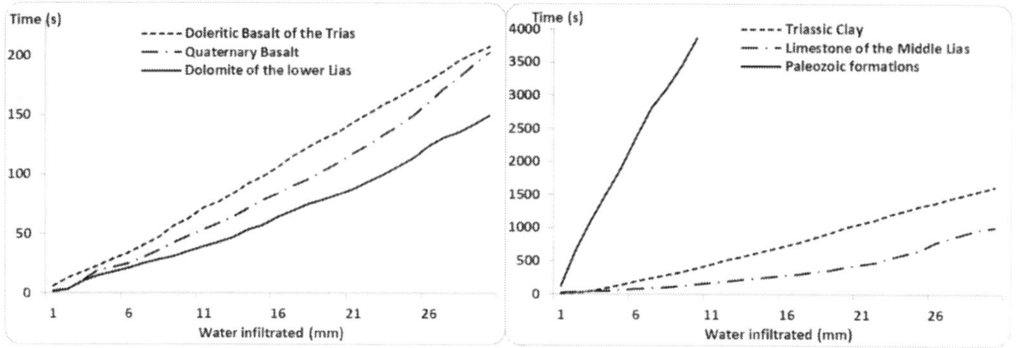

Fig. 5. Infiltration regime versus time for different soil types

In general, the specific degradation in the study area is lower than that obtained in the Rif Mountains which provide more than 60% of the sediments mobilized each year in the country [12]. In fact, if the results obtained in our case indicate values varying according to the lithology between 47.3 t/ha/yr and 6.4 t/ha/yr, the average soil loss in the Rif is 62.7 t/ha/yr in the Oued Haricha watershed [13], 61 t/ha/yr in the Oued Tlata watershed [14] and 55.3 t/ha/yr in the watershed of Oued Boussouab [9]

Conclusion

This paper presents the results of applying the Revised Universal Soil Loss Equation using a geographic information system. At the end of this study, it can be concluded that water erosion is the main source of degradation of soil capital at the level of studied watersheds. In addition, it has been possible to establish a soil loss map, taking into account the various factors at the origin of this phenomenon. This map highlights the dominance of high and very high loss classes in the Paleozoic formations that dominate the Tamelalet basin, followed by Lias limestone in the Ras Elma basin. The study made it possible to spotlight the correlation between soil loss and the factors involved in erosion as a function of the lithologies in the basins studied; it also allowed us to evaluate the infiltration regimes of the different types of soils of the studied basins. All these results make it possible to define the areas where the anti-erosive intervention is urgent.

References

1. P. Desmet, G. Govers, J. Geogr. Inf. Syst., **10**, pp. 311 – 331, (1996)
2. G. Colo, *Contribution à l'étude du Jurassique du Moyen Atlas septentrional. (*Notes et Mémoires du Service géologique du Maroc, 139, 1961)
3. A. Benslimane, *Étude hydrogéologique du bassin de Tigrigra ; apport de la géo-électrique a la reconnaissance des potentialités hydriques dans les formations basaltiques subaffleurantes*, Thèse de doctorat (Faculté des Sciences Meknès), (2005)
4. G. Beaudet, *Le plateau central marocain et ses bordures : étude géomorphologique* (Rabat, 1969)
5. WH. Wischmeier, Journal Soil and Water Conservation., **31**, pp. 5-9, (1976)
6. KG. Renard, GR. Foster, GA. Weesies, DK. Mccool, *Predicting soil erosion by water. A guide to conservation planning with the revised universal soil loss equation (RUSLE).* (Agric. Handbook 703. US Govt Print Office, Washington, DC, 1996)
7. A. Rango, H.M.J. Arnoldus, Cahiers techniques de la FAO., pp. 1-11, (1987)
8. WH. Wischmeier, DD. Smith, *Predicting rainfall erosion losses - a guide to conservation planning*. (Agriculture Handbook No. 537. US Department of Agriculture Science and Education Administration, Washington, DC, USA, 1978)

9. A. Sadiki, S. Bouhlassa, L. Auajjar, A. Faleh, J. Macaire, bulletin de l'institut scientifique., **26**, pp. 69-79, (2004)
10. A. El Garouani, H. Chen, L. Lewis, A. Tribak, M. Abahrour, Revue Télédétection., vol. 8, **3**, pp. 193-201, (2008)
11. DREF- Nord-ouest, *Étude d'aménagement du bassin versant d'Oued Beht en amont du barrage El Kansera*, (Fès, 2007)
12. B. Heusch, Ann. Recherche Forestière du Maroc., **12**, pp. 9-176, (1970)
13. M. Tahiri, H. Tabyaoui, A. Tahiri, H. El Hadi, F. El Hammichi, M. Achab, Journal of Geoscience and Environment Protection., **4**, pp. 107-119, (2016)
14. A. Tribak, A. El Garouani, M. Abahrour, Revue Télédétection., vol. 20, **4**, pp. 333-337, (2009)

Estimation of Particle Material And Dissolved Flows During Floods In The Inaouene Watershed. (Northeast Of Morocco)

Hayat Sibari[1*] , *Souad* Haida[1] and *Mohamed* Foutlane[2]

1. Laboratory : Geoscience and Naturel Ressources, Departement of Geology, Faculty of Science. Ibn Tofail Uninersity, Maamora Campus, Morocco.

2. National Office of Potable Water (ONEP), 6 Avenue Oued Akrach, 10000, Rabat. Morocco.

Abstract : This work aims to estimate the contributions of the Inaouene River during the floods. It is in this context that the dissolved and particulate matter flows were measured during the flood periods followed by the 1996/97 study year at the two hydrological stations Bab Marzouka (upstream) and El Kouchat (downstream). The specific flows of dissolved materials calculated upstream and downstream of the Inaouene watershed correspond respectively to 257 t/ km^2/year and 117 t/ km^2/year. Chlorides represent 30% and 41% respectively of the total dissolved transport upstream and downstream. The potential mechanical degradation affecting the Inaouene watershed can deliver a solid load estimated at 6.10^6 t/year corresponding to a specific flow of 2142 t/km^2/year.

Key words*:* Inaouene, watershed, Floods, Material flow, Specific flow, Morocco.

1. Introduction

The work we put forward about the Inaouene watershed relates to the Moroccan programs that aim at preventing and limiting the degradation of natural ecosystems as well as protecting dams' retention mainly in the Northern region of the country. The purpose of the present work then is to establish a quantification of flow solid matter dissolved and carried away at the level of this watershed during floods and also to study the relationships between these flows and the intake liquids.

The Inaouene watershed occupies an eastern position in the sebou watershed (fig. 1), between the last Middle Atlas foothills and the pre-Rif sector. It covers an area of 3322 km^2. The Inaouene River, second main tributary of the Sebou River after Ouergha, flows in an East-West direction, along the South-Rif corridor. The Inaouene receives the flows of the tributaries, on its right bank, which collect the runoffs of the Prerif hills and the flows of the tributaries , on its left bank, fed partly by the primary block of Tazekka, but especially by the limes of the Middle Atlas , often karstic in this region. The Sebou watershed is largely influenced by the Atlantic climate, whereas the Inaouene belongs to the western climate fringe of a Mediterranean climate with oceanic influence.

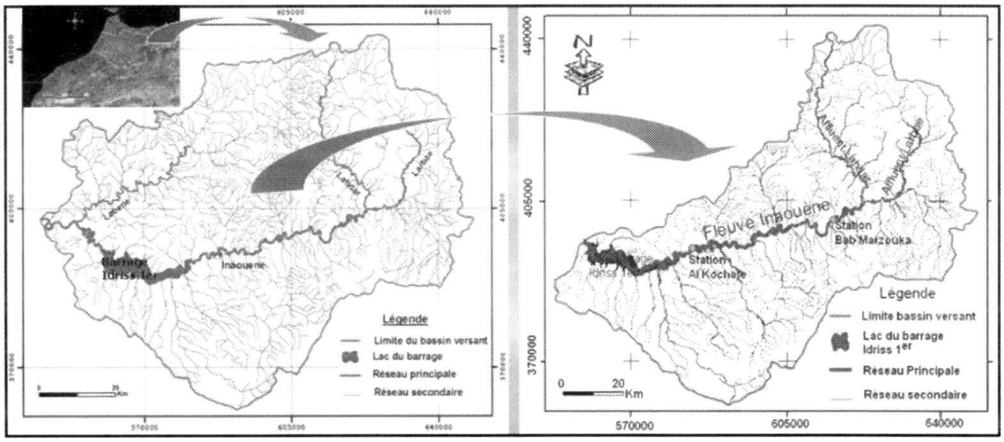

Figure.1 *: map of the geographic location of the* Inaouene watershed

2. Materials and methods

As part of this study, we carried out a sampling campaign at two hydrological stations: Bab Marzouka and El Kouchat situated respectively upstream and downstream of the Inaouène watershed. This sampling was carried out during the 1996/97 hydrological year on the basis of to two different frequencies: one weekly and the other daily during floods.

Samples have been taken approximately 1 meter deep. The collected water is placed in polyethylene bottles carefully washed with 10% nitric acid. These samples were filtered on a Millipore filter at 45 µm and the contents of suspended solids were measured by weighing the filters taken back (reported) to the volume of filtration water. As water samples were taken, water level readings were taken at the gauging stations, and using the calibration curves established by the DRPE (Research and Planing of Water Management), these water levels were transformed and corrected into liquid flows. During floods, only the water samples corresponding to the February 1997 flood upstream and downstream of the Inaouène were subjected to physicochemical analyzes (ONEP) to determine the behavior of certain major chemical elements during this period of flow. Table 1 shows the flowing volume, the liquid flow of the period during which the flood was sampled and the peak flow for each flood.

Table 1 : Sampling Période, volume water flow and the peak flow for February of Inaouene (hydrologic cycle studied in 1996/97).

Station	Beginning	End	Volume (Mm^3)	specific liquid flow ($l/s/km^2$)	peak flow (m^3/s)
El Kouchat	13/02/97	22/02/97	2820	31,94	167
Bab Marzouka	10/02/97	19/02/97	1696	35,86	109

The calculation methods of matter flow in water streams have already been dealt with by several authors such as [2, 3, 7, 10, 11]. In this case, the calculations have been achieved through the stochastic method that has already been applied to other watersheds in semi-arid areas: Tensift [4, 5] and Sebou [6]. This method makes it possible to calculate for each period (i), the average of the concentrations (Cim) weighed by the liquid flow rates. The matter flow (Fi) would be the product of this concentration (Cim) in the volume of water (Qi) obtained by integrating the flows for the period (i).

$$Cim = 1/\sum_{i=1}^{n} Qi \times \sum_{i=1}^{n} Ci \times Qi \quad ; \quad Fi = \sum_{i=1}^{n}(Cim \times \int_{t1}^{t2} Qdt)$$

3. Results and discussion

3.1. solid Flows

3.1.1. Changes in the SS rate during floods

The rate of suspended solids generally follows the variations of the liquid flow rates. There are five main SS export periods corresponding to five flood episodes observed between December 1996 and February 1997. Of these five episodes, we observe a certain proportionality between the intensity of the flood peaks. and those of suspended solids peaks. The most important flood is that of February with a maximum flow of 167 m^3 / s (59.64 l /s /km^2) at El Kouchat and 109 m^3 / s (72.66 l / s / km^2) at Bab Marzouka and suspended matter contents of 30 g / l at El Kouchat and 24 g / l at Bab Marzouka (Fig. 2).

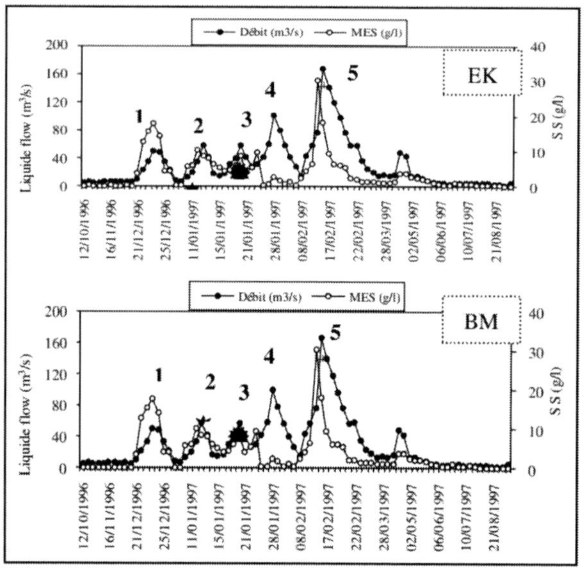

Figure 2 : Evolution of instantaneous flow rates and SS concentrations measured in the waters of Inaouène at El Kouchat (EK) and Bab Marzouka (BM) (1,2,3,4 and 5 are the main floods sampled).

In times of high water, where the transport power increases, the contents of suspended solids are very important reaching respectively upstream and downstream 10.52g / l and 6.63g/l. The highest levels of suspended matter recorded are mainly due to the mechanical erosion that predominates in the Prerif.

3.1.2. Comparison of solid flows between upstream and downstream basin

Table.2 shows the characteristics of each of the floods monitored and the calculated tonnage values of the matter exported in suspension. The flows of solid matter transported by Inaouéne upstream and downstream are respectively 4.89 10^6 tonnes and 6. 10^6 tonnes. The specific degradation corresponding to the upstream basin of an area of 1500 km^2 is 3260 t / km^2 / year and that downstream for an area of 2800 km^2 is 2141 t / km^2 / year.
As several studies have shown, erosion is a catacyclismic phenomenon, which means that most of the suspended matter is removed during a few floods [1, 2, 8, 9]. Indeed, all the major floods that occurred during the three months of the study year (December to February) are responsible for 71 to 75% of annual solid transport.

Table 2: Comparative contribution of the flood episodes to the total tonnage of suspended matter exported by Inaouène

		Annual Total	flood 1	flood 2	flood 3	flood 4	flood 5	Total flood
Bab Marzouka	period (days)	365	7	4	5	6	10	32
	Tonnage 10^6 T	6,88	0,51	0,40	0,35	0,83	2,80	**4,89**
	% year	100	7	6	5	12	41	71
El Kouchat	period (days)	365	6	6	5	6	8	31
	Tonnage 10^6 T	8,03	0,79	0,70	0,42	0,13	3,95	**5,99**
	% year	100	10	9	5	2	49	75

3.2. Dissolved flows

3.2.1. Variations of instantaneous flows during floods

The variations of the instantaneous upstream and downstream flows are shown in Figure 3. The highest dissolved loads are those of chlorides, sodium and bicarbonates. In terms of instantaneous flux, these variations show the importance of the contribution of upstream slopes in the export of chlorides and the contribution of downstream carbonate slopes in the export of bicarbonates.

 https://doi.org/10.1051/e3sconf/20183704004

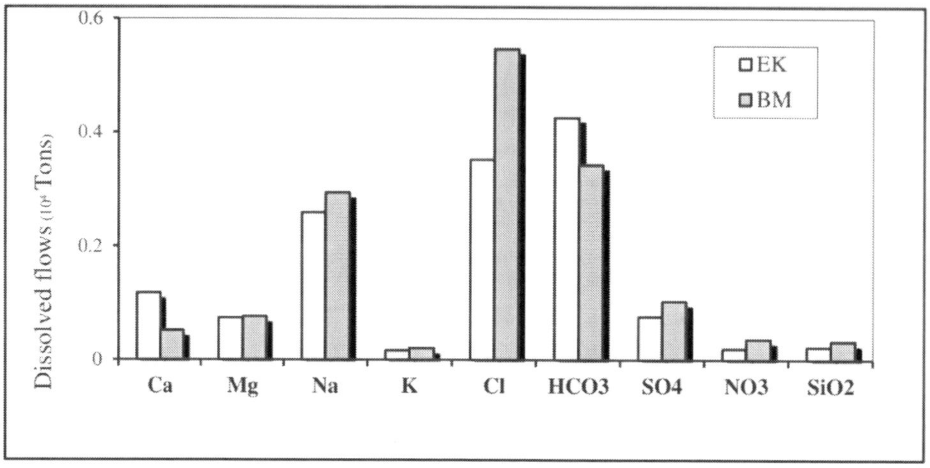

Figure 3: Average flows of major elements in Inaouene waters at the Bab Marzouka (BM) and El Kouchat (EK) stations during the February 1997 flood.

3.2.2. Dissolved flow relationship - liquid supply

Flows - liquid supply relationships show higher flows during the first flood rise than during the rest of the flood. This indicates that from the onset of the flood, rainwater and surface water strongly dragged away the elements that are highly concentrated in the soil surface horizons. The flows of the different chemical elements of Inaouene waters generally have highly significant linear relationships with liquid inputs (Fig.4).

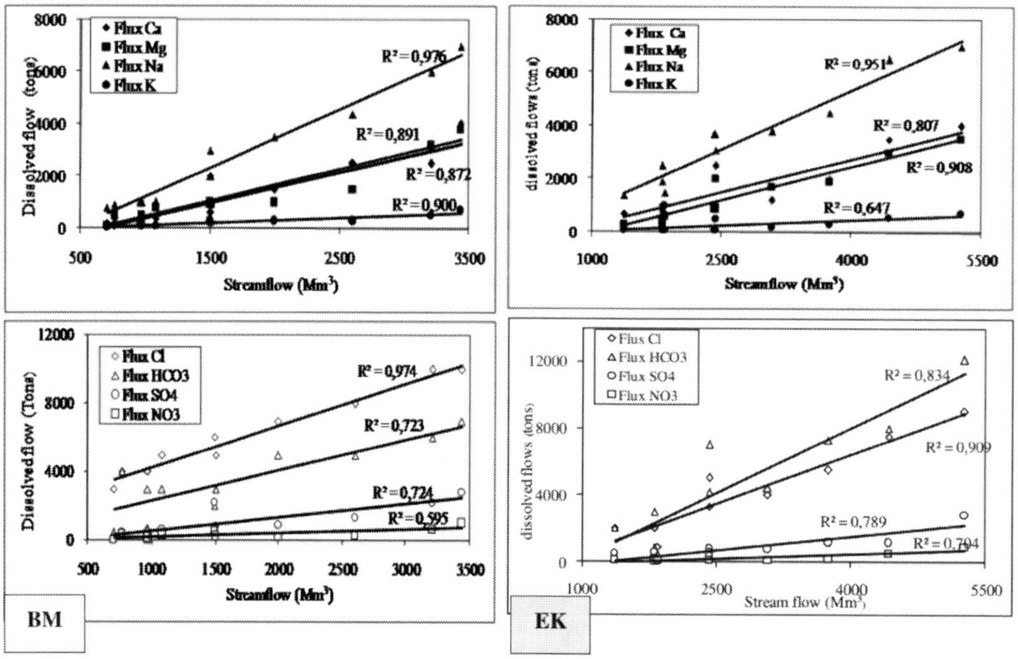

Figure 4: Relationship between the instantaneous fluxes of the major dissolved elements and the liquid inputs of the Inaouene waters to Bab Marzouka (BM) and El Kouchat (EK) during the February 1997 flood.

These relationships depend on the initial water conditions of the basin and seasonal variations. In fact, during floods following dry initial water conditions, the dissolved flows are more or less high during the rise phase of the flood due to the intense leaching of the soil surface horizons by the rainwater and by the superficial waters.

These dissolved flow - liquid supply relationships show that the different types of flow, the superficial flow and the deep flow, participate in the genesis of flows and intervene in different proportions. During the flood phase, the participation of the surface flow is greater than in the second flood phase. This flow thus causes a dilution of the deep flow rich in alteration product (Cl and Na) which participate in dominant proportions.

3.2.3. Comparison of hydrochemical flows between upstream and downstream basin

During this cycle, Inaouene exported an average of 33. 10^4 tons (117t / km^2 / year) of major chemical elements in solution. Chloride inputs downstream are about 9.6 10^4 tons, or 29.5% of total solution transport. The tonnages of bicarbonates, sodium and sulphates respectively represent 26%, 23% and 7% of the total dissolved flow. Upstream of the basin, the tonnage of the various elements in solution is on average 38. 10^4 tons (257t / km^2 / year) (Tab.3).

Table 3: Comparison of hydrochemical flows between upstream and downstream of the Inaouene basin

Inaouène in El Kouchat			Inaouène in Bab Marzouka		
kind	Annual Tonnage (10^4 tons)	Relative Contribution (%)	kind	Annual Tonnage (10^4 tons	Relative Contribution (%)
Ca^{2+}	2,20	6,7	Ca^{2+}	1,58	4,1
Na^+	7,73	23,6	Na^+	8,38	21,7
Mg^{2+}	1,30	4,0	Mg^{2+}	1,29	3,3
K^+	0,30	0,9	K^+	0,63	1,6
HCO_3^-	8,50	26,0	HCO_3^-	8,15	21,1
Cl^-	9,60	29,5	Cl^-	16,16	41,9
SO_4^{2-}	2,50	7,6	SO_4^{2-}	1,38	3,6
NO_3^-	0,33	1,0	NO_3^-	0,47	1,3
SiO_2	0,25	0,7	SiO_2	0,54	1,4
Total	32,71	100 %	Total	38,58	100 %

4. Conclusion

This study has highlighted the importance of floods in the dynamics and transport of solid and dissolved matter in the Inaouene watershed. During the 1996/97 hydrological cycle, the dissolved flux measured downstream from the Inaouene River averaged 33.10^4t / year. Chlorides, sodium and bicarbonates correspond to the bulk of this dissolved load, thus denoting the contribution of the evaporiferous pre-crystalline formations and the carbonate formations of the Middle Atlas. The potential mechanical degradation affecting the basin can deliver a solid load estimated at 6 10^6 t / year corresponding to a strong mechanical erosion (2141 t / km^2 / year). Compared to global solid flows, it is of equal magnitude as those of the Asian Hunghe River and the Colombian River Magdalena. The high intensity of the mechanical erosion revealed by the Inaouene is due particularly to the nature of the lithological formations, the climatic irregularity and the weakness of the vegetal cover favoring the erosive action of the rainwater which is generally very brutal and of short duration.

Références

1. I. DOUGLAS I: Intensity and periodicity in denudation processes with special reference of Turkey. J. of Hydrol., **14**, 233-258 (1964).
2. D. ETCHANCHU : Géochimie des eaux du bassin de la Garonne. Transfert de matières dissoutes et particulaires vers l'océan Atlantique. Thèse 3e cycle, Toulouse, 156p (1988).
3. R.I. FURGUSON: Accuracy and precision of methods for estimating river loads. Earth surface processes and landformes, n° **12**, 95-104 (1987).
4. S. HAIDA : Flux de matières issues de zone semi-aride le bassin versant de l'oued Tensift (Maroc). Estimation quantitative et qualitative : bilan. Thèse 3eme cycle, Université. Mohamed V, Faculté des Sciences. Rabat, Maroc, 147p (1991).
5. S. HAIDA., M. SNOUSSI.,C. LATOUCHE et J.L. PPRBST : Géodynamique actuelle du bassin versant de l'oued Tensift (Maroc) : Erosion et bilan des transports solides fluviaux. Bulletin des Sciences Géologiques., Strasbourg, 49, **1-4**, p. 7-23 (1996).
6. S. HAIDA. : Transports de matières et bilans de l 'érosion mécanique et de l'altération chimique sur le bassin versant de zone semi-aride : le Sebou. Thèse doctorat Es-Sciences, Université Ibn Tofail, Faculté des Sciences. Kénitra, Maroc, 253p (2000).
7. Z. KATTAN : Géochimie et hydrologie des eaux fluviales des bassins de la Moselle et de la Mossig, Transports dissous et particulaires, cycles biogéochimiques des éléments. Thèse, Université Louis Pasteur de Strasbourg., 167p (1989).
8. Z. KATTAN et J .L. PROBST : Transports en suspension et solution par la Moselle en périodes de crues. Actes Journées d'Hydrologie « Crues et Inondations », Strasbourg, 16-18 Octobre. 143-167 (1986).
9. J.L. PROBST : Géochimie et hydrologie de l'érosion continentale. Mécanismes, bilan global actuel et fluctuations au cours des 500 derniers Millions d'années. Sc. Géol, **94**, 161p (1990).
10. J.L. PROBST, R. NKOUNKOU, G. KREMPP, J.P., J.P. THIBEAUX etJ.C. OLIVRY: Dissolved major elements exported by the Gongo and Ubangui revers during the period 1987-1989. J. Hydrol. **135**,237-257 (1992).
11. J.L. PROBST et Y. TARDY: Long range streamflow and world continental runoff fluctuations since the beginning of this century. J. of Hydrology, n° **94**, 289-311(1987).

Flood risk management in the Souss watershed

Brahim Bouaakkaz[1*], *Zine El Abidine* El Morjani[1], *Lhoussaine* Bouchaou[2], and *Hamza* Elhimri [3]

1: Exploration and Management of Natural and Environmental Resources Team (EGERNE), Polydisciplinary Faculty of Taroudant (PFT), University Ibn Zohr, Morocco
2: Applied Geology and Geo-Environment Laboratory, Ibn Zohr University, Faculty of Science, Agadir, Morocco
3: Souss Massa Hydraulic Basin Agency, Av Prince My Abdellah, New Town, BP 432, Agadir, Morocco

Abstract: Flooding is the most devasting natural hazards that causes more damage throughout the world. In 2016, for the fourth year in a row, it was the most costly natural disaster, in terms of global economic losses: $ 62 billion, according to a Benfield's 2016 annual report on climate and natural disasters [1]. The semi-arid to arid Souss watershed is vulnerable to floods, whose the intensity is becoming increasingly alarming and this area does not escape to the effects of this extreme event.. Indeed, the susceptibility of this region to this type of hazard is accentuated by its rapid evolution in terms of demography, uncontrolled land use, anthropogenic actions (uncontrolled urbanization, encroachment of the hydraulic public domain, overgrazing, clearing and deforestation).), and physical behavior of the environment (higher slope, impermeable rocks, etc.). It is in this context, that we have developed a strategic plan of action to manage this risk in the Souss basin in order to reduce the human, economic and environmental losses, after the modeling of the flood hazard in the study area, using geo-referenced information systems (GIS), satellite remote sensing space and multi-criteria analysis techniques, as well as the history of major floods. This study, which generated the high resolution 30m flood hazard spatial distribution map of with accuracy of 85%, represents a decision tool to identify and prioririze area with high probability of hazard occurrence. It can also serve as a basis for urban evacuation plans for anticipating and preventing flood risk in the region, in order to ovoid any dramatic disaster.

Keywords: Flood, Disaster, Hazard, Vulnerability, Risk, Climate Change, GIS, Remote Sensing, Strategic Action Plan, Multicriteria Analysis, Watershed, Souss, Morocco.

* Corresponding author: bouaakkaz2012@gmail.com

1 Introduction

The study area (Souss River Basin) covers approximately17000 km2, in the Souss Massa, Draa tafilalt and Marrakech tensift regions, includes 128 territorials community (fig.1). It is limited by the southern mountain of the Western High Atlas, the northern mountain of the Western Anti Atlas, the Siroua massif in the east and by the Atlantic Ocean in the west. Thus, it is situated in south-west Morocco, between 29th and 31st parallels north, and between 7th and 10th degree west longitude of Greenwich [2]. Whose economic activity is mainly based on fishing, tourism, and farming (especially, vegetables, fruits and bouquets).

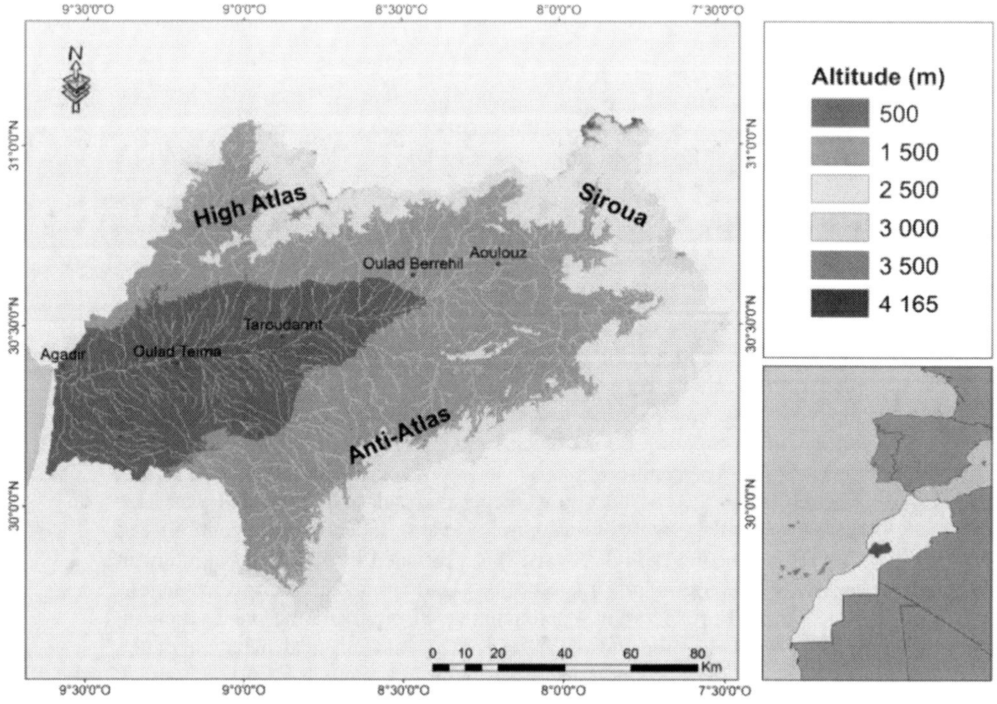

Fig. 1. Location map of the Souss watershed.

Precipitation modules exceed, in a short time, the totals typical of these regions in a normal year [2, 3].

Therefore flood hazard is a phenomena that can turn into a disaster as has been repeatedly demonstrated and is a challenge for sustainable development. Climate change will amplify existing risks and generate new ones for natural and human systems [4, 5]. Thus, the frequent overflows of Souss river (main collector of the basin) and its tributaries have caused, on several occasions, material, environmental, social and economic important damage, in particular during passage of the exceptional floods of February, November 2010 and 2014 (fig.2).

The study area was selected for its importance in the context of the regional economy and for the presence of rapidly developing anthropization, anarchic urbanization and land cover conversion dynamics; such land changes also make the territory especially vulnerable to climate change-induced flood hazard.

Fig. 2. Urban flood in Agadir city, February 2010 (ABHSMD Photo, 2010)

Although the Moroccan authority has each time made progress in the system of actions for the prevention of natural disasters (National strategy for the protection of the environment and sustainable development, water resources management strategy, national plan update study). [6], etc.), this progress was considered very modest, notably by the Hyogo Framework for Action [7]. In addition, there is no insurance system to provide financial protection against the risk of flooding [7] and there is a lack of a framework law on risk prevention. For this purpose, the damage is heavy to manage in the absence of insurance coverage.

Among the objectives of this work is the strengthening of communities' knowledge and capacities to deal with natural hazards, with a view to improving prevention systems and protection in crisis management situations. Since our society has no choice but to face this meta-risk [8] because it is an integral part of our environment.

2. Materials and methods

In this context, the determination of the areas likely to be affected flooding is essential for better management of this risk. The method adopted in this paper associates a triple approach of geomatic (remote sensing with GIS) and field, as well as the statistical treatment of data collected from the Souss Massa Hydraulic Basin Agency (ABHSMD) of the rainfall of the hydrological stations and the Frequency analysis of peak flows of Souss River.

2.1. Methodology for generating the spatial distribution of the intensity levels of flood hazard

The process used to spatially distribute flood hazard combines the extent of past flood events with the spatial distribution of causal factors influencing the flood hazard [9, 10, 11, 12], by using the Remote Sensing technique and Geographic Information Systems, as well as multicriterate analysis techniques. This combination enables the calculation of a weighted score for each individual causal factor. The spatial distribution of the weighted scores are then aggregated to derive the distribution of the flood hazard index (FHI) before being reclassified to obtain the spatial distribution of the intensity level of flood hazard.

The application of this method requires the execution of the following steps to generate the spatial distribution flood hazard map figure 4:

- Identification and generating of the causal factors at 30m resolution (Digital Elevation Model (DEM); Slope, Aspect, flow accumulation, land cover, lithology, soil type, precipitations, [9, 10] ;

- Creation and classification of the historical number of past flood events distribution map;
- Estimation of the weight of each factor (weighting);
- Standardization of factors (SWS) ;
- Production the spatial distribution map of the flood hazard index;
- Reclassification of the map generated according to five intensities of the flood hazard (very low, low, medium, high and very high) as final result (Fig.4).
- Validation of the flood hazard distribution map resulted by the Souss Massa Hydraulic Basin Agency data

For more details concerning the methodology used to generate the spatial distribution of the intensity levels of flood hazard, please refer to the [9] publication.

2.2. Frequency analysis of the peak flows of the wadi Souss

Frequency analysis of a long series of historical maximum flows (Fig.3) by the adjustment of Gumbel's law makes it possible to estimate the return time of a particular value, thus to predict the probability of occurrence of future floods, in order to anticipate the occurrence of a flood, and to alert the populations. Since risk management requires first and foremost a fair assessment of existing risks, their spatialization as well as their probability of occurrence [13].

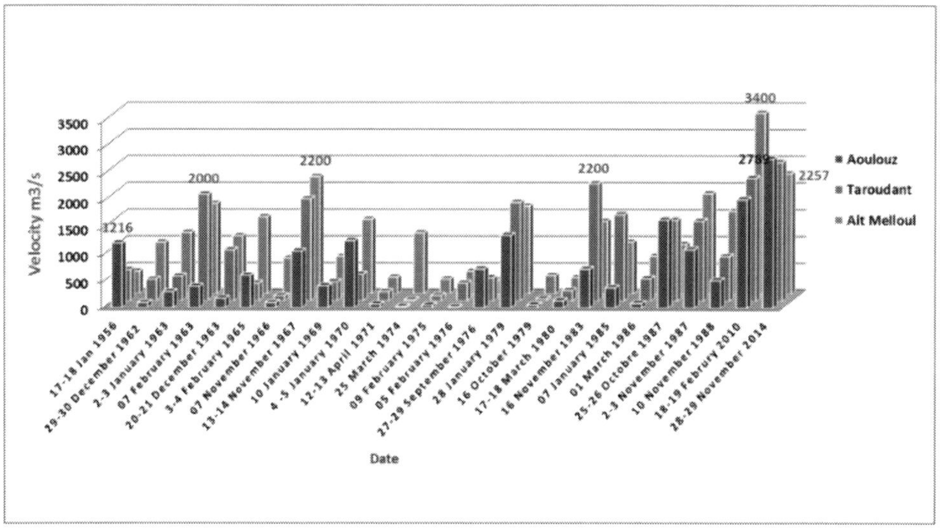

Fig. 3. Flood history of the Souss River at the stations of Aoulouz, Taroudant and Ait Mellou

2.3. Field investigation, diagnosis of the inventory and analysis of the legal framework for flood management

In addition, this method includes, exploration of the terrain through visits to vulnerable sites that can locate problems and also meet the flood victims who are the object of study and make the findings finally which are necessary. Since the increased involvement of the local populations should become essential and become obvious to improve flood risk management. As well as, the study of the different Legislative text and components of the flood risk management framework. Finally, the analytical and critical study of measures, flood protection works and direct and indirect development projects carried out by the various actors, namely: Prediction, prevention and the alert process.

3. Results and discussion

3.1. Spatial distribution of the flood hazard in the study area

The examination of the flood hazard map (Fig.4) shows that the level of flooding is very high both in the High Atlas and Siroua regions. This is due to lithology characteristics (impermeable rocks: marls, shales and crystalline rocks), highly slopes and significant annual rainfall in those areas (greater than 500 mm / year [2]). However, the level of hazard is very low in the Anti-Atlas. This is explained by the dominance of permeable limestone formations in this chain which receives only 150 to 200 mm as annual average precipitation.

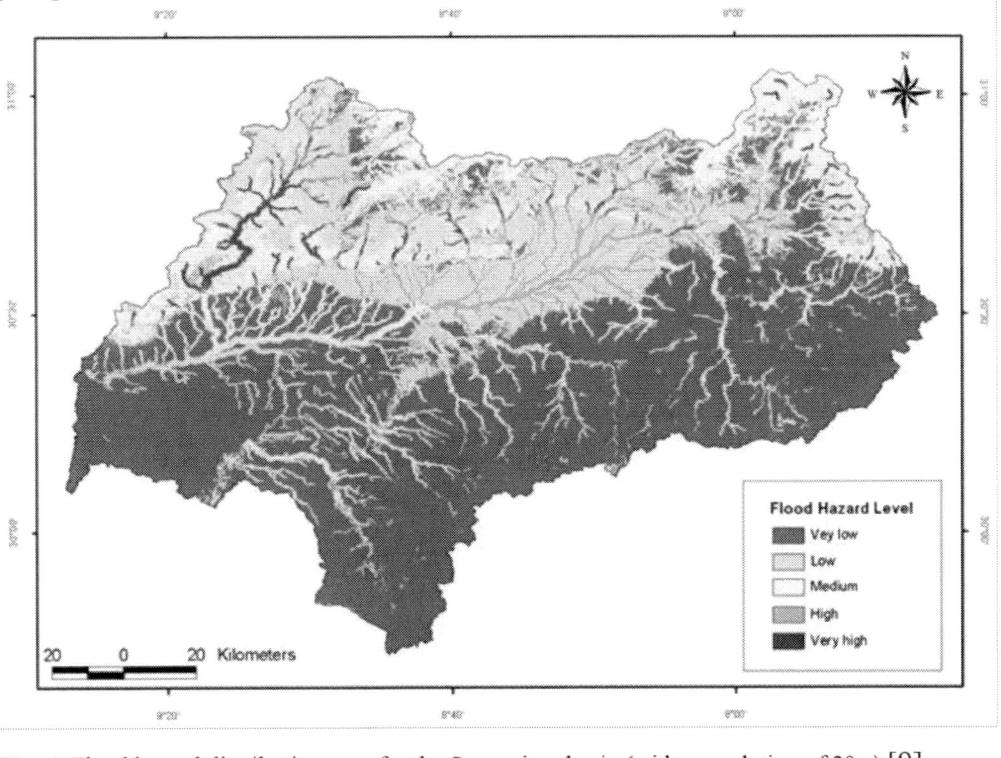

Fig. 4. Flood hazard distribution map for the Souss river basin (with a resolution of 30m) [9].

 https://doi.org/10.1051/e3sconf/20183704005

3.2. Frequency analysis of the peak flows of the Souss River

Frequency analysis of a long series of maximum flows allows to estimate the return time of a particular value. The adjustment of Gumbel's law to the floods of Souss River showed a satisfactory adequacy (Fig.5). This law describes the probability of occurrence of floods and their periods of return. Floods with a peak of at least 300 m^3/s (which is already 54 times the Aoulouz module) occur on average every year, and even floods greater than 1000 m^3/s occur every five to ten years at Ait Melloul, while those higher than 2000 m^3/s and 3000 m^3/s are respectively fifty-year and centennial floods. This relatively high frequency of floods is probably explained by the very large area drained by Wadi Souss and its many tributaries (17 000 km^2), and the low permeability of the slopes of High Atlas, which may offer response times to rather short showers

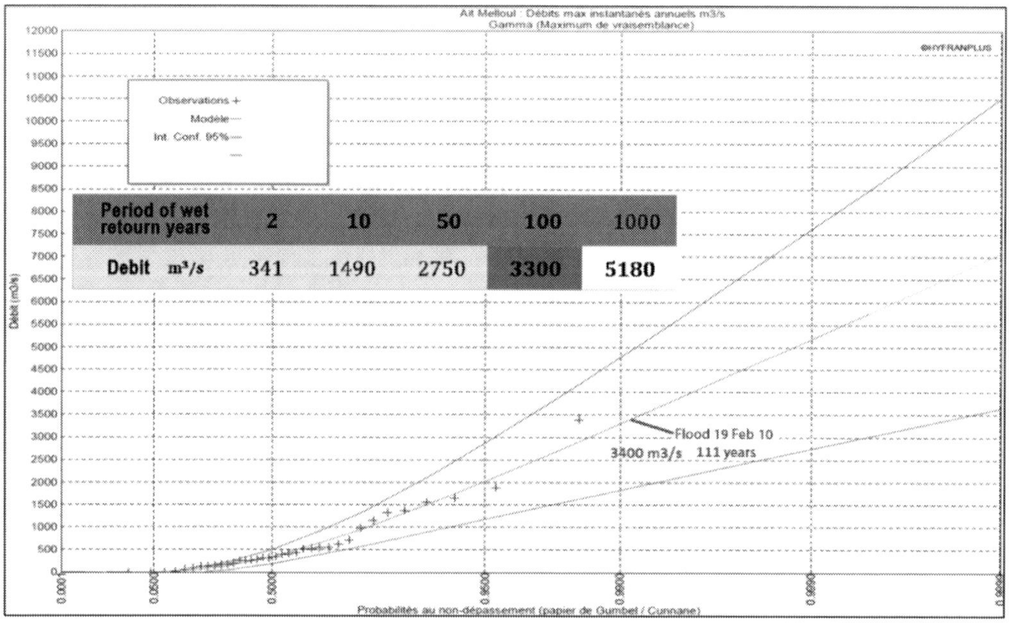

Fig. 5. Gumbel Adjustment with the floods of Souss River, at the ait Melloul station

3.3. Field investigation, diagnosis of the inventory and analysis of the legal framework for flood management

An analysis of the current situation shows that all the actors (the State, ABHSM, Council of the Region, Provincial Council) have made efforts to fight against the risk of flooding in the basin in question, but remains insufficient for the sustainable and integrated management of this risk.

Indeed, the implementation of the national strategy for sustainable development, which runs from 2015 to 2020, already developed (takes into account protection against floods) has not yet come into force.

In addition, there is a delay in the implementation of the National Program for Rural Sanitation (PNAR) and the sanitation networks in the sanitation centers have lower evacuation capacity, it happens that networks overflow before the waters reach the outlets (case of the Agadir city in February 2010.Photo.1).

As a result, the risk of storm and urban flooding is a potential hazard and necessitates upgrading of the sewerage network, in order to increase its drainage capacity due to climate change in the region.

The study of the various legislative and regulatory texts of the flood risk management framework indicates that several entanglements of skills exist because of the multiplicity of public actors and the absence of a strategy dedicated to the protection against this extreme phenomenon. The Water Act 36-15 has given diversified allocations to basin agencies including ABHSM. This multiplicity of tasks has not resulted in sufficient human and financial resources, which delays the advancement of flood risk protection work to protect the vulnerable population.

Indeed, according to the master plan study of flood protection in the province of Taroudant (97% of the basin area), the financial cost of the projects of protection against this risk is of the order of 400 Millions of dirhams. This requires the urgent intervention of all actors to treat vulnerable sites and carry out structural protection works (dams, dikes, flow diversion, etc.) in a timely manner to avoid the aggravation of the situation following the changes climate.

The analysis of the water law of 36-15 published in the official bulletin n ° 6506 of 6 October 2016 in terms of protection against floods attaches' importance to the fight against this risk of flooding. Three sub-sections of 7 articles of this legislation deal exclusively with the protection and prevention of risk, detection, monitoring and warning devices, as well as the management of flood events (Articles 117 to 123).

Nevertheless, the current institutional framework does not favor the implementation of planned actions of protection against floods. Indeed, there is a lack of regulatory and legislative framework governing the management of flood risk marked by a lack of implementing legislation for the Water Law 36-15, an absence of regulations applicable to floodplains and the risk of flooding is not integrated in urban planning documents.

On the other hand, the law 30-15 on the safety of the dams for the protection of the downstream populations is not applicable for the reinforcement of the legal arsenal in this field. Its implementation requires obligations to continue to upgrade dams, including compliance with safety standards, in order to protect as much as possible the risks of breakage or failure. In 2013, the water department updated the study of the National Flood Protection Plan (NFPP) [6] in order to draw up the state of progress of the treatment of vulnerable sites and to update data and information concerning the priority sites. However, the financial resources mobilized are also limited compared to the means provided to cover the components of the NFPP.

The latter focused on the treatment of blackheads resulting from floods, regardless of the causes and degrees of flood exposure. As a result, the approach advocated by the NFPP has meant that the protection actions undertaken in this context are often carried out piecemeal in the absence of a broader strategy for the prevention and protection of vulnerable areas.

The follow-up of the last floods that occurred in the region in February and November 2010 and 2014 revealed the unsuitable dimensioning of the crossing structures and the inadequacy of certain bridges and crossing works along roads intersecting the courses of water. These structures are impediments to flow and contribute to the raising of water levels and overflowing.

In addition, the insufficient mastery of the hydraulic public domain (HPD), which has not been defined, does not allow the ABHSM to respond to requests from stakeholders in this field (urban agencies, local authorities, etc.). What hampers the rigorous enforcement of flood-zone construction regulations and the policing role of water remains very limited in time and space.

Examination of the spatial distribution of the hydro-climatological network, shows that it is less developed on the basin on the left bank of the Souss River, its density is low (15 stations rainfall) and not automated. Indeed, the Anti-Atlas is not covered by hydrological stations, especially in the areas of high vulnerability to flooding namely wadis Arghene, Tiout and Aourgua.

In addition, high-frequency shift data transmission systems are nowadays little adapted to the needs of real-time knowledge of phenomena. In fact, this system is very often disturbed during periods of bad weather and night shift.

In this context, and in order to be able to improve the protection of people and property against floods in the study area, several technical, organizational and financial actions must be put in place to increase the resilience of the exposed territory and reduce vulnerability against this risk.

Therefore, it is desirable to promote integrated flood risk management based on three pillars: I. Prevent and develop, II. Prepare, manage the crisis and III. Act after the crisis.

Feedback is also crucial, as the information gained through this type of event must be capitalized in order to be reused in forecasting models and prevention actions.

In addition to carrying out palliative work to protect vulnerable sites, the proposed action plan provides the implementation of structural and non-structural actions based on the following main provisions (Table 1):

Table 1. Structural and non-structural measures of flood risk management

Non-structural measures	Structural measures
Strengthening meteorological and hydrological watch and creating refuge areas and evacuation traps in the most exposed areas;	Continued development of dams and hill lakes, which studies have already been finished and inventories of other potential sites for flood rolling;
Implementation of a flood forecasting and warning system and the automation of hydrological observation (rain and water level) in some flood warning stations and vulnerable wadis devoid of dams (Irguiten wadis, Tiout, Assads and Aghren). The implementation of the identification markers of the sectors at risk of flooding;	Upgrading of the 5 existing hydraulic structures to ensure that the manager implements monitoring, maintenance and management measures to ensure the safety of the structure and the protection of the residents downstream;
The reform of the legislative framework governing the flood risk management and setting up of a platform for exchange between all the stakeholders in this area to initiate the necessary political orientations in this field;	Watershed development through the construction of thresholds, the construction of dry stone terraces and reforestation with adaptive plants by the direction of water and forest to increase infiltration at the expense of runoff and reduce the rate of siltation in dams;
The updating of the delimitation of flood risk areas at territorial community level (Atlas of regions threatened with flooding), to inform the population of the risks to which they are exposed;	Reservation of a hollow volume to absorb the floods at the levels of dam reservoirs and mitigate the brutal pulsations of the main collector (Souss river);
Development of the awareness of the populations to risk by the sensitization, the development of the	Development of water diversion structures and flood protection dams in upstream;

memory of the risk and the diffusion of the information;	
The updating of land-use planning documents to control the use of areas in risk of flooding;	Study the possibility of raising the height of dams in operation, to composite the rate of siltation and to reduce the floods;
The development and strengthening of financial mechanisms (insurance and disaster funds), consultation and coordination between different actors;	Cleaning and recalibration of the sections of wadis filled by erosion deposits, to increase flow and avoid overflow;
Perform technology transfer through the training of flood risk management personnel of the various actors.	Update studies and analyzes of all aspects related to the phenomena of floods: Climate, hydrology, flow, topography, morphology, geology, geomorphology, and hydraulics aspects to determine the thresholds of warning in case of extreme precipitation and flows.

To identify this problem in its entirety, the establishment of real sustainable financing is a crucial element to the success of any program and action plans of intervention works to increase capacity and reduce the vulnerability of local residents. It is desirable to set up a joint flood commission which will be a place for dialogue and exchange between the various actors involved in flood prevention (water management, urban planning, etc.)

4 Conclusions and recommendations

The map of the spatial distribution of the flood hazard with a high resolution of 30m, and accuracy 85% is a tool to identify areas where decision makers can intervene. As it can serve as a basis for urban evacuation, anticipation and prevention plans in case of flood risk.

By studying the peaks of various floods calculated in the Souss valley since 1956, only those of February 2010 and November 2014 are the most devastating. The precipitation that triggered these catastrophic floods have largely exceeded, in a short time, the typical totals of these regions during a normal year.

This methodological process could be transferred and implemented in other basins or regions of Morocco by exploiting their own precise and reliable data and information. In order to study the risk of flooding in a global and integrated ways, the field opens up innovative perspectives for the future use of the very high resolution models and the assessment of social and economic vulnerability within this extreme event environment. The knowledge of the various aspects constituting the vulnerabilities of a territory subject to flood risk represents an important contribution to reduce and control the consequences on already existing sensitive equipment.

This study also demonstrated the need to improve the institutional framework for climate change, including flood prevention and control through the introduction of a compensation system. In addition, updating the laws governing the management of this risk and their applications is desirable.

In order to prevent the damage of future floods, the mapping of the dynamic evolution of this risk, is desirable by means of radar remote sensing which ensures at all times the availability of the information despite the meteorological conditions.

Finally, promote research, innovation and technology transfer in the field of adaptation to climate change, specially the risk of flooding.

Acknowledgments

We would like to thanks the Souss-Massa Hydraulic Basin Agency (ABHSM) for providing us with the complete database needed for this study.

References

1. A. Benfield (2016). *Analytics Impact Forecasting : Annual Global Climate and Catastrophe Report.* 73 p;

2. B. Bouaakkaz (2015). *Le bassin versant de l'oued Souss : caractéristiques physiques, climatiques et comportement hydrologique : prospectives territoriales*, «Mémoire d'accomplissement des modules du master spécialisé : Aménagement, Développement, Gestion et Prospectives territoriales dans les zones arides et semi-arides ». Université Ibn Zohr, Agadir. 137 p.

3. L. Chakir, A. Aït Hssaïne, B. Bouaakkaz (2015). *Les crues de l'oued Souss et le risque de l'inondation dans le tronçon aval de la vallée (Maroc),* Laboratoire: Géoenvironnement et développement des zones arides et semi-arides (GEDEZA) Département de géographie, Université Ibn Zohr, B.P. 29/S, Agadir, Maroc.

4. GIEC, (2014). *Changements climatiques 2014: «Rapport de synthèse» : Contribution des Groupes de travail I, II et III au cinquième Rapport d'évaluation du Groupe d'experts intergouvernemental sur l'évolution du climat.*GIEC, Genève, Suisse, 161p

5. S. Rannow, M. Neubert (Eds.). (2014). Managing protected areas in central and eastern Europe under climate change. Springer, 322p.

6. Royaume Du Maroc, Ministère Délégué Auprès Du Ministre De L'énergie, Des Mines, De L'eau Et De L'environnement, Charge de L'eau (2013). *Étude d'actualisation du plan national de protection contre les inondations, Sous-Mission I-1 : «Rapport d'Analyse des études et travaux de protection réalisés pour la lutte contre les inondations»*.112p

7. Département De L'environnement Du Ministère De L'énergie, Des Mines, De L'eau Et De L'environnement (2013). *Rapport national de suivi sur la mise en œuvre du Cadre d'action de Hyogo «Rapport d'évaluation»*,41p

8. E. Rioust (2013). *Gouverner l'incertain : adaptation, résilience et évolutions dans la gestion du risque d'inondation urbaine : les services d'assainissement de la Seine-Saint-Denis et du Val-de-Marne face au changement climatique.* «Thèse en Architecture, aménagement de l'espace». Université Paris-Est, 429p

9. Z.E.A. El Morjani, M. Seif Ennasr, A. Elmouden, S. Idbraim, B. Bouaakaz, Et A. Saad (2016). *Flood Hazard Mapping and Modelling using GIS applied to the Souss River watershed. R. Choukr-Allah et al. (eds.), the Souss-Massa River Basin, Morocco. «Chapter part of the series The Handbook of Environmental Chemistry pp 1-30», DOI 10.1007/698_2016, Springer International Publishing Switzerland*

10. Z.E.A. El Morjani, (2011a). *«Preparation of the dataset. Methodology document for the WHO e-atlas of disaster risk. Volume 1. Exposure to natural hazards. Version 2.0»*. Taroudant poly-disciplinary faculty of the Ibn Zohr University of Agadir, Morocco. ISBN: 978-9954-0-5397-3.

11. Z.E.A. El Morjani, (2011b). *Methodology document for the WHO e-atlas of disaster risk. «Volume 1. Exposure to natural hazards Version 2.0: Flood hazard modeling»*. Taroudant poly-disciplinary faculty of the Ibn Zohr University of Agadir, Morocco. ISBN: 978-9954-0-5393-5.

12. United Nations International Strategy For Disaster Reduction (ISDR), *Basic terms of disaster risk* reduction at http://www.unisdr.org/eng/library/lib-terminology-eng%20home.htm [Accessed Nov. 23, 2017]

13. A. T. Ngo (2014). *Evaluation environnementale du risque d'inondation dans le delta du fleuve ha Thanh (centre Viêt-Nam).* « Thèse en Géographie-Aménagement-Environnement ». Université d'Orléans, France, 429p.

Application of electrical geophysics to the release of water resources, case of Ain Leuh (Morocco)

A. Zitouni[1*], *A.* Boukdir[1], *H.* El Fjiji[2], *W.* Baite[1], *V.R.* Ekouele Mbaki[1], *H.* Ben Said[1], *Z.* Echakraoui[1], *A.* Elissami[3] et *M.R.* El Maslouhi[4]
[1]Laboratory of Industrial engineering, Faculty of Sciences and Technics, Sultan Moulay Slimane University, Beni Mellal, Morocco.
[2] Méditerranée Travaux Etudes Et Conseils D'ingénierie (MEDITEC-IN), Beni Mellal, Morocco.
[3] National Office of Electricity and Drinking Water (ONEE), Rabat, Morocco.
[4] Oum Er-Rbia Hydraulic Basin Agency (ABHOER), Beni Mellal, Morocco.

Abstract. Being seen needs in increasing waters in our contry for fine domestics, manufactures and agricultural, the prospecting of subterranean waters by geologic and hydrogeologic classic method remains inaplicable in the cases of the regions where one does not arrange drillings or polls (soundings) of gratitude (recongnition) in very sufficient (self-important) number.
In that case of figure, the method of prospecting geophysics such as the method of nuclear magnetic resonance (NMR) and the method of the geophysics radar are usually used most usually because they showed, worldwide, results very desive in the projects of prospecting and evaluation of the resources in subterranean waters.
In the present work, which concerns only the methodology of the electric resistivity, we treat the adopted methodological approach and the study of the case of application in the tray of Ajdir Ain Leuh.
Key words: electric resistivity, methodology, hydrogeology, prospection, ressources in subterranean waters, tray of Ajdir Ain Leuh.

1 Introduction

The mountains are generally called "water towers of the planet". This qualifier is based on the precipitation potential associated with terrain and snow-ice retention. The barrier effect (according to moist air masses) and the altitudinal effect result in more abundant precipitation and snow-ice storage (depending on altitude) of precipitation; storage returned to surface and underground flows during the summer seasons. The mountain water appears in this scheme as an abundant and inexhaustible resource renewing from one year to another [14].

However, access to drinking water, especially in mountainous areas, is a major factor in economic development, and in improving the standard of living of populations and their stabilization in rural areas.

This study is part of securing the drinking water supply for the mountainous area at the Ajdir Ain Leuh plateau.

It is in this context, we present in this work the contributions of the application of the method of electrical soundings to the recognition of different aquifer formations. This will make it possible to define underground water potential and quantify the groundwater table in order to decide on the most appropriate option for mobilizing groundwater resources,

* Corresponding Author: abdelhamidgat@gmail.com

taking into account productivities, quality as well as the constraints linked to the influence of water mobilization on water resources already exploited, including sources [5].

2 Description of the study area

2.1 Geographical setting

The Ain Leuh and Mrirt zones are located southeast of the Sebou basin between Ifrane and Khénifra on the main road n° 24 (**Fig. 1**), they belong to the Middle Atlas tabular plateau, framed on the north by the South-West corridor-Rifain, to the south by the High Atlas and the valley of Upper Moulouya, to the east by the valley of the Middle Moulouya and to the west by the Moroccan Meseta.

Morphologically, the Ain Leuh-Mrirt area is characterized by a tabular structure, more faulted than folded, by a monotonous relief implying the Middle Atlas pleated with accentuated folds, high mountains and deep marly depressions. It is a monotonous lithology of weakly folded liasic limestones which is responsible for this platitude.

The Causse, south of Ain-Leuh presents a typical landscape poljés adapted to the network of faults, this karst plateau, like all those of the same kind, has a very underdeveloped hydrographic network.

Water resources in these areas are used as sources by local people for irrigation of small irrigated areas and water supply. Generalized groundwater is used mainly for intensive and modern irrigation.

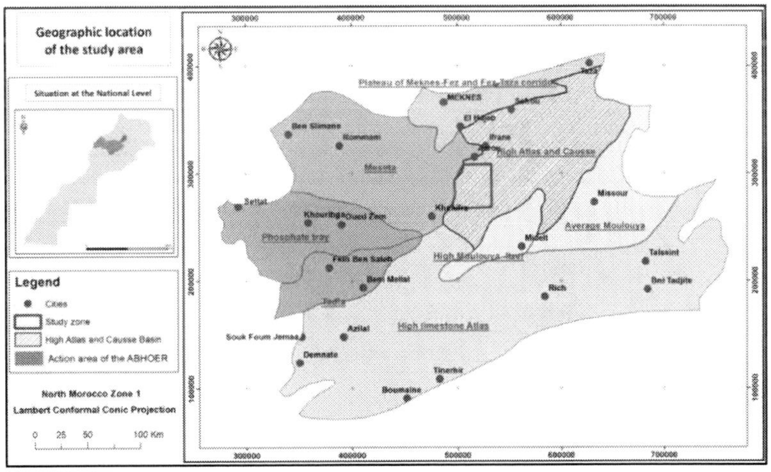

Fig. 1. Situation of the study area

2.2 Climate context

The climate prevailing in the Middle Atlas is Mediterranean Type Mountain; it is characterized by a wet and cold climate. This particular climate of the Middle Atlas is due mainly to its altitudinal position, its geographical situation and its exposure to marine influences [15]. The clouds coming from the west give abundant precipitations (rain, snow) in contact with the Middle Atlas. The natural barrier that forms the chain Atlasique creates a dissymmetry on the climatic plan: the Atlantic facade exposed to the winds coming from the NW is more watered; as for the SE facade which is subject to the influence of the Saharan climate [16].

In general, the climate of the causses is a Mediterranean climate of mountain, cool and wet; it snows in abundance on the heights and the quantities of rain exceed 1100mm / year in Ifrane, but as we decent towards the south, the rains become less abundant.

2.3 Hydrogeological context

The Moroccan average atlasic causse has been the subject of several geological studies. Il est constitué de plateaux d'altitude comprise entre 1000 et 2200 m, constitués par la prédominance de formations carbonatées du Lias inférieur et moyen, découpées en blocs t consists of plateaus of altitude between 1000 and 2200 m, consisting of the predominance of carbonate formations of the lower and middle Lias, cut into fault blocks along several northeast-trending faults (**Fig. 4**). The Causse is limited to the west by the primary lands of Wadi Beht (Moroccan primary Meseta) and those of Tazzeka to the northwest. To the north, the limit is determined by Tertiary and Quaternary overburden in the South Rifain Corridor [17].

The dominant facies consists of dolomite and limestone and liasic, locally covered quaternary basaltic flows, resulting from recent volcanic activities (**Figure 4**). The dolomitic layers of the lower Lias lie in probable concordance on the Triassic (Colo, 1961). The middle Lias is characterized by limestones of different colors, in fact, the northern part of the Causse is essentially Dolomitic, that of the southern section is essentially limestone.

The Atlas Mean Causse includes two main aquifers of unequal interest; these are the aquifer of the quaternary basalts (Dolerites) and that of the dolomitic limestones of the Lias.

2.3.1 Basaltic aquifer

The basaltic aquifer of the Middle Atlas Causa can be subdivided into two hydrogeological sub-basins: Tigrigra (area of approximately 546 km²) and Timahdite (area of approximately 435 km²). These two compartments are separated by a watershed SSW-NNE, passing between the centers of Ain Leuh and Timahdite.

The approximate balance of the Quaternary basalt water table of the Middle Atlas is summarized in the table below.

Table 1. Assessment of Quaternary Basalt Water Resources in the Middle Atlas Causa

Inputs (Mm^3/year)	
Infiltration of rainwater	57
Childbirth with the tablecloth of Lias	20
Return of irrigation water	9
Total Entries	**86**
Outputs (Mm^3/year)	
Drainage of sources	31 à 46
Drainage of wadis Tigrigra and Derboura	16
Agricultural levies	15
Rural AEP levies	2
Total Outputs	**64 à 79**

2.3.2 Aquifer of Middle Lias

This water table circulates mainly in dolomitic formations (compact, sandy or bedded) of the lower and middle Lias. In the southern section of Causse, the aquifer consists essentially of limestone formations. The accident line of Tizi N'Tretten marks the boundary between the two facies constituting the aquifer of the Middle-Atlas Causse.

Water circulates in cracks, fractures and karst channels. These discontinuities favor the infiltration of rainwater, which is the main source of food for the Liasique aquifer.

The only feeds in the basin are meteoric; the emissions are constituted by the drainage of the waters of Ain Aicha-Hammad (n ° IRE 49/30), the Wadi Ain-Leuh and the sources Aioun-Akadous and toufstelt.

In the absence of piezometric monitoring of the Aïn Leuh-Azrou basin and large withdrawals from wells and boreholes, the water balance (of this basin) is assumed to be balanced. The drainage flow of Ain Leuh wadi and quaternary basalts (Tioumliline and Tagounite) is estimated (from this assessment) at about 41 Mm^3 / year (1.3 m^3 / s), which is of the same order of magnitude than that evaluated by the Sebou Basin Hydraulic Agency (ABHS) in 2003 **(Table 2)**.

Table 2. Assessment of water resources in the basin Lias

Inputs (Mm^3/year)	
Infiltration of the rain	36
Total Entries	**36**
Outputs (Mm^3/year)	
Drainage of sources	1.7
Drainage of Ain Leuh wadi and Quaternary basalts (Tioumliline and Tagounite)	34.3$^{(*)}$
Total Outputs	**36**

(*) : Derived from the hydraulic balance of the basin.

2.3.3 Water point data IRE located in the study area

The study of drilling datasheets showed us that the following N ° IRE points (ABHOER), are the most representative for the hydrogeological recognition of the study area:

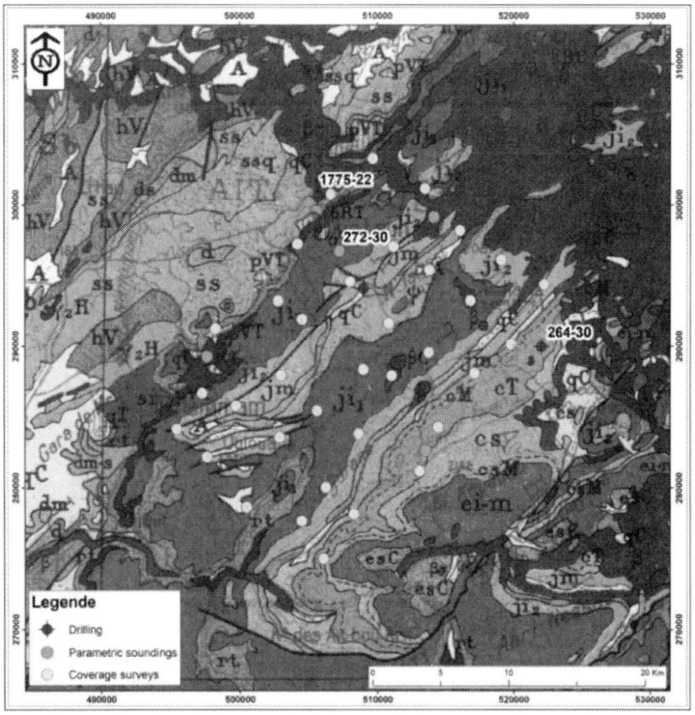

Fig. 2. Geological map of the study area (extracted from the geological map of Morocco at 1: 500,000)

<u>N°IRE: 1775-22</u> (Located northwest of the study area).

- ***Nature:*** *Drilling.*
- ***Situation map:*** *Ain Leuh au 50 000 th.*
- ***Lambert coordinates (in m) :*** *X = 505 600, Y= 300 600 et Z= 1330.*
- ***Measurable depth:*** *150 m.*
- ***NP / sol :*** *40.98m.*
- ***Electrical conductivity*** *(at 25°C):* *1042 mho/cm.*
- ***Water temperature:*** *19.4°C.*
- ***Dry residue at 105°C :*** *729 mg/l.*

<u>N°IRE : 272-30</u> (Located west of the study area).

- ***Nature:*** *Drilling.*
- ***Situation map:*** *Al Hammam au 50 000 th.*
- ***Lambert coordinates (in m) :*** *X = 509 000, Y= 294 100 et Z= 1840.*
- ***Measurable depth:*** *130 m.*
- ***NP / sol:*** *41.20m.*
- ***Electrical conductivity*** *(at 25°C):* *426 mho/cm.*

- ***Water* temperature:** *12.4°C.*
- ***Dry residue at 105°C:*** *298 mg/l.*

N°IRE : 264-30 (Located east of the study area).

- ***Nature:*** *Drilling.*
- ***Situation map :*** *Timahdit au 50 000 th.*
- ***Lambert coordinates (in m)*** *: X = 522 000, Y= 290 000 et Z= 1920.*
- ***Measurable depth:*** *160 m.*
- ***NP / sol:*** *77.07m.*
- ***Electrical conductivity*** *(at 25°C):* *743 mho/cm.*
- ***Water* temperature:** *11.1°C.*
- ***Dry residue at 105°C*** *: 520.1 mg/l.*

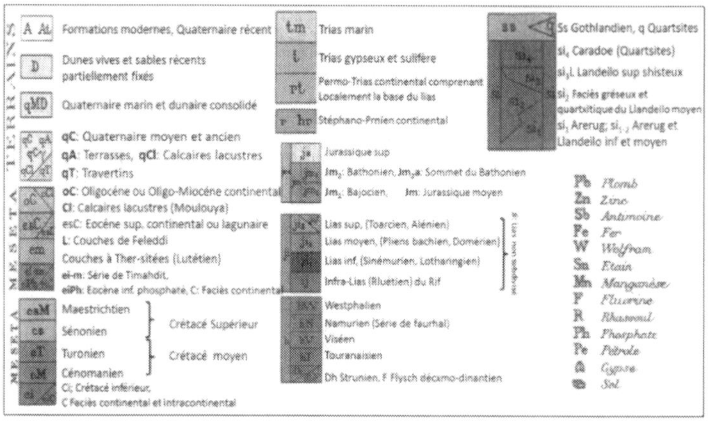

Fig. 3. Legend of the geological map of Morocco at 1/500 000

3 Materials and methods

3.1 Methodology

3.1.1 Principle of the ES method (fig. 4)

The electrical survey allows to study the variation of the resistivity of the soil with the depth [1-9-10-11]. Data acquisition in the field is as follows: A current of intensity (I) is sent through two electrodes A and B by means of batteries or a generator. Using a potentiometer (direct reading) or a recorder, the potential difference (ΔV) between two measurement electrodes M and N is measured by increasing the length of line AB each time. The values of apparent resistivities obtained represent deeper and deeper depths [1-3-5-6-7-9].

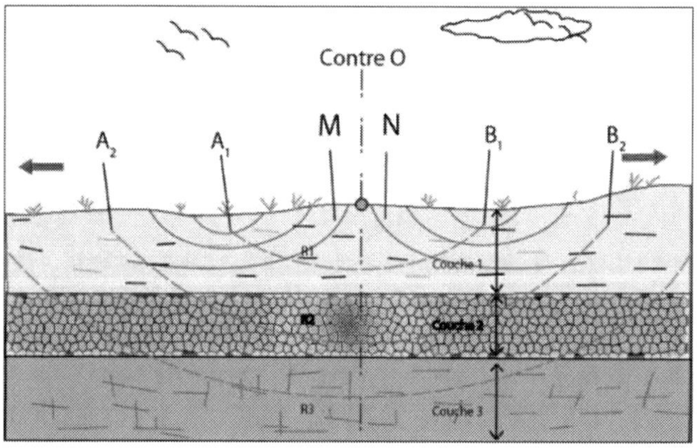

Fig. 4. Schematic diagram of the ES method (Schlumberger device)

The **Schlumberger** quadrupole, which has been applied in the context of the present study, is characterized by a small distance MN in front of AB in order to introduce the notion of electric field thus facilitating the theoretical calculations. The application of the ohm law makes it possible to calculate for each line lenner AB an apparent resistivity value (R) defined by the following formula [1-9-13]:

$$R = K.\ \Delta V/I$$

K being a coefficient that depends on the AMNB geometry.

3.1.2 Processing of ES diagrams

The electrical sampling diagram (ES) is obtained by plotting on a grid with a bilogarithmic scale the values of AB / 2 (in m) on the abscissa and the apparent resistivities (in ohm.m) on the ordinate [1-9-12-13].

The study of these diagrams, their comparison with each other and with abacuses as well as their analysis using computer programs make it possible to determine, in the majority of the cases, the vertical succession of "electric layers" as well as the true resistivity of each. This requires, of course, that the contrasts of resistivities between the different formations are sufficient [1-9-12].

The quantitative interpretation of the electrical soundings conducted in the study area was done on a microcomputer using specific interpretation and processing software. These allow an analysis of smoothed curves with theoretical charts to eliminate the effects of possible taps that could harm the computer analysis [1-5-9-13].

3.2 Geophysical measurements made

3.2.1 Number of surveys

The geophysical measurements collected in the field were measured by a GRM 3000 resistivity meter. A total of thirty-nine vertical electrical soundings (39 ES) which shows the following distribution (**Fig. 5**):

- 35 electrical cover surveys carried out on a grid of approximately 4Km × 4km;
- 4 parametric electrical soundings carried out on the different outcrops of the geological formations.

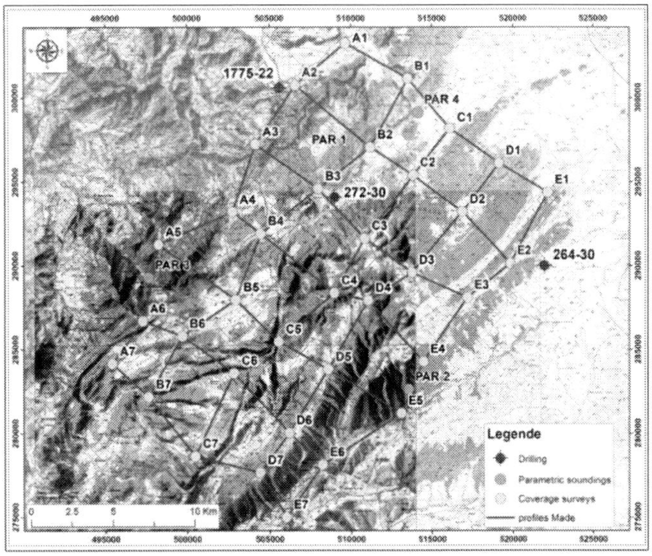

Fig. 5. Situation Map of Electrical Surveys and Geo-Electrical Profiles (Extract of topographic maps of Morocco at 1/50 000)

The 35 coverings were distributed over 7 profiles of 5 electrical soundings according to a 4Kmx4Km mesh (**Table 3**).

Ce maillage n'a pas été respecté totalement sur toute la zone d'étude vu les difficultés et les obstacles rencontrés lors de réalisation des mesures électriques sur le terrain (terrain accidenté, clôtures des fermes, oueds…).

Table 3. Distribution of electric soundings by profile

Profile	Electrical survey
Profile A	A1, A2, A3, A4, A5, A6 et A7
Profile B	B1, B2, B3, B4, B5, B6 et B7
Profile C	C1, C2, C3, C4, C5, C6 et C7
Profile D	D1, D2, D3, D4, D5, D6 et D7
Profile E	E1, E2, E3, E4, E5, E6 et E7

3.2.2 AB line length

The totality of these electrical soundings was carried out with variable current transmission line lengths AB ranging up to 3 km depending on the depth and thickness of the geological formations traversed.

3.3 Results

3.3.1 Electrical soundings diagrams

The analysis of the diagrams of the electrical soundings made it possible to highlight the existence of a deep resistant substratum noted Rp and a heterogeneous cover formed of alternation of conductive and resistant grounds.

The comparative study of the lithological layers revealed by the rigging drilling with the horizons detected on the diagrams of the electrical soundings, made it possible to establish geoelectric correlations. The results obtained led to distinguish three main families of electrical soundings (**Table 4**):

Table 4. Families of electrical soundings

Family number of electrical soundings	IRE number of calibration drilling
Family 1	1775-22
Family 2	272-30
Family 3	264-30

a) *Family 1:*

Table 5 shows the geoelectric characteristics of electrical soundings belonging to this family:

Table 5. Geoelectric characteristics of the family 1

	Electrical levels	Resistivity (Ohm.m)	Thicknesses (m)	Geological formation
RT level coverage	R1	102	10	Whitish dolomitic limestones
	C1	75	24	Red clay + sandy yellow marl
RT Resistant	RT	1420	12	Basalts of the Permo-Triassic
Deep conductor Cp	Cp	2.5	18	Yellowish Marl
Deep Resistant Substrate	Rp	5000	-	Black shales (Primary)

These electrical soundings were carried out on the outcrops of the lower Jurassic carbonate formations (lower Lias). The analysis of the diagram of these electrical soundings made it possible to highlight two deep resistant levels having resistivities of the order of 1420 Ohm.m and 5000 Ohm.m (**figure 6**).

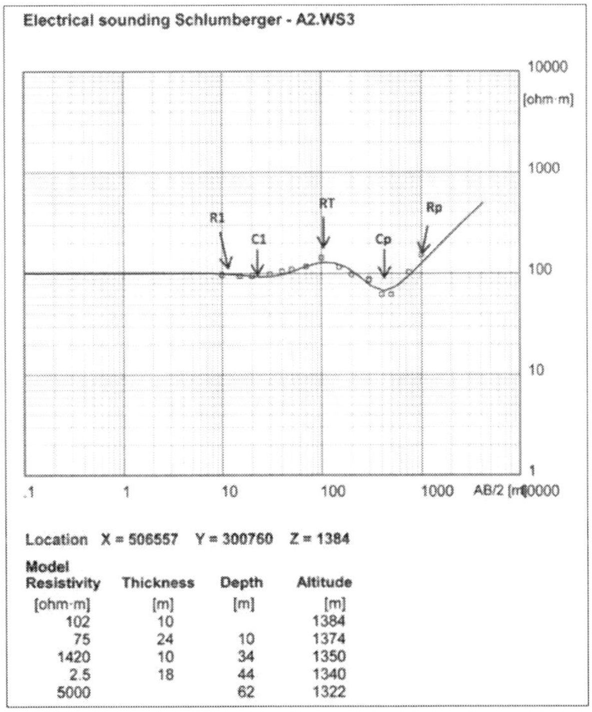

Fig. 6. Electrical resistivity curve of the electrical sounding No.IRE 1775-22.

b) Family 2:

This family is characterized by the thickening of the level coverage. It consists essentially of alternating conductive and resistant levels of the lower and middle Jurassic whose total thickness is of the order of 150m (**Table 6**).

Table 6. Geoelectric characteristics of the family 2

	Electrical levels	Resistivity (Ohm.m)	Thicknesses (m)	Geological formation
RT level coverage	R0	70-150	26	Yellowish marl + marno-limestone + limestone fractured dolomitic limestone + gray marl (lower and middle lias)
	C0	2	5	
	R1	120-200	20	
	C1	35-500	34	
RT Resistant	RT	1000	12	Basalts of the Permo-Triassic
Deep conductor Cp	Cp	50	22	Gray marl
Deep Resistant Substrate	Rp	2000	-	Black shale (Primary)

The analysis of the diagram of this electrical survey allowed to individualize a succession of electric grounds (RT, Cp and Rp). The deep resistive RT substrate with an electrical resistivity of 2000 Ohm.m corresponds according to the geology of the region and the lithological cut of the rigging drilling to shale formations of the Primary. The latter is based on a conductive level Cp (50 ohm.m) whose thickness is 22m.

The set of Rp and Cp lands is surmounted by a resistant level noted RT, the value of the electrical resistivity is of the order of 1000 Ohm.m. This resistant level is attributed to Permo-Trias basalts.

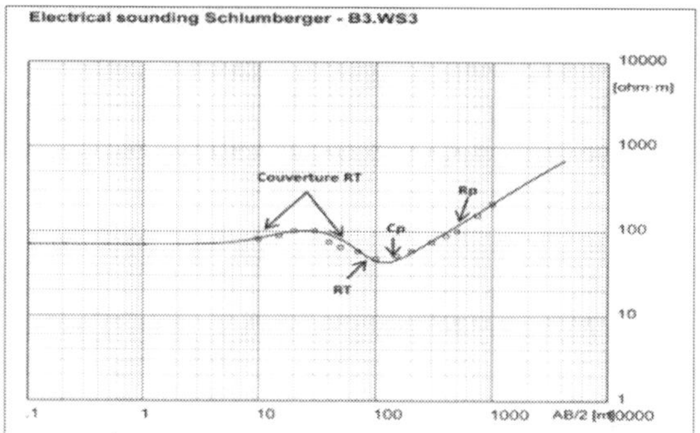

Location X = 511166 Y = 297066 Z = 1823

Model Resistivity [ohm·m]	Thickness [m]	Depth [m]	Altitude [m]
70	6		1823
150	20	6	1817
2	5	26	1797
120	5	31	1792
30	4	36	1787
200	10	40	1783
50	5	50	1773
500	51	55	1768
35	24	106	1717
1000	12	130	1693
50	22	142	1681
2000		164	1659

Fig. 7. Electrical resistivity curve of the electrical sounding No.IRE 272-30.

c) Family 3:

These electrical soundings have the characteristic that they do not show the resistant level RT attributed to basalts of Permo-Trias. The Jurassic cover lies directly on an impervious substratum consisting of triassic clays and marls **(Table 7)**.

Table 7. Geoelectric characteristics of the family3

	Electrical levels		Resistivity (Ohm.m)	Thicknesses (m)	Geological formation
RT level coverage	R0	c_0	15-62	56	Yellowish marl + marno-limestone + limestone fractured limestone gray marl plastic + hard limestone (lower and middle lias)
		r_0	160	31	
	C0		18	5	
	R1		100	43	
	C1	c_1	4	11	
		r_1	200	7	
Deep conductor Cp	Cp		25	22	Gray marl (Permo-Triassic)
Deep Resistant Substrate	Rp		550	-	Oil Shale (Primary)

The interpretation of these electrical soundings and their correlations with the lithological succession given by the corresponding calibration holes made it possible to distinguish **(figure 8)**:

- A deep resistant substrate Rp whose resistivity is equal to 550 ohm.m. Its roof is located at a depth of 157m; and which, according to the description of the drilling, corresponds to oil shale of primary age.
- A conductive level Cp, attributed to gray marls of Permo-Trias whose resistivity value is of the order of 25 Ohm.m.
- A conductive level cover Cp formed of a succession of conductive and resistant levels.

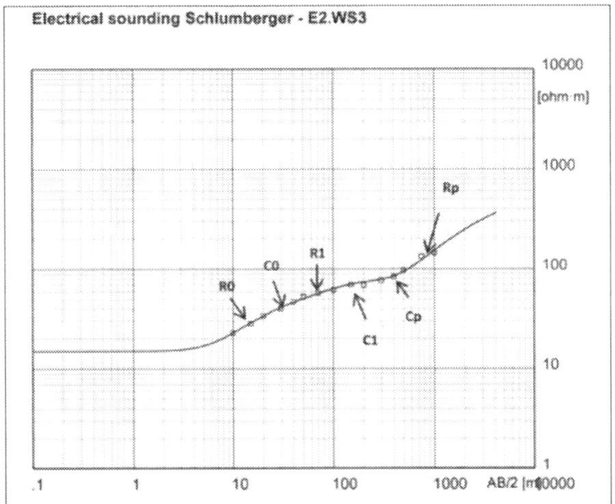

Location X = 522206 Y = 294384 Z = 1920

Model

Resistivity [ohm·m]	Thickness [m]	Depth [m]	Altitude [m]
15	5		1920
62	51	5	1915
160	31	56	1864
18	5	87	1833
100	43	92	1828
4	11	135	1785
200	7	146	1774
25	4	153	1767
550		157	1763

Fig. 8. Electrical resistivity curve of the electrical sounding No.IRE 264-30.

3.3.2 Geoelectric cuts

On the one hand, the geoelectric sections make it possible to visualize in a clear manner the structure in depth of the study area and to follow the lateral and vertical variation of the geological formations, and on the other hand to determine the roof and the wall of all formations likely to constitute aquifers.

Two types of geoelectric cutting were carried out as part of this work, the North-South longitudinal cross-sections and the West-East cross-sections. In order to better follow in depth the variation of the different electrical levels detected, two longitudinal NE-SW geoelectric cuts and two NW-SE cross sections were made. (**figure 9**):

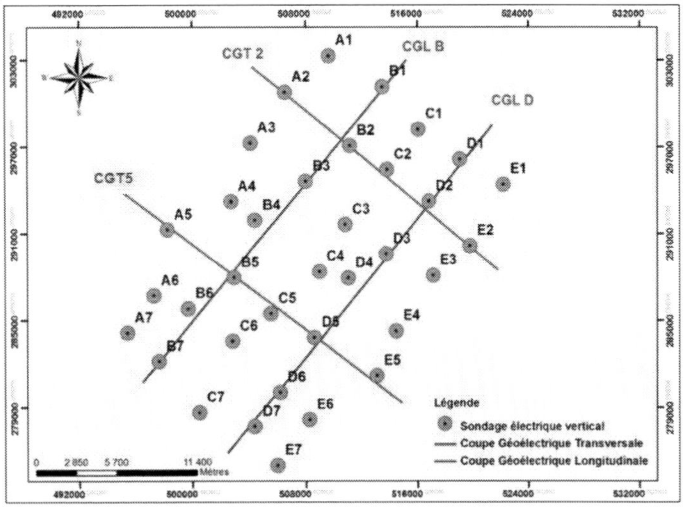

Fig. 9. Plan de situation des coupes géoélectriques.

- the longitudinal geoelectric section B comprises the electrical soundings B1, B2, B3, B4, B5, B6 and B7;
- the longitudinal geoelectric section D comprises the electrical soundings D1, D2, D3, D4, D5, D6 and D7;
- the transverse geoelectric section 2 comprises the electrical soundings A2, B2, C2, D2 and E2;

 https://doi.org/10.1051/e3sconf/20183705001

- the transverse geoelectric section 5 comprises the electrical soundings A5, B5, C5, D5 and E5.

a) Longitudinal geoelectric section B (CGL B)

This section was made on the outcrops of carbonate formations of Middle Lias (Middle Jurassic) over a length of about 25 Km (**figure 10**).

The analysis of the geoelectric section shows the following information:

- The roof of the aquifer levels (R0, R1, RT and Rp) shows a wave structure over the entire section.
- The deep resistant substratum noted Rp attributed to Paleozoic schistose formations is surmounted by a conductive level (Permo-Triassic clay loams), whose thickness varies between 18m and 25m, and its resistivity oscillates between 30 Ohm.m and 55 Ohm.m.
- The roof of the four aquifer levels shows a deepening from the B3 electrical survey towards the South-West.
- Lower Lias aquifer in calcare dolomitic formations (noted R1) shows deepening and thickening to the southwest.

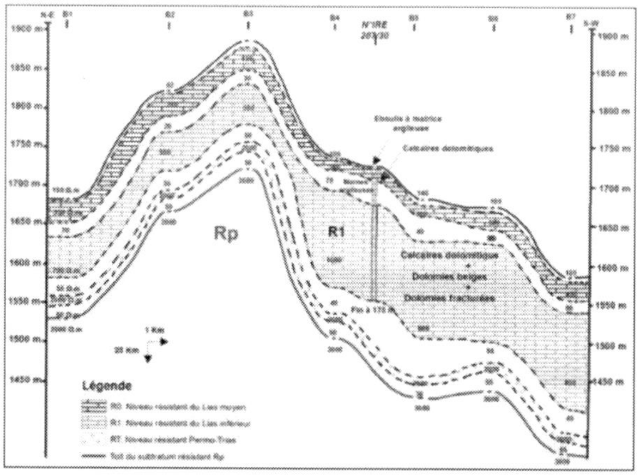

Fig. 10. Longitudinal geoelectric section B.

b) Transverse geoelectric section 2 (CGT 2)

This cut was made in the south of the study area over a length of about 17km. It crosses the outcrops of the lower and middle Lias (**figure 11**).

The analysis of this section made it possible to deduce the following information:

- Resistant level R1 outcrops to the northwest. It shows a significant burial in the electrical sounding A2.sa electrical resistivity varies between 100 Ohm.m and 300 Ohm.m ;
- The roof of the resistant bedrock Rp shows a progressive burial going from the South-East towards the North-West towards the outcrops of the Lower Lias where the depth of its roof is of the order of 64m ;
- An electrical discontinuity at the contact of the RT level and the deep conductor level noted Cp was highlighted. This discontinuity could also be interpreted in terms of fault.

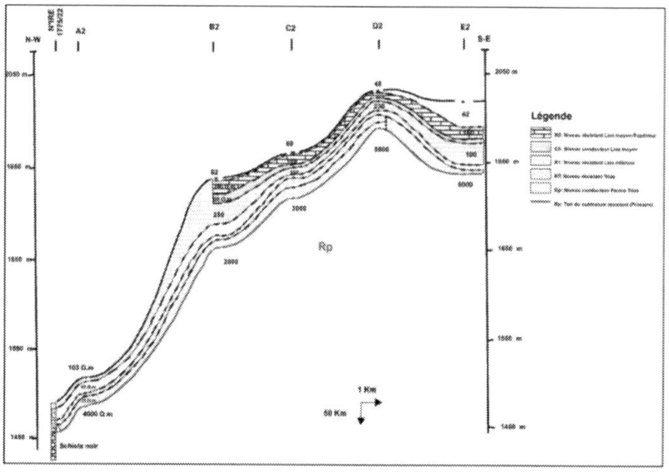

Fig. 11. Transverse geoelectric section 2.

3.3.3 Geoelectric maps

The isopach maps are based on the thickness values of the main aquifer levels recorded on the geoelectric sections: the resistant levels R0; R1 and RT.

This map (**figure 12**) tracks the variation of the first level aquifer power R0 attributed to fractured carbonated and dolomitic lands of the middle and upper Lias.

The analysis of this map makes it possible to distinguish:

- A shallow area located northeast and southwest of the study area;
- An area of high thickness values located mainly in the extreme south of the study area where the thickness has a maximum value of 52 m.

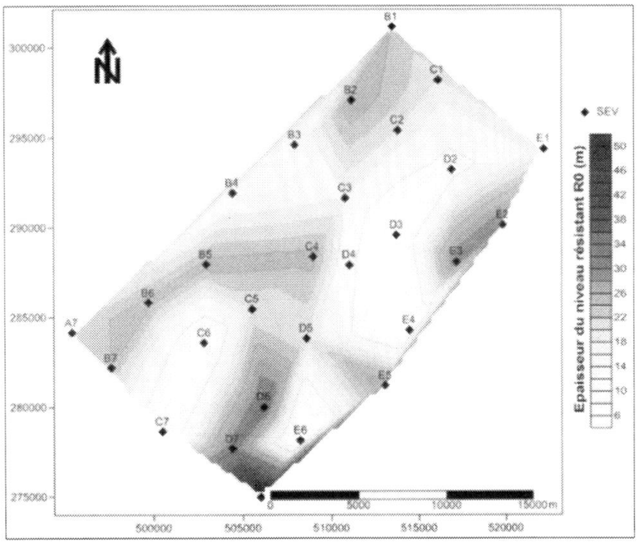

Fig. 12. Isopach map of resistant level R0.

The isopach map of aquifer level R1 (**figure 13**) shows that the thickness of this level varies between 10m and 110m.

Examination of this map made it possible to distinguish:

- A high-power area, located at the northeastern and southwestern end of the study area.
- A low to medium area in the northern and central parts of the study area.

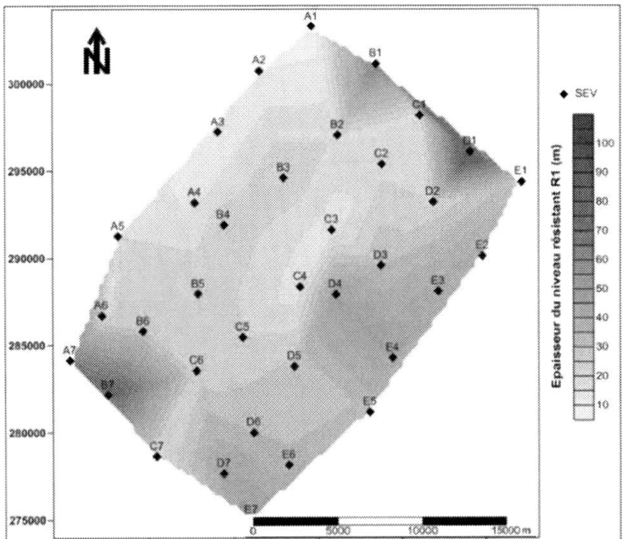

Fig. 13. Isopach map of the resistant level R1

Examination of this map shows that the ranges of the thickness values of this aquifer level RT oscillate between 6m and more than 56m. The maximum thickness of this reservoir is observed at the southwestern end of the study area, while the low thicknesses (thickness ≤ 22m) are distributed north and west of the study area.(**figure 14**).

Overall, aquifer level R1 shows a thick zone located in the southwest and a thin zone located northeast of the study area.

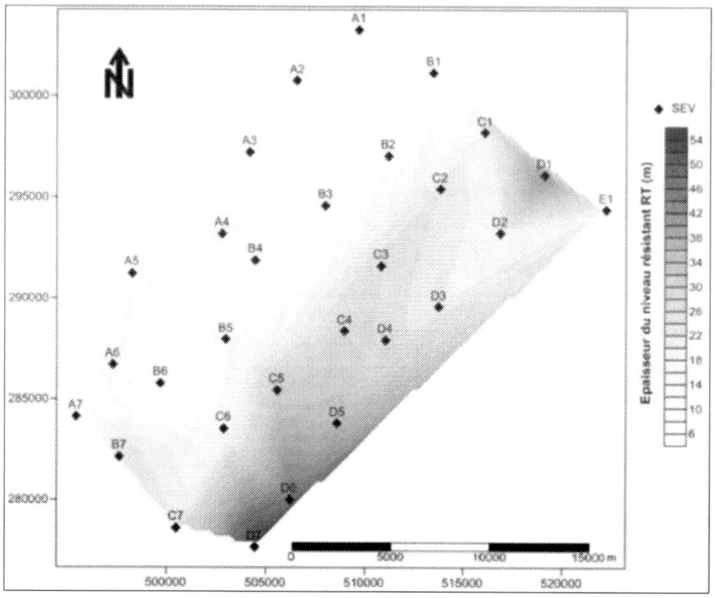

Fig. 14. Carte des isopaches du niveau résistant RT

The isohypse maps were developed from the roof elevation values of the aquifer levels found on the geoelectric sections (R0, R1, RT and Rp).

This map is based on the values of the aquifer-level wall coast recorded on geoelectric sections (**Figure 15**).

Examination of this map allowed showing:

- that the roof moves progressively from the northwest to the southeast of the study area;
- that the high altitude values of the roof recorded in the south-east of the study area reflect the raising of the roof of this level towards the surface at this location;
- a rise of the roof of this aquifer R0 towards the South-East in the direction of the outcrops of dolomitic-limestone formations of Lias medium-superior.

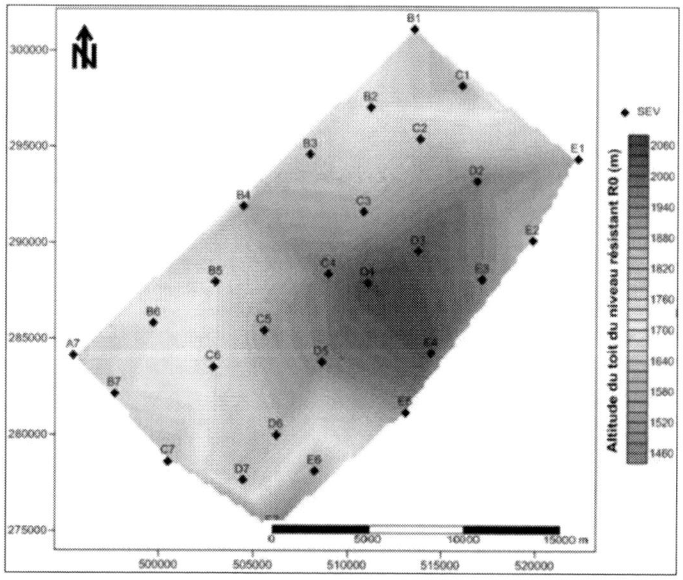

Fig. 15. R0 aquifer roof roof isohypsis map

The following map shows that the roof of this aquifer level varies from 1350 m and more than 2050m. This resistant level corresponds according to the geoelectric correlation with limestone and dolomitic formations of the lower Lias. The analysis of this map (**figure 16**) shows:

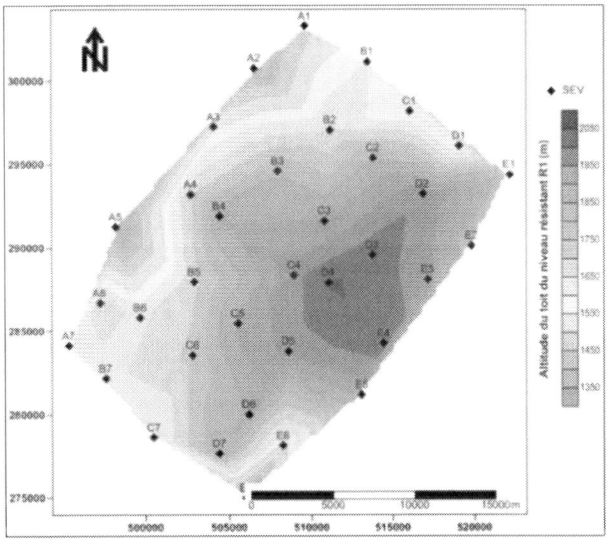

Fig. 16. Map of the isohypse of the roof of aquifer level R1

- That the high altitude values are recorded in the north-west of the study area while the low values are observed in the south-east of the study area ;
- The roof of this aquifer level shows a rise from north-west to south-east of the study area ;
- The high values recorded in the South-East reflect the rising of the roof of this aquifer level in these places.

This map (**Figure 17**) is based on the values of the roof coast of the aquifer level RT recorded on the geoelectric sections. Examination of the map of the map made it possible to highlight:

- A low elevation area (≤1600), located in the northwestern part of the study area ;
- An area located in the south-east of the study area, where the roof of the aquifer reaches more than 1950 m altitude;
- The roof of this aquifer level RT is gradually rising towards the south-east of the study area.

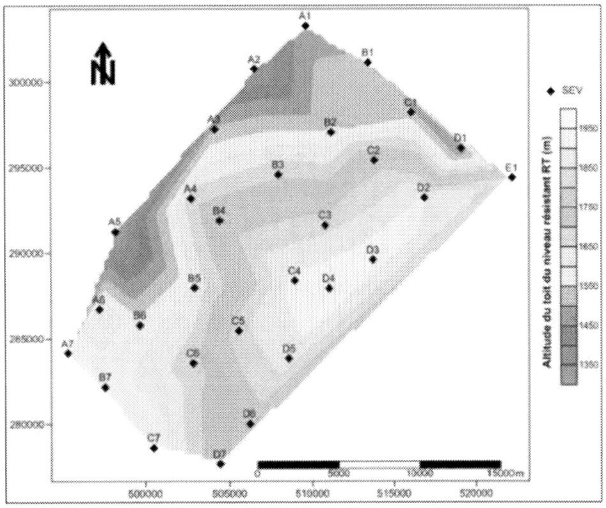

Fig. 17. Map of the isohypse of the roof of resistance level

The following map shows the elevation variation of the roof of the Rp resistant bedrock attributed to Paleozoic black shale. This variation extends from 1300 m to a maximum value of 2000 m (**figure 18**).

Areas of low altitude are observed south-west of the study area, these areas correspond to areas where the roof of the resistant bedrock is close to the surface, ie they correspond to the low-power areas of the area blanket.

Overall, the resistant level Rp shows a progressive deepening from the North-West to the South-East of the study area.

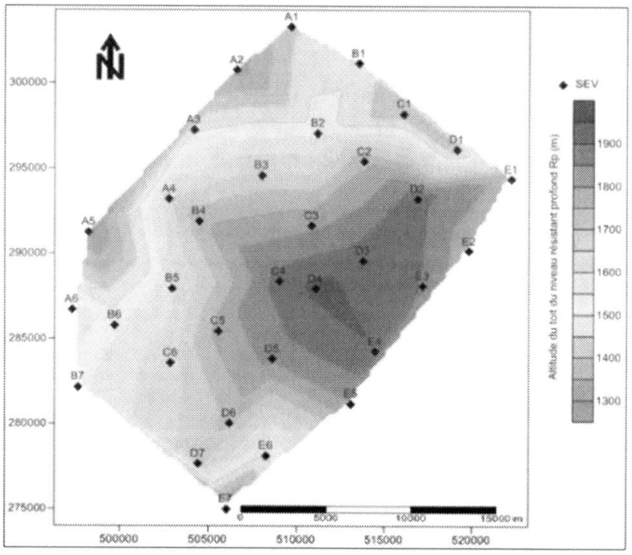
Fig. 18. Map of the isohypse of the Rp aquifer level

It appears from the analysis of the results obtained that:

- For the aquifer level of the lower-middle Lias (the first two potentially aquiferous levels R0 and R1 forming the deep-bedrock cover) :
  - ⇒ The aquifer-level roof of the Lias shows a wave structure that rises to the south-east of the study area.
  - ⇒ This aquifer level shows a thickening towards the south-east of the study area, in the direction of the outcrops of the limestone and dolomitic formations generally fractured of Lias inferior-average.
  - ⇒ These outcrop zones constitute the natural recharge zones of this layer.
  - ⇒ Cracking and karstification of limestone formations forming this aquifer favoring the infiltration and circulation of rainwater and subsequently its feeding.

- For the RT aquifer level:
  - ⇒ This second aquifer level corresponds to altered basalts Permo-Trias.il shows a thickening towards the South-East of the study area where its power holds a maximum value of 60m.
  - ⇒ The roof of this level has a wave structure that plunges to the northwest of the study area.

- For the RP aquifer level:
  - ⇒ This deep aquifer level, attributed to Paleozoic schistose formations.
  - ⇒ This resistant substratum shows a progressive deepening from the North-West to the South-East of the study area (towards the Lias and Turonian outcrops).

4 Conclusions

The electrical resistivity method applied to the recognition of underground aquifers is a method that can be adapted to both regional and local studies. In fact, in the regional case, it makes it possible to follow the evolution of petro-physical, geometrical characteristics, etc. in addition to the fact that it allows to locate the areas with potential in waters. In the case of local prospecting, it allows at least to define the vertical succession of existing aquifers in addition to the local petrophysical characteristics that can be compared with those of the closest rigging holes.

The comparison of drilling data with the different geophysical surveys carried out on our study area makes it possible to draw the following conclusions:

- Reveal the existence of four resistant electric horizons R0; R1; R2; RT and Rp.
- Define the geometry of these aquifer levels by drawing the isopach maps ;
- Follow laterally and vertically the structure and extension of these aquifer levels.
- To monitor the importance and hydrogeological productivity of these aquifer levels by mapping their transverse resistance RT.

Thus, we confirm by our results that the electrical resistivity method is an important guide for the recognition of underground aquifers and for the implantation of mechanical soundings of recognition by increasing their chance of success and thus minimizing the cost of recognition.

5 acknowledgments

We thank the management of the Oum Er-Rbia Hydraulic Basin Agency for providing us with the data used in this article as well as for their help and constructive suggestions. We also thank all those who contributed directly or indirectly to the development of this work.

6 References and notes

1. ASTIER J.L. Géophysique appliquée à l'Hydrogéologie. Masson et Cie, Editeurs, Paris, 1971.
2. Abderahmane BENTAYEB et Claude LECLERC, Le causse moyen atlasique. Ressource en eau du Maroc, Tome 3 : Moyen Atlas et Causse, 1977.
3. BOUKDIR A., Etude hydrogéologique en vue d'une modélisation du système aquifère du Tadla-Plateau des phosphates. CEA, Fac. Scien. Semlalia Marrakech, 1990.
4. BOUKDIR A., Etude géothermique du bassin du Tadla-Plateau des phosphates. DES, Fac. Scien. Semlalia Marrakech, 1994
5. BOUKDIR A. et Al, Prospection des eaux souterraines par méthode des résistivités électriques-Approche méthodologique et etude de cas dans le bassin de l'Oum Er R'Bia. Rencontre national des géomorphologues marocains, Faculté des Lettres et sciences humaines, Béni Mellal, Maroc, 2002.
6. DRPE, Etude par sondages électriques – plaine de Tadla. GEOATLAS SA, 1992.
7. DOBRIN M.B., Introduction to geophysical prospecting. 3rd ed.: New York, Mc Graw Hill Book Co., Inc, 1976.
8. LAGABRIELLE R., TEILHAUD S., Prospection de gisements alluvionnaires en site aquatique par profils continus de résistivité au fond de l'eau, 1981, Bull. liaison labo. P. et Ch. 114:17-24.
9. MEYER DE STADELHOFEN C., Application de la géophysique aux recherches d'eau. Tec. & Doc. Lavoisier, Paris, 1991.
10. TELFORD W.M., GELDARD L.P., SHERIFF R.E., KEYS D.A., Applied geophysics. New York, Cambridge University Press, 1976
11. MECHLER P., Les méthodes de géophysique. Dunod, Bordas, Paris, 1982.
12. ZHODY A. A. R., A new method for the interpretation of schlumberger and wenner curves, 1989, géophysics, vol. 54, pp. 249-253.
13. G Marquis. Prospection électrique. 2005,5-11.
14. Berengere Charnay, Pour une gestion intégrée des ressources en eau sur un territoire de montagne. Le cas du bassin versant du Giffre (Haute-Savoie). 2010 pp 18.
15. MARTIN J. - Le Moyen Atlas central étude géomorphologique. Notes et Mémoires du Service géologique, 1981 , 1-447.

16. BAALI A. Genèse et évolution au Plio-Quaternaire de deux bassins intramontagneux en domaine carbonaté méditerranéen. Les bassins versants des dayets Afourgagh et Agoulmam (Moyen Atlas, Maroc). Thèse d'État, Université Sidi Mohamed Ben Abdellah, Fès, 1998.
17. Gentil, 1916 ; Russo, 1939 ; Termier, 1936; Dubar, 1954 ; Golo, 1961

E3S Web of Conferences **37**, 05002 (2018) https://doi.org/10.1051/e3sconf/20183705002
EDE6-2017

Contribution of gravity data and Sentinel-1 image for structural mapping. Case of Beni Mellal Atlas and Beni Moussa plain (Morocco).

Ikram Boutirame[1,*], *Ahmed* Boukdir[1] , *Ahmed Akhssas*[2] , *Fatima* Boutirame[3] , Ahmed Manar[4] and *Brahim Aghzzaf*[5]

[1]Laboratory of Industrial engineering, Faculty of Sciences and Technics, Sultan Moulay Slimane University, Beni Mellal, Morocco.
[2]Laboratory of Applied Geophysics, Geotechnics, Engineering Geology and Environment, Mohammadia School of Engineers, Mohammed V University, Rabat, Morocco.
[3]Faculty of Medicine and Pharmacy, Mohammed V University, Rabat, Morocco.
[4]Ministry of Energy, Mines and sustainable development, Rabat, Morocco.
[5] Oum Er'Rbia Hydraulic Basin Agency, Beni Mellal, Morocco.

Abstract. The present work is a combined study of gravity and Sentine-1 data for fracture mapping in the karstic massif of Beni Mellal Atlas and the adjacent plain of Beni Moussa. In order to locate the various faults that contribute to the study area structuring, the gravimetric contacts analysis method, based on the joint use of the horizontal gradient and the upward continuation at different altitudes, has been applied to the gravity data. To optimize the structural mapping in the study area, the gravimetric lineaments obtained were completed and correlated with the lineaments got from Sentinel-1 image. Four faults families of NE-SW; E-O; N-S and NW-SE directions have been highlighted. There fault families are perfectly combined with the studied area's surface water network, moreover, they corroborate with the previous geological and structural studies.

* Corresponding author: i.boutirame@usms.ma

1 Introduction

The northern border of Beni Mellal Atlas is a complex zone where two different geological domains cross via thrusts with various compositions [1]: Beni Mellal Atlas and Tadla plain. This junction zone has been an abject of numerous geological studies [2, 3, 4, 5, 6] which remain incomplete to this date, especially in structural mapping; because of the rugged landform that dominates this zone as well as a thick vegetal cover boosted by a semi arid climate. In such conditions of difficult field mapping, the new techniques of remote sensing and geophysical data allow to highlight geological structures and to realize the mapping of fractures, usually, hidden by vegetations.
The remote sensing with its possibilities of synoptic view, represent a key tool for lineaments detection [7-8]. The use of passive remote sensing has been described in morocco by many authors in different geographical contexts [9, 10, 11, 12, 13, 14, 15, 16, 17, 18]. However, this technique remain strongly influenced by atmospheric conditions which affect the quality of acquired images and make hard the lineaments detection, in contrast with radar images -independent of atmospheric conditions[19]- which provide important informations in structural mapping. Geophysical prospecting methods are widely used in litho-strucural mapping studies; they allow to decrypt different geological structures signatures and localize tectonic accidents in surface that define structural architecture of studied area.
The present study represent a combined analysis of data provided by sentinel 1A radar image and gravity to optimize lineaments network mapping which cross the study area. Gravity data has been treated and analyzed in order to extract gravimetric lineaments which will be correlated and completed by the discontinuities deduced from sentinel 1A radar GRD type (Ground Range Detected) image analysis and filtering. This type of image is recognized as having a great potential, generally in geological applications and particularly in structural mapping [20, 21, 22]. Numerous examples in literature demonstrate the effectiveness of geophysical and satellite data coupling method in fractures detection and litho structural maps updating [23, 24, 25, 26, 27, 28].
The study area belongs to the superior basin of Oum Er'Rbia, it covers a surface of 2850 km2, its width includes the zone between El Ksiba et ait attab and crosses dozen of kilometers of Beni Mellal atlas mountains in south-east and Beni Moussa plain in north west (left edge of Tadla plain). It covers two different geological domains (Fig.1):

- In the south, Beni Mellal atlas is represented by the north occidental edge of central High Atlas [29, 2, 30]. It is located in the south of Tadla plain and the outskirts of the junction between high and middle atlas. The altitudes decrease from the south east to the North-West, They pass from 1000 m (in el Ksiba) in the atlas border to less than 500 m in Tadla plain. Its highest point is Jbel R'Nim that reaches 2411 m;

- The Tadla plain with altitudes limited between 400 and 700 m is shown like an asymmetric wide depression with ENE-WSW direction [31]. It extends to the phosphate plateau in the north and Beni Mellal atlas in the south. With its monotonous landscape, the Tadla plain is drained by Oum Er'Rbia river which divides it into two asymmetric sectors. Beni Amir in the right side and Beni Moussa in the left one.

Between the mountainous zone of Beni Mellal Atlas and the Tadla plain, there is a slim band of fertile lands named piemont zone or Beni Mellal Dir [32-33]. This rough contact between a glacis in slight slope and steep terraces which dominate it occupies fifteen kilometers of width in the south-west and become narrows in the north east. Along the northern karstic massif border, many springs row along the thrust fault of Tadla, they emerge from Lias contact at the level of tertiary thrusts and travertine of Dir (Fig.4a and b). Finally, they provide an abundant resource of good quality water which is exploited for drinking water supply and agricultural irrigation.

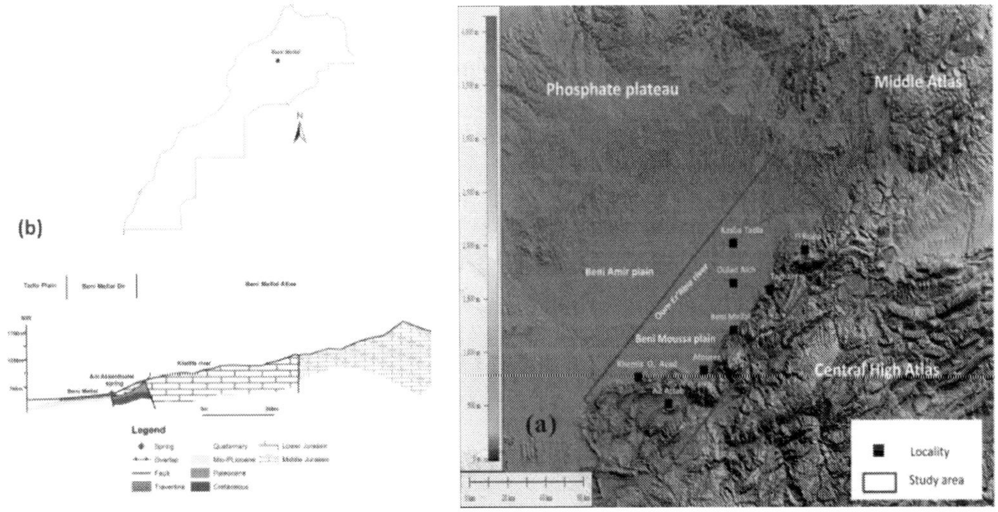

Fig.1. Location of study area. **a**) Digital Elevation Model of the study area , **b**) NW-SE geological cross-section showing the Atlas overlap on the Dir and Beni Moussa plain formations.

2 Geological and hydrogeological setting

2.1 Geological setting

Beni Mellal Atlas is essentially constituted by a dolomitic massif facies from Middle and Lower Jurassic which is strongly fractured and karstified (Fig.4c and d). Facies lateral variations are observed when moving from the basin (marly limestones) to the platform (dolomitic limestones) [3-6]. Stratigraphic column starts with the Permo-Triassic and ends by the quaternary deposits. Its current structure is the result of tectonic movements occurring in hercynian and alpine phases [4, 34, 35, 36]. Therefore, the carbonate slab has been folded to the North-West, then it overlapped plio quaternary series of the plain [2, 37, 38, 39] (Fig.2).

To the north, Beni Mellal High Atlas dominates abruptly Beni Moussa plain because of faults system. The most important one is the north atlasic fault (Tadla overlap). Armed with a thick liassic limestone series, it leads up atlasic structures to Tadla plain (Fig.1c). Tadla plain is a wide dissymmetric depression that is oriented WNW-ESE [40, 41, 31]. Its axis is located in atlas border and support strong series of deposits moving from Triassic to Quaternary.

At the structural level, Beni Mellal karstic massif is fractured by a faults system oriented towards two main directions: NE-SW to ENE-WNW (N40-70°) and N-S (N150-180°). This system of faults facilitates atlas waters drainage to the subatlasic formations of the plain and allows their contact, at some points, with permeable facies. This contact makes easy hydraulic connections either by underground circulation, or by re infiltration of Beni Mellal complex atlas-dir waters springs [39].

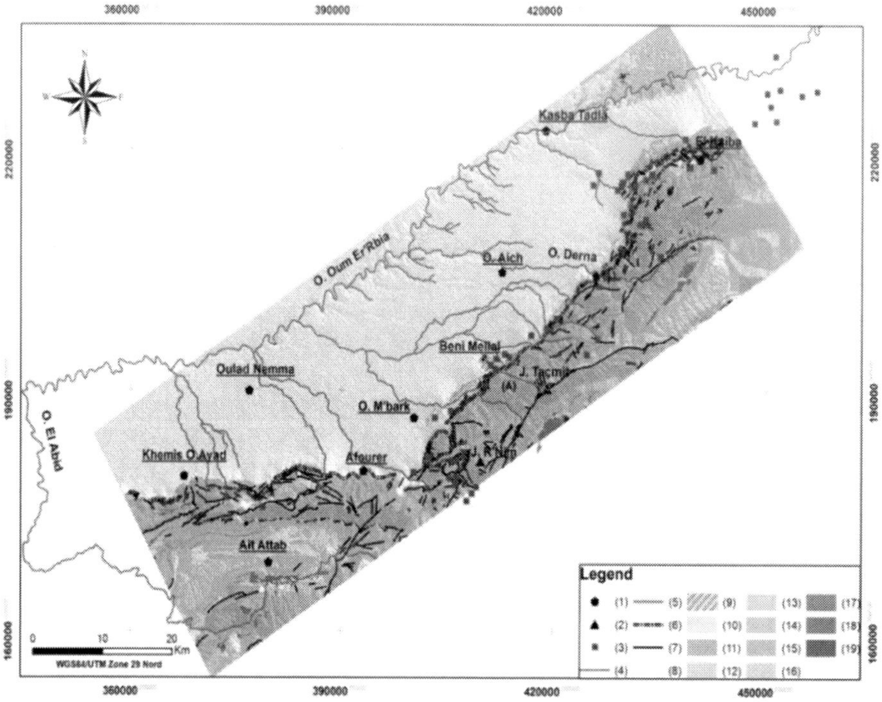

Fig.2. Geological map simplified of studied area [3, 5, 42].
(1) locality; (2) mountain ; (3) spring; (4) hydrographic network; (5) geological cross-section; (6) Tadla overlap; (7) fault; (8) Quaternary (alluviums + loams); (9) travertine; (10) Mio-Pliocene; (11) Paleocene-Eocene; (12) Cretaceous; (13) Middle Jurassic (Dogger); (14) Toarcian-Aalenian; (15) Lias (dolomitic and marly limestones); (16) Triassic (clay and pink marl); (17) Cretaceous basalts and dolerites ; (18) Middle Jurassic Igneous rocks; (19) Upper Triassic altered basalts.

2.2 Hydrogeological setting

In Beni Mellal atlas, the aquifer is represented by a karstic groundwater which is constituted by dolomitic limestones from Lower Lias [43]. It is drained by many springs which emerge from Tadla overlap. The most important one is ain Asserdoune spring (1,1 m^3/s) (Fig.3a). This groundwater is supplied by rain waters, snow melting, waters originated from upstream Oued Derma basin and waters of Tasmit hinterland [38].
This karstic aquifer overlaps Dir thrusts that are fragmented into many little reservoirs, and thus, the transition can be possible with other Tadla plain aquifers which have been a subject of many hydrogeological studies [31, 38, 43, 44, 45, 46].
Those different aquifers observed in Tadla plain have a special organization. It consists in a multilayer system that is represented by four aquifers with an important hydrogeological potential:

- Turonian aquifer;
- Senonian aquifer;
- Eocene Aquifer;
- Tadla phreatic groundwater (Mio-Plio-Quaternary aquifer).

Turonian underground water is a reservoir with an important hydrological potential. It contains water resources strongly used in drinking water supply and industry [48-49].The thickness of Turonian formations increase by moving from north to south. They measure 20 m in their outcrop and reach 80 m nearby the atlas.

Fig.3. Pictures showing the fractured limestones and the emergence context of some springs. a) Ain Asserdoune spring; b) Tamegnounte spring; c) Liassic fractured limestones; d) El Ksiba thrusts.

3 Materials and methods

3.1 Gravity data

The gravity data used in this study result from a measurement campaign carried out in 1961 by the African Company of Geophysics (CAG), for the Mines and Geology department (Ministry of Energy, Mines, and Water). The data are available in the form of a 1/200000 scale Bouguer anomaly map with a contour interval of 5 mGal and a reduction density of 2.2 g / cm3. This map was scanned and digitalized by using ArcGIS software (version 10.3) before proceeding with the various filtering techniques using Geosoft's Oasis Montaj software (version 8.2).

The Examination of Bouguer anomaly map shows local variations in the gravity field moving from -98 mGal to -47 mGal, which are clearly displayed as an increasing regional gradient from North-West to South-East (Fig.4). This regional anomaly, assimilated to a plane, was estimated by using the polynomial regression method and subtracted from the Bouguer anomaly in order to calculate the residual anomaly (Fig.7).

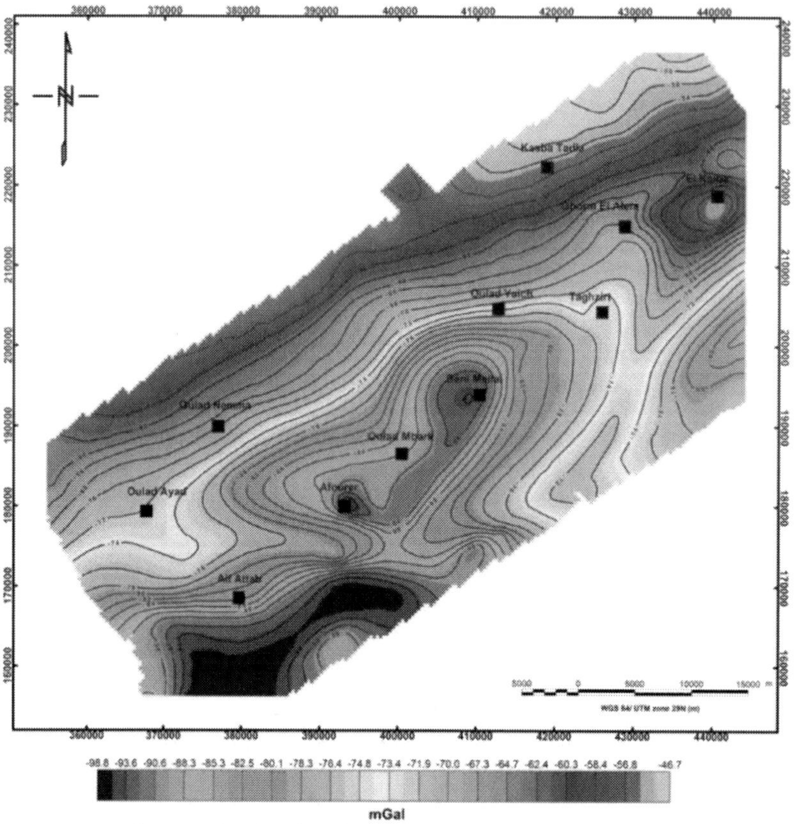

Fig.4. Bouguer anomaly map (reduction density d=2.2; interval = 2 mGal).

The residual anomaly map shows several gravity gradient areas (high and low trend gravity) that may correspond to contacts or discontinuities (faults, flexures, etc.). In order to make better use of this gravity data and to bring out more informations compared to those can be drawn from the residual map, different filtering techniques have been applied to the gravity data to highlight the major structural axis that affect the study area [50].

We have applied the analysis contacts method based on the joint use of the horizontal gradient and the upward continuation (Fig.6). The interest of this method has been mentioned for a long time [51]. The success of this method's application has been demonstrated by numerous previous studies [51, 52, 53, 54, 55, 56, 57]. The principle of this technique is to compute the horizontal gradient of the residual anomaly and its upwards continuations at different altitudes. This treatment is very useful for localizing the geological contacts because the limit between two blocks of different densities corresponds to this gradient maxima [58, 59, 60, 61]. Indeed, above a vertical contact between two rocks of different density, the gravity field is manifested by a level's change marked by the passage of low values above low density rocks to higher values at the aplomb of high density rocks. The inflection point which marks the transition between the two zones is located at the vertical between the two types of rocks. Local maxima of the horizontal gradient are narrow lines above geological contacts marked by density contrasts [62-63] (Fig.5)

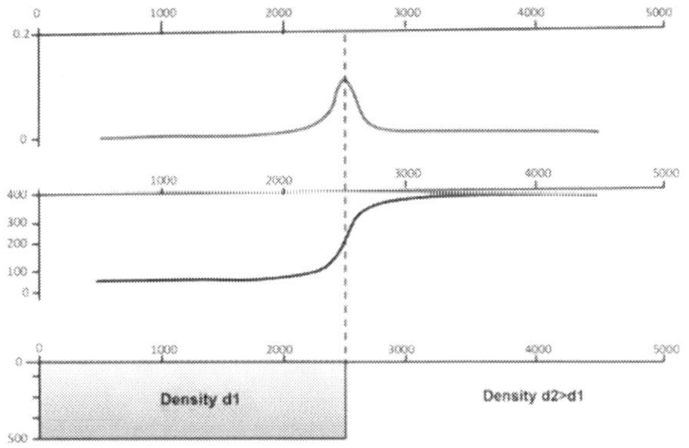

Fig.5. Representation of gravity anomaly caused by a vertical contact (black curve) and its horizontal gradient (red curve).

3.2 Sentinel-1A processing and analysis

In order to bring more precision to the structural mapping and to optimize the results obtained from the gravity data processing, we relied on the analysis of Level-1 Ground Range Detected (GRD) product of Sentinel. The radar image used in this study was acquired in 2 April 2016 in descending orbit and dual polarization VV and VH, with twenty meters (20 m) spatial resolution recorded in the band C (Table 1). It is available free on the website of the European Space Agency ESA (https://scihub.esa.int/dhus/).The different processing and filtering of the Sentinel-1A radar image were performed using SNAP ESA Sentinel-1 toolbox (S1TBX).

Table 1. Senytinel-1A data Characteristics

Specifications	Sentinel-1A data
Acquisition time	02 April 2017
Acquisition orbit	Descending
Imaging Mode	IW (Interferometric Wide)
Imaging frequency	C-band (5.4GHz)
Polarization	VV-VH
Data product	Level-1 GRD
Resolution mode	20 m (Full resolution)

A set of processing and filtering operations were applied to the Sentinel-1 radar image (Fig.6) to reduce uncertainty of the data, mainly due to atmospheric disturbances, as well as the topographic effects of the target [64] and improve the possibility of the lineaments extraction[65, 66, 67].

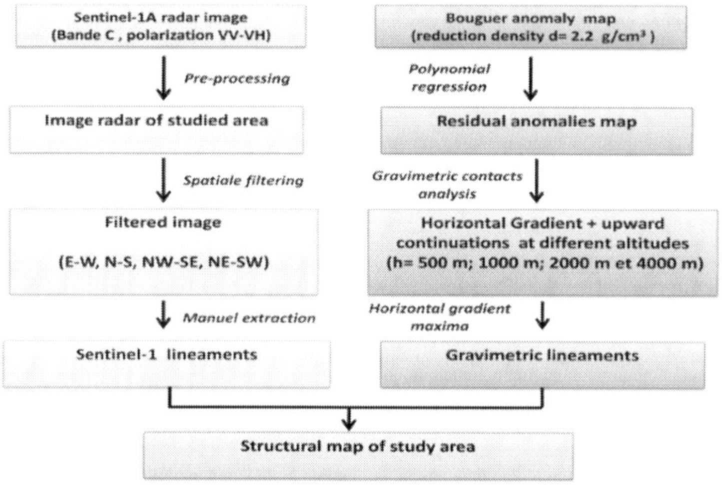

Fig.6. Flowchart of adopted methodology for lineaments extracted

In this study, the manual extraction by visual interpretation was adopted for the lineaments identification in the study area [18, 68, 69, 70, 71, 72, 73]. Those lineaments are materialized by the limits formed by the dark and clear zones lines [74-75] more or less regularized and the hydrographic network disturbances [76].Their signature in the field is very diverse, they may correspond to contacts between formations of different lithology, outcrop limits or to fractures or faults lines [77-78]. The figure 7 shows a map of lineaments extracted from the Sentinel-1A image processing of the study area.

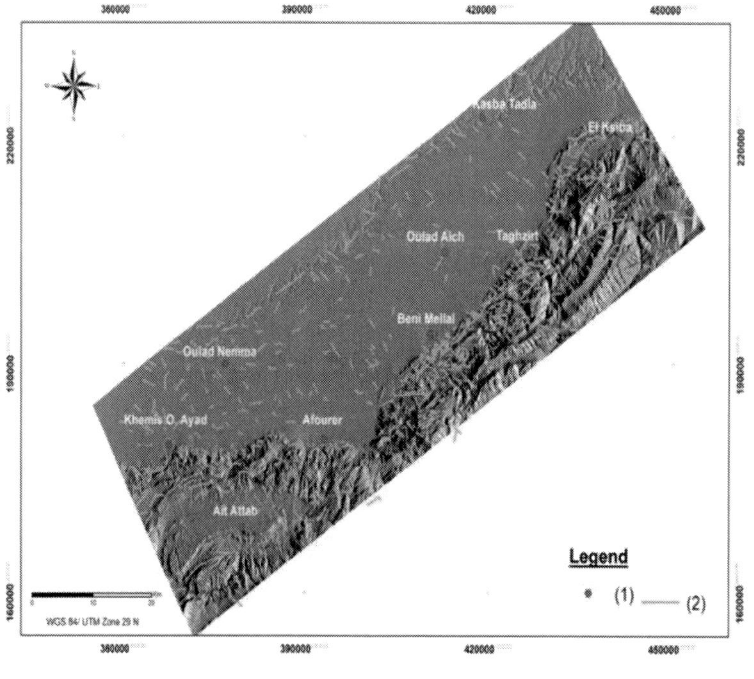

Fig.7. Map of lineaments extracted from Sentinel-1 image. (1) locality; (2) lineaments Sentinel-1.

4 Results and discussion

Residual anomalies map obtained from the regional trend removal, show values that vary between -20 mGal and more than 6 mGal (Fig.8a). Those variations are associated with underground density variations. Many positive anomalies (high trend gravity) and negative ones (low trend gravity) have been highlighted by this map. To understand their origin, we overplayed residual anomalies and study area geological maps [3, 5, 42], then we correlate those anomalies with existing borehole logs. By analyzing all this data, we can deduce the following principal structural traits:

- **Low trend gravity**

  - N1 negative anomaly: located in the map center with strong amplitude of 10 mGal and NE-SW direction, it corresponds actually to the lengthening direction of Tadla basin. It is a zone of subsidence that affects a steeply dipping socle, so, it allows a very important deposition of light tertiary and quaternary formations;
  - N2 negative anomaly: located in the north of Afourer locality, it is less important than precedent anomaly. It shows a continuation to NE (where it extends to N1 anomaly). It corresponds probably to the effect of sedimentary fill of Tadla basin with a displaced depocenter to the south nearby the piedmont of atlasic chain;
  - The two N3 and N4 anomalies, with low amplitude, correspond respectively to Middle Jurassic and cretaceous outcrops in the atlas.

The alignment of N1 and N2 negative anomalies represents Tadla basin axis, where the sedimentary cover is thick (Fig.8b). This result is consistent with the general asymmetric structure of Tadla basin. Correlations trials between existing boreholes (Fig.9) show clearly the syncline shape of Tadla plain [79-80] and the thickening of post-Palaeozoic cover towards the northern border of Beni Mellal Atlas [41, 46, 80]. The northern flank of N1 anomaly shows a strong gradient that can be explained by the presence of faults which lead the Liassic slab to overlap Dir northern and Tadla quaternary formations [30-81].

- **High trend gravity**

  - P1 and P2 positive anomalies

In the northern border of area study which corresponds to Beni Moussa plain, gravimetric axis linking P1 and P2 positive anomalies correspond to conglomerates formations represented in Tadla plain by lentils drowned in plio quaternary cover. This gravimetric axis would present a hydrogeological interest because it can play a role of underground waters accumulation and circulation seat.

  - P3, P4, P5 and P6 anomalies:

The southern part of area study is characterized by the existence of positive anomalies series (P3, P4,P5 and P6) which constitute pink to red coloured band. Those four anomalies are approximately oriented from NS-SW to ENE-WSW, they are probably caused by the liassic slab elevated roof which overlaps tertiary and quaternary adjacent plain (Beni Moussa plain)

- P3 positive anomaly : caused by the anticline effect of jbel Imouzzer located in Afourer atlas;
- P4 positive anomaly: reflects gravimetric signature of jbel Tazergount's anticline overlapping [30];
- P5 positive anomaly: lengthened following a NE-SW direction with high amplitude. It is bordered by jebl Ighnayene and jbel Tasmit, it opens up towards the south where it should continue to Ouaouizeght high atlas basin.
- P6 positive anomaly :located in the north-east of study area, it corresponds to the double effect of two anticlines respectively originated from jbel Izerfane and jbel Antar (El Ksiba Atlas)

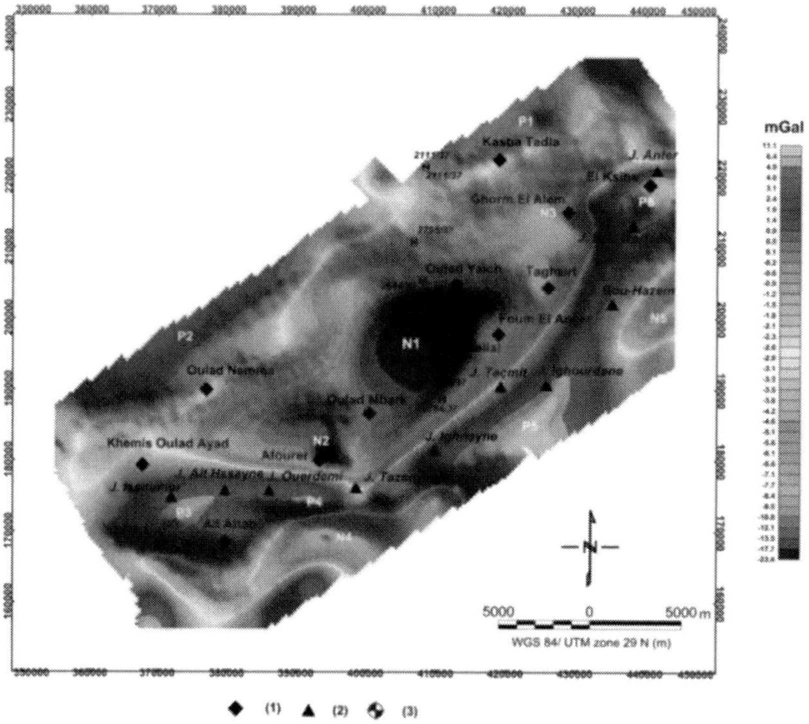

Fig.8a. Residual anomalies map of studied area. (1) locality; (2) mountain; (3) borehole.

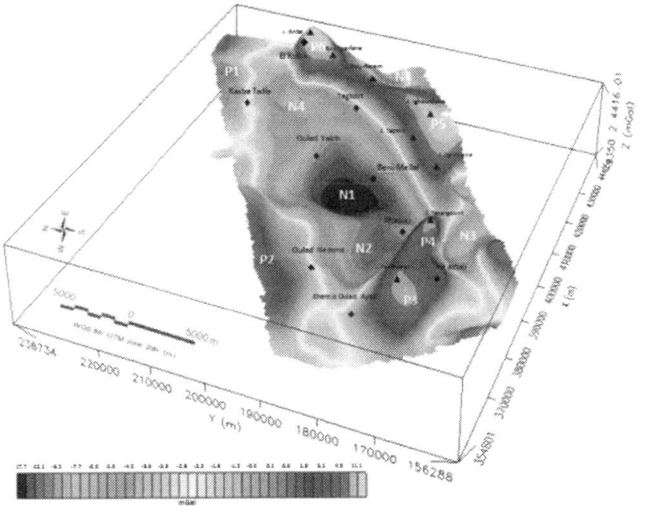

Fig.8b. 3D view of residual anomalies map of studied area.

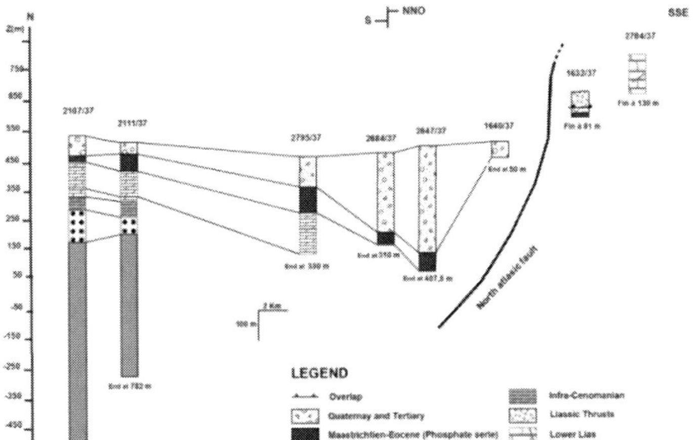

Fig.9. Correlation between boreholes BJ101, 2795/37, 2684/37, 2647/36, 1640/37, 1632/37 and 2784/37.

Indeed, this high trend gravity is consistent with Beni Mellal atlas general morphology. This mountainous chain is derived from a very open anticline [38-39] which overlaps in north-west direction. The progressive decrease of gravity anomalies values observed from South-East to North-West (towards Beni Moussa plain) is correlated with main study area structural traits (**Fig.10**). The Liassic slab that dominates Tadla plain underwent a displacement in north-west direction: a geological fact confirmed by the decrease altitude that we note in the same direction [29, 30, 38, 39, 40].

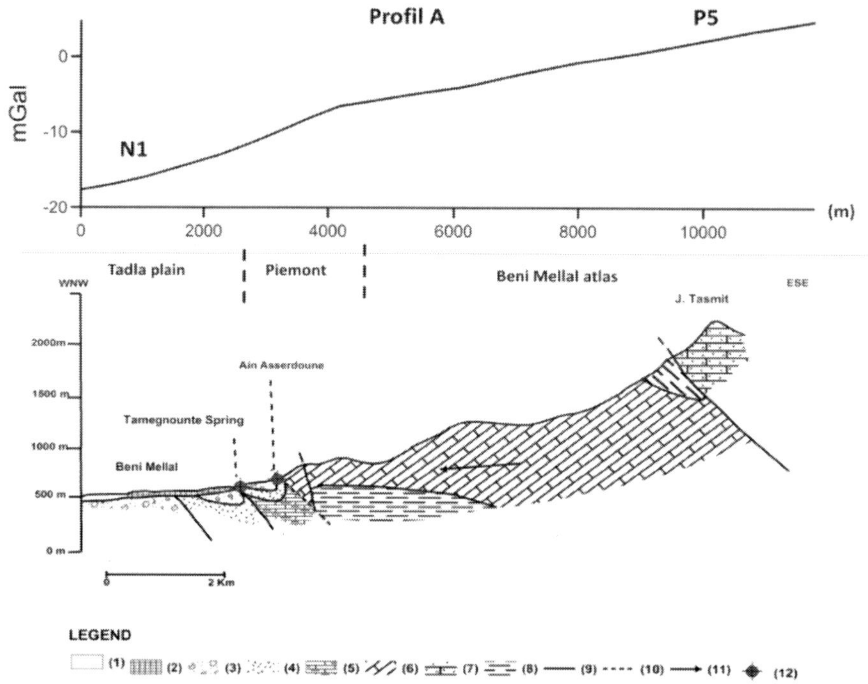

Fig.10. Correlation between the WNW-ESE geological cross-section [38] and the gravimetric profil crossing of central study area.

The results obtained from analyzing gravimetric contacts method, previously described, allow to highlight different faults which participate to structure the study area. The figure 11 summarizes all horizontal gradient maxima of the residual anomaly as such as its upward continuations at different altitudes represented by different colours. Linear contacts can correspond to faults while circular ones can represent diapers limits or intrusive bodies [60, 82, 83, 84].

Gravimetric lineaments show a polymodal directionnel distribution that is similar to the one revealed by lineaments deduced from Sentinel-1 data (Fig.12). The main lineaments directions are, preferably, oriented in NE-SW à ENE-WSW (N20-40° to N40-70°); N-S (N150-180° and N20°) ; E-O (N90°) and NW-SE (N120-140°) directions. The first two directionnel ranges are dominant in the study area, they are mainly identified in the mountainous zone that correspond to Beni Mellal karstic massif [2, 38, 39], while the latter two directions are revealed by previous geophysical studies that have been realized in Tadla plain [55]. The two directions NW-SE and NE-SW guide in surface the majority of hydrographic network as well as the main springs distribution which emerge from the northern atlasic border of Tadla plain. Indeed, [16-85] had demonstrated the influence of the brittle tectonic on the drainage network organization and the spring resurgence because it accelerates the erosion and alteration processes and therefore the rivers development

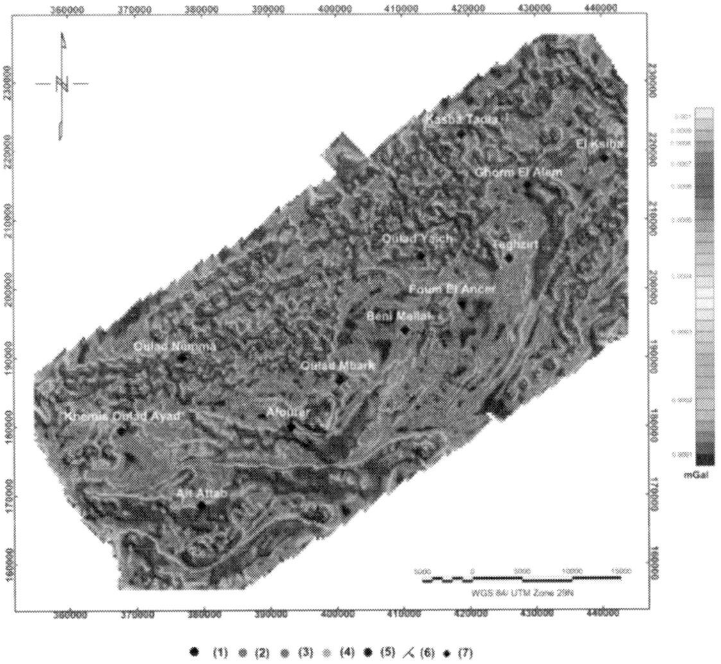

Fig.11. Superposition of horizontal gradient anomaly and its local maxima obtained from residual anomalies map (1) and its upward continuations at different altitudes: 500 m (2); 1000 m (3); 2000 m (4) et 4000 m (5); (6) Dipping direction ; (7) locality.

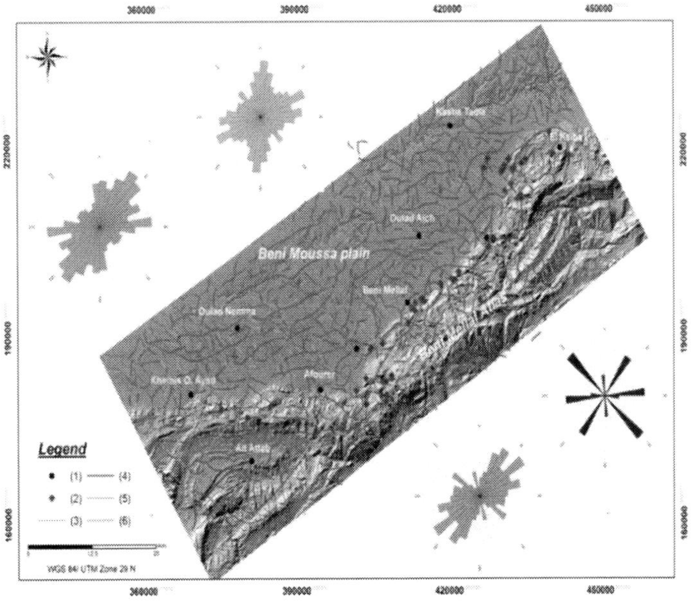

Fig.12. Synthetic lineaments map of study area and rose diagrams for highlighted directions. (1) locality; (2) spring; (3) hydrographic network; (4) gravimetric lineaments; (5) Sentinel-1 lineaments; (6) mapped faults from geological map of study area [3, 5, 42].

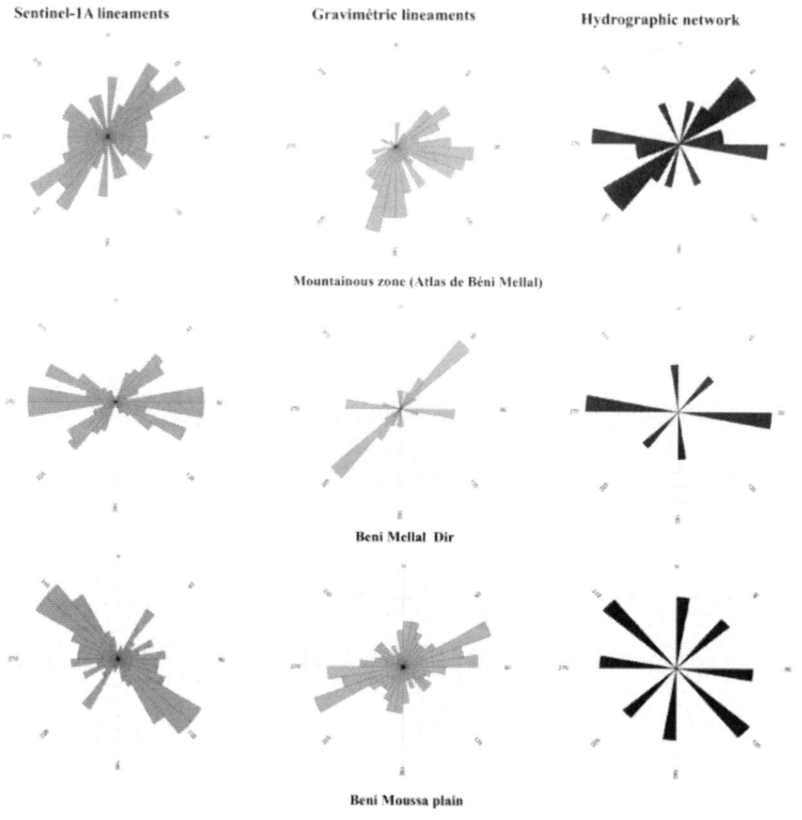

Fig.13. Rose diagrams of hydrographic network (in the right); gravimetric lineaments (in the center) and lineaments deduced from Sentinel-1 image (in the left) in the different sectors of study area.

For a better results analyzing and in order to rule on strucural origin [70-86] of deduced lineaments from gravity data and radar images combination, we proceeded with a breakdown of the study area into tectonic sectors. Gravimetric lineaments and the ones deduced from the Sentinel-1image were correlated and compared for different parts of the study area (Figu.13). Indeed, the structural map obtained (Fig.12) shows a clear difference in the distribution and fractures density in the study area. The southern part of the study area (mountainous area of the Beni Mellal Atlas); dominated by NE-SW to ENE-WSW and N-S directions; shows a high fracturing density which decreases towards the Beni Moussa plain (northern sector). At this level, quaternary alluvial deposits make difficult the lineaments identification. The latter has a NW-SE trend in its orientation. Those two sectors are separated by an intermediate zone that corresponds to the Beni Mellal piemont. In turn, this zone presents a NW-SE and E-O trend in its orientation. The analysis of rose diagrams for each sector (Fig.13) revealed the following important points:

- The major directions NE-SW to ENE-WSW and NS distributed in the Beni Mellal Atlas (southern sector) bear witness to the importance of the structural heritage [87, 88, 89, 90] noticed by several authors [4, 40, 91, 92]. These directional peaks characterize the Atlas chain evolution during Mesozoic and Cenozoic periods [81-93].The direction NE-SW to ENE-WSW is parallel to the Atlas chain structural direction [39] and also to the Tadla basin lengthening. While the direction N-S;

orthogonal and posterior to the previous one; corresponds to the alpine orogenesis phase [6-88].These fractures are responsible for the Beni Mellal carbonate reservoir karstification, they favor karsts evolution and underground caves and cavities formation by increasing the carbonates dissolution [94, 95, 96]. [97] carried out a speleological exploration of Beni Mellal Atlas where he deduced that the majority of the underground cavities (mainly avens, caves and absorbent holes) are developed on the N60 ° and N150- 180 ° directions faults resulting from the atlasic tectonics;

- The gravimetric lineaments in E-O direction (N80-90 ° to N100 °) are timidly represented in the Dir of Beni Mellal compared with those identified from the Sentinel-1 image. The faults associated with this direction have guided the Mesozoic sedimentation in order to be reactivated during the Plio-Quaternary period into dextral strike-slip faults. Dominate the Dir sector, they are the seat of an intense compression linked to the overlap movement (The northern atlasic border) [2, 88, 91, 98].These faults are carried by the overlaps that mark the north-east of the Beni Mellal Dir (Taghzirt-El Ksiba alignment), they rarely allow the sources emergence unlike those of NE-SW direction. The spatial arrangement of these faults is responsible for the Beni Mellal piemont thrust pronounced particularly in the El Ksiba Atlas hence the name of "El Ksiba thrusts" [2] (Fig.3c);
- The family N120-140 ° direction is very widely represented in the Beni Moussa plain. It is mainly detected by the radar image, in contrast to the one identified by the gravity data analysis whose frequency is relatively low. The associated faults facilitate the water drainage from the Atlas to the Beni Moussa plain (southern part of the plain), which confirms the decrease of local waters salinity in the plain [31, 45, 55, 99]. This direction is related to the compressive constraint results due to the two African and European continents rapprochement [36, 40, 87]. As we approach the Beni Mellal Atlas, the NW-SE direction becomes a minority or even absent to give place to another faults family oriented NE-SW. This latter controls the drainage network orientation in the mountainous area of the Atlas.

The comparison between the different rose diagrams of the lineaments obtained and the hydrographic network made it possible to establish a correlation between the principal fractures directions and those runoff waters circulations. Indeed, in the mountainous area of the Beni Mellal Atlas the hydrographic network borrows the major fractures of NE-SW to ENE-WSW and E-O directions, characterizing the said zone. In Beni Mellal Dir, the rivers – of atlasic origin- coincide perfectly with the major thrust faults oriented E-W. They also borrow the minority directions NE-SW and N-S that imposes a circulation in these two directions.

In the Tadla plain, the surface water drainage network is oriented mainly, in order of importance, in the NW-SE, E-O, NE-SW and N-S directions. All of the atlasic origin rivers converge towards the Beni Moussa plain which corresponds to a subsidence zone where a streams accumulation occurs. Those latter end up joining the Oum Er'Rbia river in the Tadla plain. The revealed directions perfectly conjugate with the major faults directions affecting the Beni Moussa plain. They represent weakness areas used by rivers to minimize the energy required for their flow [100].

Conclusion

The results obtained in this study provide informations on the study area structuring and highlight a major accidents mapping that affect it. The results obtained by the integration of the gravity data and the Sentinel-1A radar image, show that four fractures directions were highlighted in the studied area: NE-SW to ENE-WSW; E-O; N-S and NW-SE. In the three identified tectonic sectors, the lineaments are predominantly oriented NE-SW to ENE-WSW and E-W. The directions N-S and NW-SE, in turn, characterize respectively the mountains zone of the Beni Mellal Atlas and the northern sector of the Beni Moussa plain. The directional correlation between the resulting lineaments and the surface hydrographic network shows a general parallelism between the preferential directions of the runoff circulation and the fracturing. The NE-SW and E-O directions are the most frequent in the study area in general and particularly in the mountain zone of the Beni Mellal Atlas, they control the drainage network and allow the springs emergence. This general coincidence between the hydrographic network directional distribution and that of fractures makes it possible on the one hand to confirm that these identified lineaments correspond to real fractures continuing in depth and on the other hand to validate the methodology adopted for lineaments mapping based on the combination of gravity data and radar imagery.

References

1. L. Bouchaou, *Contribution à la connaissance de l'aquifère karstique de l'Atlas de Béni Mellal (Maroc)* - Karst Hydrology (Proceedings of Workshop W2 held at Rabat, Morocco, IAHS Publ. no. **247**, **(1997)**.
2. M. Benzaquen, *Bordure septentrionale de l'Atlas de Béni Mellal. Contribution à l'étude géologique de la région d'El Ksiba*. Notes et Mém. Serv. Géol. Maroc **22**(170), 20-45,**(1963)**.
3. J.P. Rolley, *Carte géologique du Maroc au 1/100.000 : feuille d'Afourer. Notice explicative*. Notes et Mémoires du Service Géologique du Maroc, **247**, 247 bis: 1-103, **(1978)**.
4. E. Laville, *Rôle des décrochements dans le mécanisme de formation des bassins d'effondrement du versant nord du haut atlas central au cours du temps triasiques et liasiques*. Bull. Serv.Géol. France (7), t. **XXIII, n°3**, pp : 303-312, **(1981)**.
5. M. Monbaron, *Carte géologique du Maroc au 1/100 000ème, feuille Béni Mellal*. Notes et Mém. Serv. Géol. Maroc, n°**341**, **(1985)**.
6. M. Monbaron, *Précisions sur la chronologie de la tectogénèse atlasique*. C.R. Acad. Sc. Paris : t. **294**, série II pp : 883-885, **(1982)**.
7. M.S. Abdou Babaye, *Evaluation des ressources en eau souterraine dans le bassin de Dargol (Liptako-Niger). Thèse en cotutelle*, Univ. de Liège et Univ. Abdou Mamouni, Niger, 265p **(2012)**.
8. A. Koudou, T.V. Assoma, B. Adiaffi, Ta.M. Youan, K.F. Kouamé, T. Lasm, *Analyses statistique et géostatistique de la fracturation extraite de l'imagerie Asar Envisat du sud-est de la cote d'ivoire*. Larhyss Journal, ISSN 1112-3680, n°**20**, Décembre 2014, pp. 147-166, **(2014)**
9. L. Ait Brahim, *La zone de fracture trans-marocaine d'Agadir-Nékor: critères géophysiques, données de terrain et analyse de documents Landsat*. Bull. Inst. Scient, Rabat, **10**, 27-40, **(1986)**.
10. A. Chaouni, *Apport des données Landsat M.S.S., Radar ERS1-SAR ; modèles numériques de terrain ; données morphométriques et structurales de terrain à la compréhension de la*

cinématique des principales failles de la péninsule de Tanger depuis l'Oligocène supérieur jusqu'à l'Actuel (Rif septentrional, Maroc). Thèse D.E.S., Univ. Mohammed V, Rabat, 295 p, **(1996)**.

11. H. Tabyaoui, L. Ait Brahim, P. Chotin & J.P. Gelard, *Exemples d'artefacts à l'utilisation de l'imagerie radar SAR ERS-1. Effet de rabattement au sommet : faille de l'oued Za. Monts de Taourirt-Oujda.* Actes du 13ème Coll. Bassins sédimentaires marocains, Marrakech, p. 192, **(1996)**.
12. H. Tabyaoui, *Apport des données satellitaires (Spot-XS, Radar SAR-ERS, Landsat-MSS) à la cartographie des structures géologiques du Maroc nord-oriental. Tectonique cassante, cinématique et contexte géodynamique du Trias à l 'Actuel.* Thèse Doct. Nationale, Mohammed V, Rabat, 338 p, **(2000)**.
13. A. Emran, J. Chorowicz, B. Cervelle, N. Lyberis, G. Tamain & E.M. Alem, *Cartographie géologique et analyse de la fracturation du sud de l'Anti-Atlas central (Maroc) à partir d'une image Landsat MSS*, Photo Interprétation, **2**, 1-7, **(1988)**
14. S. Himyarim, C. Hoepffner, M. Benzakour & D. El Hadani, *Étude structurale du Haut-Atlas oriental (Maroc) à l'aide de l'analyse linéamentaire des images HRV (XS) de SPOT.* Télédétection, **2**, 243-253, **(2002)**.
15. A. Saidi, *Etat de contraintes et mécanismes d'ouverture et de fermeture des bassins permiens à la reconnaissance des faciès et des réseaux de failles.* Doctorat d'état. Faculté des sciences de Rabat, Université Med 5, 222 p, (2005)
16. M. Tahiri, A. Emran, M. Hadkaoui, F. Sossey-Alaoui, M. Achab, H. El Hadi, *Rôle de la fracturation dans la répartition du réseau hydrographique du bassin de Tahaddart (Rif nord occidental, Maroc). Apport du SIG et de la télédétection.* ScienceLib Editions Mersenne, **5**, 131211, **(2013)**
17. Z. Adiri, A. El Harti, A. Jellouli, L. Maacha, E.M. Bachaoui, *Lithological mapping using Landsat 8 OLI and Terra ASTER multispectral data in the Bas Drâa inlier, Moroccan Anti Atlas.* Journal of Applied Remote Sensing **10**(1), 016005 14, **(2016)**.
18. Z. Adiri, A. El Harti, A. Jellouli, R. Lhissou, L. Maacha, M. Azmi, M. Zouhair, , E.M. Bachaoui, *Comparison of Landsat-8, ASTER and Sentinel 1 satellite Remote Sensing data in Automatic Lineaments Extraction: a case study of Sidi Flah-Bouskour inlier*, Moroccan Anti Atlas, Advances in Space Research **(2017)**.
19. D. Casasent & R. Shenoy, *Synthetic aperture radar detection and clutter rejection minace filter.* Pattern Recognit. **30,** 151–161, **(1996)**
20. A. El-Harti, A. Bannari, M. Bachaoui, E.M. Aarab & A. El-Ghmari, *Etude spectroradiométrique des roches des Jebilet centrales (Maroc) : perspective d'utilisation de la télédétection hyperspectrale pour la cartographie géologique.* Télédétection, 2004, vol. **4**, n°3, p. 251–262, **(2004)**
21. D. Amitrano, G. Di Martino, A. Iodice, F. Mitidieri, M.N. Papa, D. Riccio & G. Ruello, *Sentinel-1 for Monitoring Reservoirs: A Performance Analysis.* Remote Sens. 2014, **6**, 10676-10693, **(2014)**.
22. S. Abdikan, F.B. Sanli, M. Ustuner, F. Calò, *Land cover mapping using Sentinel-1 SAR data. The International Archives of then Photogrammetry*, Remote Sensing and Spatial Information Sciences, Volume **XLI-B7**, 2016 XXIII ISPRS Congress, 12–19 July, Prague, Czech Republic **(2016)**.
23. S. Rustami, M. Saadi Nureddin, E. Aboud, K. Yonezu, and K. Watanabe, *Investigating Structural and Tectonic Evolution of Central Afghanistan using Remote Sensing and Gravity Data.* J Geol Geophys 2017, **6**:6 DOI: 10.4172/2381-8719.1000313, **(2017)**
24. T. Lasm, K.F. Kouamé, N. Soro, J.P.R. Jourda & J. Biémi, *Analyse géostatistique de la fracturation extraite de l'imagerie spatiale aéroportée et satellitaire. Application à la région de Man-Danané (ouest de la Cote d'Ivoire)*, Revue Ivoirienne des Sciences Technologiques, n°**5**, 135-154. **(2004)**.

25. S. Boutaleb, F. Hammichi, H. Tabyaoui, L. Bouchaou & K. Dindane, *Détermination des écoulements préférentiels en zone karstique (Tafrata, Maroc), Apport des données satellitaires SAR ERS-1 et Landsat ETM+ et de la prospection géophysique*. Revue des sciences de l'eau, **22**(3), 407–419, (**2009**).
26. A. Yassaghi, *Integration of Landsat imagery interpretation and geomagnetic data on verification of deep-seated transverse fault lineaments in SE Zagros, Iran*. Int J of Remote Sensing **27**: 4529-4544, (**2006**)
27. Bilim, *Investigations into the tectonic lineaments and thermal structure of Kutahya-Denizli region, western Anatolia, from using aeromagnetic, gravity and seismological data*. Physics of the Earth and Planetary Interiors **165**: 135- 146, (**2007**).
28. B. Redhaounia, *Contribution of geophysics approaches to the hydro-geological characterization of fractured limestone karst of Amdoun region (North-Western Tunisia)*. PhD Thesis, 196p, **(2016)**
29. G. Dubar, *Haut Atlas central (Maroc)*. 19ème congrès géol.. Alger, série mai n°**4**, 74p, **(1952)**
30. J.P. Rolley, *Etude géologique de l'Atlas d'Afourer - Haut-Atlas central - Maroc. Stratigraphie*. Thèse Doctorat 3ème cycle, Univ. de Grenoble, 100 p, (**1973a**).
31. Y. Hsissou, *L'aquifère des calcaires du Turonien (bassin du Tadla, Maroc). Alimentation locales et lointaines à partir de l'Atlas*. Doct. Univ. Franche Comté, Besançon, 196p, **(1991)**
32. Ph. Le Noir, *Rapport d'activité sur les sources du Dir de Béni Mellal. D.R.H de Béni Mellal*. (**1978**).
33. A. Bouguenouch, *Etude de la nappe des Béni Moussa et des sources du Dir*. Mém. De fin d'étude, Ecole Mohammedia des Ingénieurs (EMI), Rabat, (**1978**).
34. M. Monbaron, *Sédimentation, tectonique synsédimentaire et magmatisme basique l'évolution paléogéographique et structurale de l'Atlas de Béni Mellal (Maroc) ay cours du Temps Mésozoïque, ses incidences sur la tectonique atlasique*. Ecologea Géol. Helv, Vol. **74/3** pp. 625-638, (**1981**).
35. Dercourt et al, *Geological Evolution of the Tethys Belt from the Atlantic to the Pamirs since the Lias*. Tectonophysics, 123, 241-315, (1986)
36. A. Piqué et al, *Le poinçon maghrébin: contraintes structurales et géochimiques*. C. R. Acad. Sci. Paris, **326**: 575-581, **(1998)**
37. J.P. Rolley, *Sur quelques paléoreliefs récifaux du Lias de l'Atlas de Béni-Mellal (Maroc)*. Notes et Mémoires du Service Géologique du Maroc, **34/254** : 113-120, (**1973b**).
38. L. Bouchaou, *Fonctionnement des aquifères atlasiques et leur relation avec les aquifères de la plaine de Tadla : Cas de l'Atlas de Béni Mellal et de la plaine de Tadla (Maroc)*. Doctorat d'Etat, Univ. Cadi Ayyad, Marrakech, Semlalia, 154p, (**1995**).
39. L. Bouchaou, *Hydrogéologie du bassin des sources karstiques du complexe calcaire haut atlasien du Dir de Béni Mellal (Maroc)*. Doct. Univ. Franche Comté, Besançon, 182p, (**1988**).
40. Choubert & A. Faure Muret, *Evolution paléogéographique et structurale des domaines méditerranéens et alpins d'Europe « Tome 1 ». Evolution du domaine atlasique marocain depuis les temps paléozoïques*. Mém. hors série, Soc. Géol. France, n°**1**, 1, pp. 447-527, **(1960-1962)**
41. C. Archambault, *Compte rendu des travaux du marché 41/70DH et étude hydrogéologique de l'extension nord est du périmètre irrigué des Béni Moussa*. CRTMB/ DRE/DH, Béni Mellal, (**1971**).
42. Y. Verset, *Carte géologique du Maroc au 1/100 000ème, feuille Kasba Tadla*. Notes et Mém. Serv. Géol . Maroc, n°**340**, (**1985**).
43. M. Combe, *Ressources en eau du Maroc. Domaines atlasiques et sud atlasiques. Tome 3*, Notes et Mém. Serv. Géol, n°231, Rabat, Maroc, **(1977)**

44. C. Archambault, *Piézométrie des aquifères du plateau des phosphates et de la plaine du Tadla*, (**1972**).
45. S. Dirrhami, *Contribution à l'étude hydrogéologique et hydrogéochimique de la plaine de la Tassout aval : Bahira Oriental (Maroc).* Thèse 3ème cycle Univ. Cadi Ayyad, Fac. Sc. Marrakech, 230 p, (1990)
46. B.R.G.M, *Etude du système aquifère multicouche de la plaine de Tdla ; mission 1 : description des aquifères et analyse du système multicouche du Tadla ; annexe : synthèse* ***géologique***, (**1993**).
47. A. Boukdir, *Fonctionnement hydrogéologique de l'aquifère du Turonien du bassin du Tadla-Plateau des Phosphates-Tassaout Aval (Maroc).*Thèse d'Etat. Eole Mohammedia des Ingénieurs, 128p, (**2007**).
48. Agence du Bassin Hydraulique de l'Oum Er'Rbia, Béni Mellal, *Appui à la préparation de contrat de nappes dans le Bassin de l'Oum Er'Rbia en s'appuyant sur une approche inclusive. Mission 3 : Revue et actualisation du modèle hydrodynamique du complexe de Tadla*, rapport interne, 176p, (**2016**).
49. Agence du Bassin Hydraulique de l'Oum Er'Rbia, *Plan Directeur d'Aménagement Intégré des Ressources en Eau du Bassin de l'Oum Er'Rbia et des bassins côtiers atlantiques*, Béni Mellal rapport interne, 179p, (**2012**)
50. M. Everaert & J.L. Mansy, *Le filtrage des anomalies gravimétriques ; une clé pour la compréhension des structures tectoniques du Boulonnais et de l'Artois (France).* Bull. Soc. géol. Fr., **172**(3), 267-274, (**2001**).
51. A. Gérard, & P. Griveau, *Interprétation quantitative en gravimétrie ou magnétisme à partir des cartes transformées de gradient vertical*. Geophysical Prospecting **22**, 460-481, (**1972**)
52. N. Debeglia & C. Weber, *Geological mapping of the basement of the Paris basin (France) by gravity and magnetic data interpretation*. In: The Utility of Regional Gravity and Magetic Anomaly Maps, 154–163. SEG Publication, (**1985**).
53. D. Khattach, P. Keating, M. Mili, T. Chennouf, P. Andrieux & A. Milhi, *Apport de la gravimétrie à l'étude de la structure du bassin des Triffa (Maroc nord-oriental) : implications hydrogéologiques*, C. R. Geoscience 3361427–1432. (**2004**).
54. D. Khattach, H. Mraoui, D. Sbibih, T. Chennouf, *Analyse multi-échelle par ondelettes des contacts géologiques : application à la carte gravimétrique du Maroc nord-oriental*. CR Geoscience ; **338** : 521-526. (**2006**)
55. A. Najine, M. Jaffal, K. El Khammari, T. Aifa, D. Kattach, M. Himi, A. Casas, S. Badrane, et H. Aqil, *Contribution de la gravimétrie a` l'étude de la structure du bassin de Tadla (Maroc): Implications hydrogéologiques*. C. R. Géoscience, **338** : 676–682, (**2006**).
56. B. Farhat, R. Benassi, C. Jallouli & A. Ben Mammou, *Contribution de la gravimétrie à l'étude de la structure de la plaine de Mornag (nord est de la Tunisie): implications hydrogéologiques*, Hydrological Sciences Journal, **55**:8, 1396-1404, (**2010**)
57. S. Rochdane, A. El Mandour, M. Jaffal, M. Himi, A. Casas, M. Amrhar & M. Karroum, *Géométrie de l'aquifère du Haouz oriental et Tassaout amont, Maroc occidental: approche géophysique et hydrogéologique*, Hydrological Sciences Journal, **60**:1, 133-144, (**2014**)
58. L. Cordell & V. J. S Grauch, *Mapping basement magnetic zones from aeromagnetic data in San Juan basin, New Mexico. In: The Utility of Regional Gravity and Magnetic Anomaly Maps*. 181–197. SEG Publication, (**1982**).
59. R.J. Blakely & R.W. Simpson, *Approximating edges of source bodies from magnetic and gravity anomalies*. Geophysics **51**, 1494-1498, (**1987**).
60. L.T.A. Vanié, D. Khattach, M.R. Houari, *Apport des filtrages des anomalies gravime'triques a` l'e'tude des structures profondes du Maroc oriental*. Bulletin de l'Institut Scientifique, Rabat, section Sciences de la Terre 2005 ; **27** : 29-40, (**2005**).

61. Aquil, D. Khattach, R. El Gout, A. El Mandour & O. Kaufmann, *Contribution de la gravimétrie à la reconnaissance de l'aquifère profond de la plaine des Angad (Maroc nord-oriental)*. Sécheresse 2010 ; **21** (4) : 252-256, (**2010**).

62. P. Keating, M. Pilkington, P. Sailhac & L. Nadeau, *Locating magnetic contacts and source characteristics from magnetic data: application to the Canadian Shield*. EAGE 66th Conference & Exhibition, Paris, (**2004**)

63. J.D. Phillips, R.O. Hansen and R.J. Blakely, *The use of curvature in potential-field interpretation: Exploration Geophysics*, v.38, p.111-119, (**2007**).

64. D.W. Eckhardt, J.P. Verdin and G.R. Lyford, *Automated Update of an Irrigated Land GIS Using SPOT HRV Imagery*. Photogrammetric Engineering and RemoteSensing, **59** (11): 1515-1522 ; (**1990**).

65. M.L. Süzen & V. Toprak, *Filtering of Satellite Images in Geological Lineament Analyses: An Application to a Fault Zone in Central Turkey*. International Journal of Remote Sensing **19**(6): 1101-1114, (**1998**).

66. S. Corgne, R. Magagi, Yergeau M. & D. Sylla, *An integrated approach to hydro-geological lineament mapping of a semi-arid region of West Africa using Radarsat-1 and GIS*. Remote Sensing of Environment **114**, 1863–1875, (**2010**)

67. Amer, T. Kusky & A. El Mezayen, *Remote sensing detection of gold related alteration zones in Um Rus area, Central Eastern Desert of Egypt*. Advances in Space Research, **49**, 121–134, (**2012**).

68. Savane, B. Goze & H. Gwyn, *Evaluation de la productivité des ouvrages dans le socle par l'étude des fractures et le SIG dans la région nord-ouest de la Côte d'Ivoire*. IAHS-Publications, Vol. **241**, pp: 103-111, (**1997**)

69. Kouamé, *Méthode de cartographie des discontinuités-images extraites d'images satellitales: exemple de la région semi montagneuse à l'ouest de la Côte d'Ivoire*. Télédétection 1, 139–156, (**1999**)

70. J.P. Jourda, E.V. Djagoua, K. Kouamé, M.B. Saley, C.C. Gronayes, J.J. Achy, J. Biémi & M. Razack, *Identification et cartographie des unités lithologiques et des accidents structuraux majeurs (nord de la côte d'ivoire): Apport de l'imagerie ETM+ de Landsat*. Revue Télédétection, **6**(2), pp. 123- 142, (**2006**).

71. Ta. M. Youan, T. Lasm, J. P. Jourda, K. F. Kouamé & M. Razack, *Cartographie des accidents géologiques par imagerie satellitaire Landsat-7 ETM+ et analyse des réseaux de fractures du socle précambrien de la région de Bondoukou (nord est de la Côte d'Ivoire)*, Télédétection, **8**(2): 119–135, (**2008**)

72. Anwar, N. Shawki, & G. Abdoh, *Landsat ETM-7 for Lineament Mapping using Automatic Extraction Technique in the SW part of Taiz area, Yemen*. Global Journal of HUMAN SOCIAL SCIENCE Geography, Geo-Sciences, Environmental & Disaster Management Volume **13** Issue 3 Version 1.0, (**2013**).

73. S. Sedrette & N. Rebaï, *Automatic extraction of lineaments from Landsat Etm+ images and their structural interpretation: Case Study in Nefza region (North West of Tunisia)*. Journal of Research in Environmental and Earth Sciences, **04**, 139-145, (**2016**).

74. G. Lachaine, *Structures géologiques et linéaments, Beauce (Québec). Mémoire de maîtrise, Département de géographie et télédétection, Université de Sherbrooke*, 83 pp. (**1999**)

75. M. Ta, T. Lasm, J. Jourda, K. Kouamé & M. Razack, *Cartographie des accidents géologiques par imagerie satellitaire Landsat-7 ETM+ et analyse des Réseaux de fracture du socle Précambrien de la région de Bondoukou (Nord-Est de la Cote d'Ivoire). Revue Télédétection, **8***(2), 119-135, (**2008**).

76. M.J. Lefèvre, *Détection aéroportée de phénomènes géologiques de sub-surfaces révélées par une anomalie de température de brillance (Coat en Noz, Côtes du Nord) France*. Revue photo interpretation, **6** (3a), (**1979**)

77. W.H. Hobbs. *Earth features and their meaning: an introduction to geology for the student and the general reader*. Macmillan, New-York, 506 pp, (**1912**).
78. D.W, O'Leary, J.D. Friedman & H.A. Pohn. *Lineament, linear, lineation: Some proposed new standards for old terms*. Geological Society of America Bulletin, **87**: 1463–1469, (**1976**).
79. E. Bolleli, *Hydrogéologie du Maroc (Béni-Amir, Béni Moussa)*. Notes et Mémoires du Service Géologique du Maroc, **97,** pp. 205-215, (**1952**).
80. C. Graf, *Synthèses géologiques du bassin de Quasbat-Tadla, Béni Mellal, Tanhaset (d'après les données géophysiques et de forage)*. Rapport B.R.P.M., 89p, nombreuses coupes, profils et pl.h.j. Rabat, (**1976**)
81. El. Zouine, *Géodynamique récente du Haut Atlas. Evolution de sa bordure septentrionale et du Moyen Atlas sud-occidental au cours du Cénozoïque*. Doct. Etat, Es-Sc. Nat. Univ. Mohamed 5 Fac. Sc. Rabat (Maroc), (**1993**).
82. H. Gabtni, C. Jallouli, K. Mickus, H. Zouari, M.M. Turki, *The location and nature of the Telemzan High - Ghadames basin boundary in southern Tunisia based on gravity and magnetic anomalies*. J. Afr. Earth Sci. **44**, 303–313, (**2006**).
83. N. El Goumi, M. Jaffal, D. Kchikach & A. Manar , *Apport de la gravimétrie à l'étude de la structure du bassin du Haouz (Maroc)*. Estudios Geológicos, **66**(2), Julio-diciembre 2010, 181-191, (**2010**)
84. M. Dhaoui , H. Gabtni, C. Jallouli, A. Jleilia, K.L. Mickus & M.M. Turki, *Gravity analysis of the Precambrian basement topography associated with the northern boundary of Ghadames Basin (southern Tunisia)*, Journal of Applied Geophysics, **111**, 299–311, (**2014**).
85. C. Bonnet, Interactions entre tectonique et processus de surface dans l'avant pays alpin. Apport de la modélisation analogique et analyse de la fracturation récente. Thèse de doctorat, Univ. Montpellier II, Sciences et Techniques du Languedoc (France), 194 p, (**2007**).
86. M. Razack, *Application de méthodes numériques et statistiques à l'identification des réservoirs fissurés carbonatés en hydrogéologie*. Thèse Doct. Es Sci., Univ. Languedoc, 384p, (**1984**)
87. A. Michard, *Elément de géologie marocaine*. Note & Mém, **252**. Rabat : éditions du service géologique du Maroc, (**1976**).
88. M. Mattauer, P. Tapponier, F. Proust, *Sur les mécanismes de formation des chaines intracontinentales L'exemple des chaines atlasiques du Maroc*. Bull. Soc. Géol. France **19**:521-526, (**1977**).
89. A. Charrière, *Héritage hercynien et évolution géodynamique alpine d'une chaine intracontinentale : le moyen atlas au sud-est de Fès (Maroc)*. Thèse Doct. Sci. Fac. Sc. Univ. De Toulouse III, 589 p, (**1990**).
90. Piqué, A. Soulaimani, C. Hoepffner, M. Bouabdelli, E. Laville, M. Amrhar, et A. Chalouan, *Géologie du Maroc*. Editions GEODE, Marrakech. (**2007**).
91. R. Du Dresnay, *Influence de l'héritage structural tardi-hercynien et de la tectonique contemporaine sur la sédimentation jurassique, dans le sillon marin du Haut- Atlas, Maroc*. 9ème Congrès international de Sédimentologie (Nice), thème 4 : 103-108, (**1975**)
92. Laville, *L'évolution synsédimentaire et tectonique de la couverture jurassique de la région d'Imouzzer-Marmoucha (Moyen Atlas, Maroc)*. Bull. Soc. Géol. France, **7**, XIX, pp. 1151-1158, (**1977**).
93. D. Sadki, *Le Haut Atlas central (Maroc) – stratigraphie et paléontologie du Lias supérieur et du Dogger inférieur : dynamique du bassin et des peuplements*. Thèse Doct. D'Etat ; Univ. Cadi Ayyad, Marrakech, 331 p, (**1992**).
94. A. Mangin, *Contribution à l'étude hydrodynamique des aquifères karstiques. Doct. Etat, Univ. Dijon,* Ann. Spéléo. 1974, **29**(3), p.283-332 ; **29**(4) p. 495-601, **30**(1°P ; 21-124), (**1975**)

95. M. Bakalowicz, *Les processus de la karstification et les différents types de karst associés. Mémoires Société Géologique de France*, 169, 363-37, (**1996**).

96. W. Al-Fares, *Caractérisation des milieux aquifères karstiques et fracturés par différentes méthodes géophysiques*. Thèses Univ. Montpellier II, 221p,(**2002**).

97. P. Bienfait, *Note relative aux cavités dans la région de Béni Mellal (Maroc)*. Direction de la région Hydraulique de Béni Mellal (D.R.H), 27p, (**1978**).

98. J. Ferrandini, J.J. Cornée, H. Saber & J. Aubouin, *Mise en évidence d'une compression subméridienne d'âge permien probable dans le massif ancien du Haut Atlas occidental (Maroc)*. C.R. Acad. Sc. Paris, t. 304, série II, n°**20**, pp. 1243-1248, (**1987**).

99. Y. Hsissou, P. Chauve, &, J. Mania, *L'aquifère des calcaires turoniens (bassin du Tadla (Maroc).Alimentations locales et lointaines à partir de l'Atlas*. Hydrol. **183**, 433-443, (**1996**)

100. B. Deffontaines, *Développement d'une méthodologie morpho néotectonique et morphostrucurale. Analyse des surfaces enveloppes, du réseau hydrographique et des modèles numériques de terrain ; application au Nord-Est de la France*. Thèse nouveau doctorat, Paris VI, n 90 PA06 6740, 260 p, (**1990**).

APPLICATION OF THE ELECTRICAL (GEOPHYSICAL) METHOD FOR THE IDENTIFICATION OF THE SEALED BEVEL IN THE COASTAL REGION OF WALIDIA PROVINCE OF EL JADIDA – MOROCCO

H. ELFJIJI[*1], *A.* BOUKDIR[1] et *A.* ZITOUNI[1]

1 : Faculté des Sciences et techniques, Université Sultan Moulay Slimane, Béni Méllal, Morocco

ABSTRACT. The Walidia area knows a great development of agricultural, industrial, maritime activities and a very great demographic expansion. This development is accompanied by a growing demand for water. On the other hand, any intense abstraction of fresh water at a coastal aquifer capture field causes a decrease in groundwater flow and a lowering of the piezometric surface. There is therefore a danger of advancing towards the land of the saltwater bevel and salinization of groundwater which is gradually loaded into mineral salts and can reach abnormally high levels.

This work deals with the application of geophysics by electrical sounding method to the in-depth study of the freshwater-salt water interface (salted bevel) on the one hand and the study of the lateral and vertical evolution of the geometry of the different aquifers.

The principle of the method is, when performing an electrical survey, to research how varies, at a given point on the surface, the resistivity of the subsoil vertically. For this one executes in a same place a succession of measurements, increasing each time the dimensions of the device and thus the depth of investigation which is proportional to them. At this site, we explore a slice of ground increasingly thick and thus highlights the changes in geological constitution along the vertical.

The electrical soundings were carried out with the classic Schlumberger device.

The interpretation and the exploitation of the data of the electric soundings carried out made it possible to :

- Follow the evolution of the salted bevel roof,
- Draw the upstream limit (marine invasion) of salted bevel Hauterivian, on the one hand, and Valanginian and Jurassic on the other hand,
- Follow the roof structure of Dridrat limestone (Hauterivien) and limestones of Valanginien and Upper Jurassic,
- Draw the map of the marl roof of Safi (Hauterivian) constituting the impermeable substratum of Dridrat limestones.
- **KEYWORDS :** Geophysical prospecting, electrical soundings, groundwater aquifers, saltbee, freshwater, saltwater, piezometric surface, limestones, marls, coastal region of Walidia, El Jadida, Morocco

* haelfjiji@yahoo.fr

I- INTRODUCTION

In recent years, the mastery of irrigation has become synonymous with safety and success for the farm. In the Doukkala, the use of groundwater for irrigation, in particular, has become increasingly important. The main aquifer of the Walidia Sahel region (Lower Cretaceous limestones) is subject to subterranean marine invasion downstream. From where the will of the managers to know, with precision, the importance of the invaded zone, with a view to a reasoned exploitation.

The present work highlights the contribution and contribution of geophysical prospecting by the electrical survey method applied to the study of the coastal region of Walidia. The use of these geophysical methods for the detection of water, whether in continuous aquifers or in fissured environments, contributes, in the same way as the geomorphological analysis and the examination of aerial photos or satellite images carried out in advance. , to select one or more drill sites with a maximum chance of success.

II- PRESENTATION OF THE STUDY AREA

II-1 Geographic location (fig. 1)

The study area is located 23 km north-east of the city of Walidia in the province of El Jadida. With an area of about 170 km2 (28 km x 6 km), this area is located, more exactly, in the Walidia sahel in the form of a low plain formed of a set of plio-quaternary platforms.

II-2 Geology (fig. 2)

Plio-Quaternary deposits consist of detrital and marine formations:

- Sandstone-limestone slabs;
- coarse bio-detritic limestones;
- Shell sands, more or less indurated;
- Clay sands.

The Plioquaternary ensemble, whose thickness is between a few meters and more than 100 meters, rests on the Lower Cretaceous formations. The following stratigraphic levels are observed from top to bottom:

- Upper Hauterivian: red plastic clays and at the base a level of calcareous sand, sometimes conglomeratic;
- Middle Hauterivian: crystalline limestone, sometimes with clayey pastures (Dridrat limestone);
- Lower Hauterivian: marls, gray or green clays with limestone levels (marl of Safi),
- Valanginian: limestone crystalline, sometimes marly ;
- Valanginien : calcaire cristallin, parfois marneux,
- Berriasian: gray to blue plastic clays;
- Upper Jurassic: anhydrite with limestone levels, sometimes dolometic.

II-3 Hydrogeology

The recent formations of the Plioquaternary contain a shallow water table, exploited by a battery of shallow wells. The piezometric surface is at absolute elevations between +5 and +15m. The alimentation is ensured by the infiltration of the rains. Locally, this aquifer can feed the underlying aquifers, when the impervious level of the Upper Hauterivian has disappeared by pre-quaternary erosion.

An underground aquifer develops in the limestone of Dridrat, with a power of about 30 m. This aquifer is fed, at the same time, by outcrops to East and South-West Zemmamra, Cape Cantin), and by vertical drainage and karst conduits from the plio-quaternary aquifer system. The two tablecloths are sometimes confused near the shore.

The aquifer of Dridrat, sometimes communicates in depth, with valanginian limestones (10 to 50m). However, the marls of Safi, more or less impervious and variable thickness (2 to 30m) can constitute a substratum to the aquifer of Dridrat.

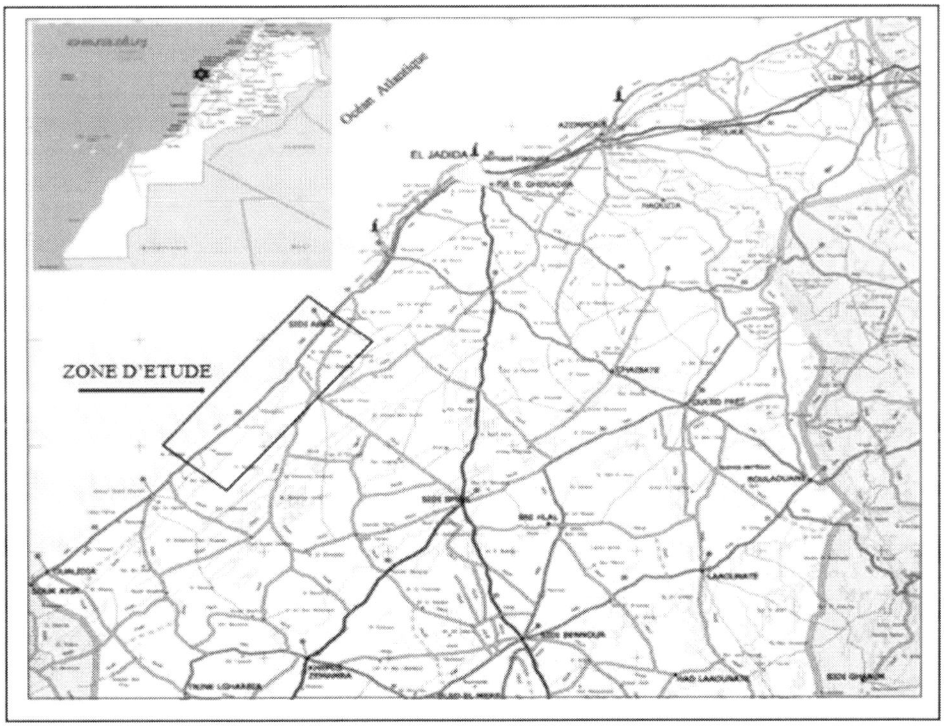

Fig.1 . Situation of the study area

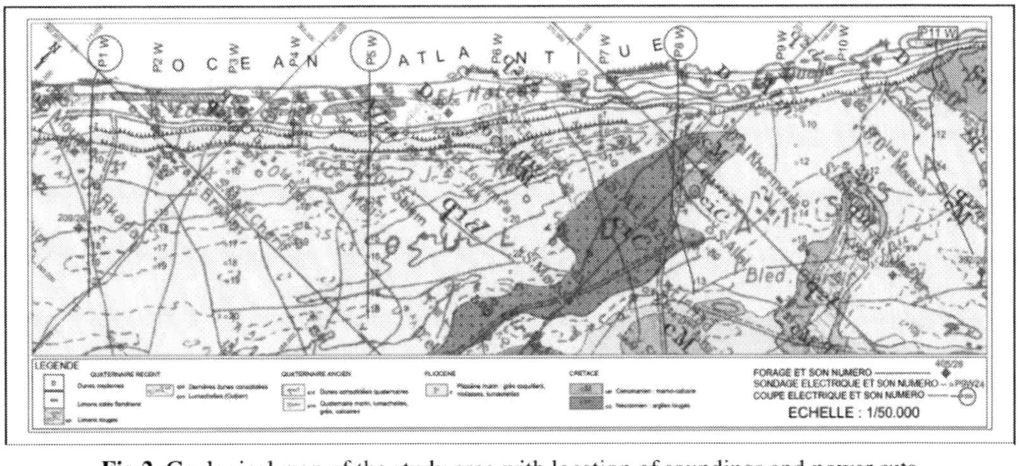

Fig.2. Geological map of the study area with location of soundings and power cuts

The sketch below (Figure 3) presents in a synthetic way the distribution of aquifer geological formations and their mode of operation.

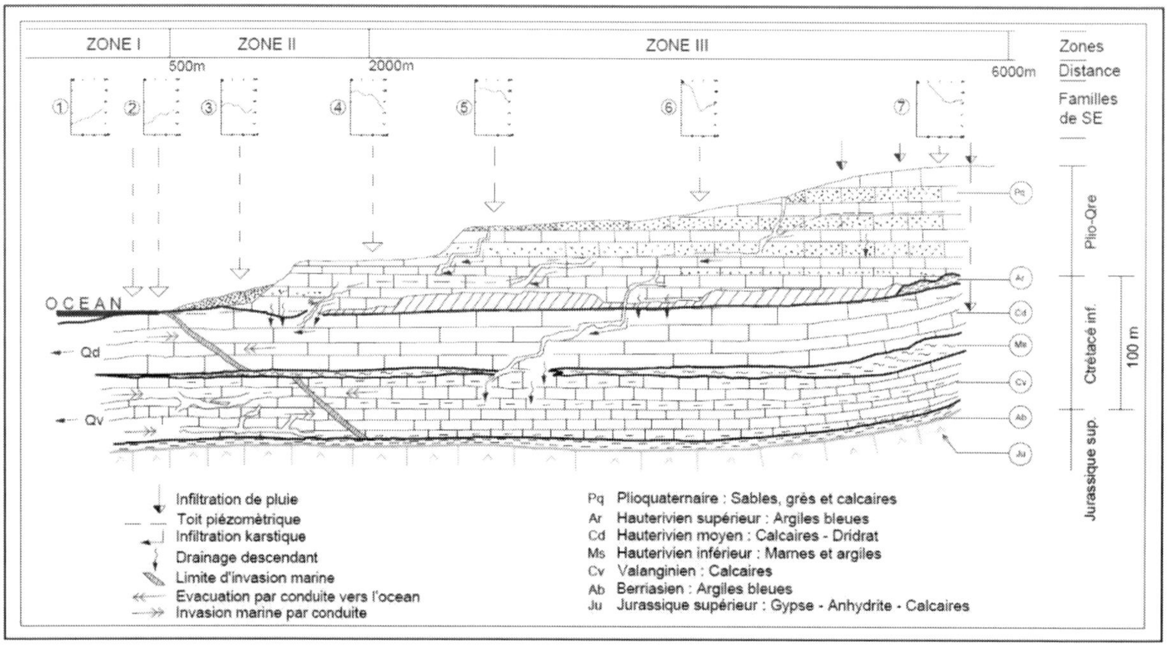

Fig.3. Hydrogeological diagram of the coastal basin of Walidia and the different families of Electrical Surveys

III- METHOD OF ELECTRICAL SURVEYS (SE)

III-1 Purpose of the method of SE

The use of electric prospecting method, allows to "visualize" the invaded areas in salt water and their lateral and vertical extension. In addition, this study results in the location and stratigraphic identification of two aquifer systems (Dridrat and Valanginien) and their hydrogeological boundaries.

Ultimately, the combination of geophysics and hydrogeological synthesis allows a good definition of the geometry of the Lower Cretaceous aquifer and a better understanding of the freshwater / saltwater system of the Walidia sahel.

III-2 Principle of the method of SE

The principle of the method is, when conducting an electrical survey, to research how varies, at a given point on the surface, the resistivity of the subsoil vertically. For this one executes in a same place a succession of measurements, increasing each time the dimensions of the device and thus the depth of investigation which is proportional to them. At this site, we explore a slice of ground increasingly thick and thus highlights the changes in geological constitution along the vertical. The electrical soundings were carried out with the classic Schlumberger device.

We obtain the apparent resistivity (**ρa**) of a certain volume of ground:

$$\rho a = \Delta V / I \, [2\Pi \, (1/AM - 1/AN - 1/BM + 1/BN)] \quad ,$$

We have : $\boldsymbol{\rho a = (\Delta V/I).K}$ *with* $K = 2\Pi\,(1/AM - 1/AN - 1/BM + 1/BN)$

I : current intensity injected (electrodes A and B)

ΔV : the d.d.p. (potential difference between M and N)
K : geometric coefficient f(AMNB)

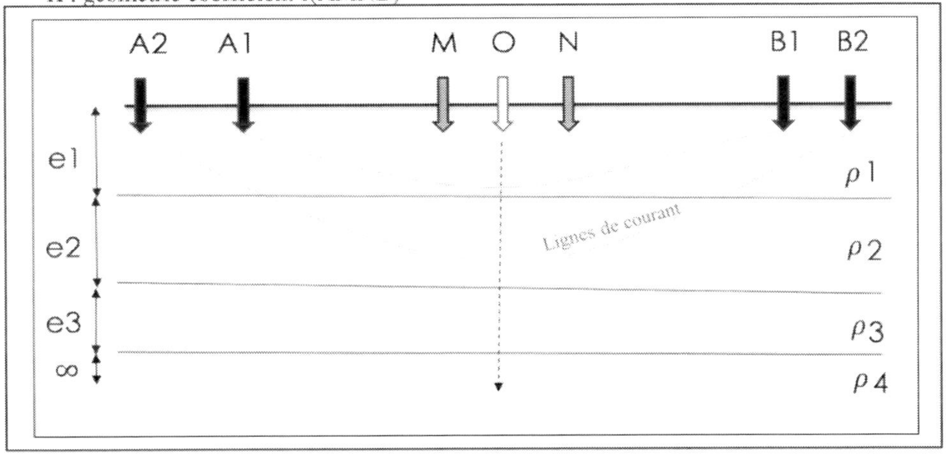

Fig.4. Diagram of the principle of electrical sounding in Schlumberger device

III-3 Processing of diagrams of SE

The quantitative interpretation of the measurements, previously carried out on a bilogarithmic diagram, begins with the smoothing of the curves, then by semi-automatic method, one carries out the inversion of the data. The model obtained is then corrected by successive approximations until the identification of the geological layers. The comparison of these results with those of boreholes, if any, allows a good estimate of true resistivities.

It should be noted, however, that the theoretical scheme of interpretation thus presented considers that the geological formations explored are tabular and isotropic.

IV- RESULTS OBTAINED

IV-1 Analysis of diagrams of SE

The mapping of the electrical soundings carried out with the established hydrogeological diagram makes it possible to group the SEs into a "typical" family set. Each family is characteristic of a specific electrical situation.

This correlation was based on drilling data, where available. Figure 5 summarizes the reference ES diagrams to illustrate the electrical evolution observed from shore to inland.

Table n°1. The main electrical levels highlighted by electrical soundings

N°	Electrical levels	Resistivity (ohm.m)	Thickness (m)	Nature	Marine invasion (near the shore)
1	C0, R0, C1, R1	< 4	variable	Plio-quaternary Salty Dridrat Salted Valanginian	Yes
2	C0, R0, C1, R1	30-130		Plio-quaternary	No
	Rd	140-600		Dridrat Limestones (Middle Highlander)	No
4	Rv	100-250		Limestones marl (Valanginien)	No
5	Cv	2,6		Limestones marl (Valanginien)	Yes
6	Rp	>40		Blue clays (Berriasian) and gypsum formations (Jurassic sup.)	No
7	Cp	32			
8	Rj	>75		Limestone anhydrite (Jurassic sup.)	

It should be noted that formation lithology is no longer the only factor conditioning the value of resistivity.

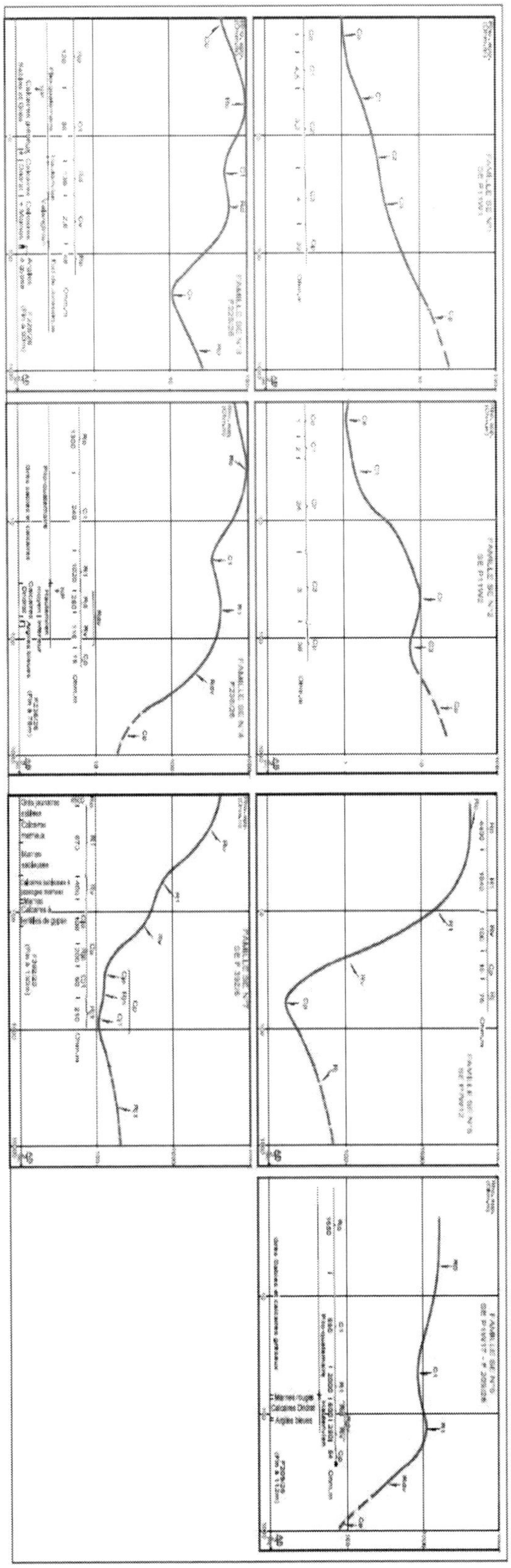

Fig. 5. The different families representative of electrical soundings

IV-2 Geoelectric cuts

Four profiles or geoelectric sections were sent perpendicular to the shore. Their direction is approximately NO-SE. (Fig. 6).

These sections were drawn at the same horizontal and vertical scale (/5.000) so that the dipping of the layers and the shape of the salted bevel are presented without any deformation.

The examination of geoelectric sections presents, in a schematic way (Fig. 7), the following information:

- The marine invasion is wider in the South-West (2000-3500m) than in the North-East (500-900m);
- The highlighting of two stepped salt bevels. The upper bevel, having a gentle slope, corresponding to the limestone of Dridrat (Middle Hauterivian) and the lower bevel with very steep slope, corresponding to the marly limestones of Valanginien;
- The very important rise of the Jurassic formations towards North and North-East accompanied by a reduction, or an erosion of the formations hauteriviennes;
- Plio-Quaternary formations may be in direct contact with Valangian limestones.

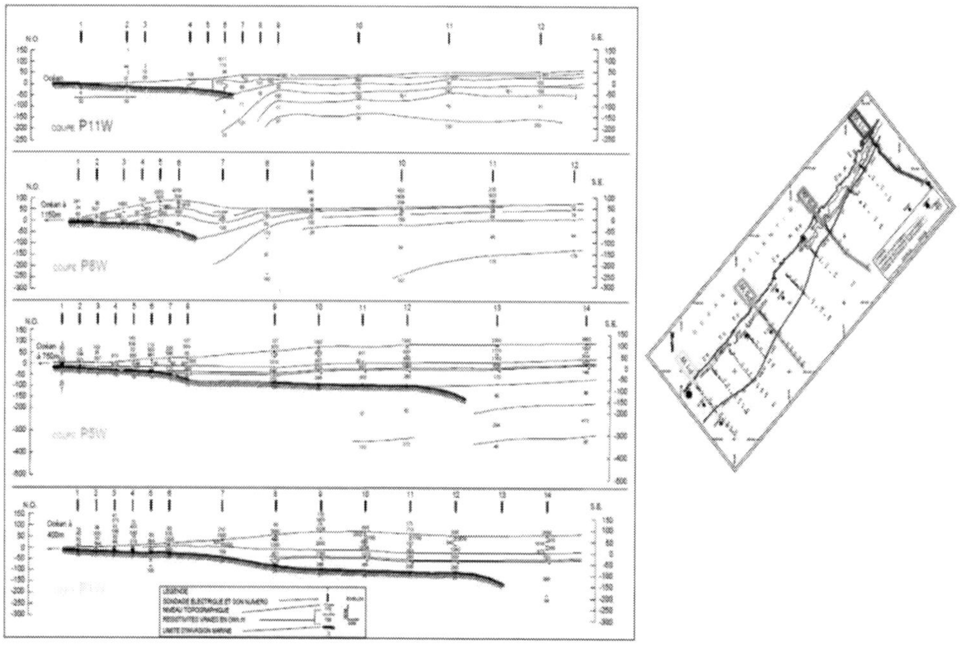

Fig. 6. Geoelectric cuts at the invasion zone marine

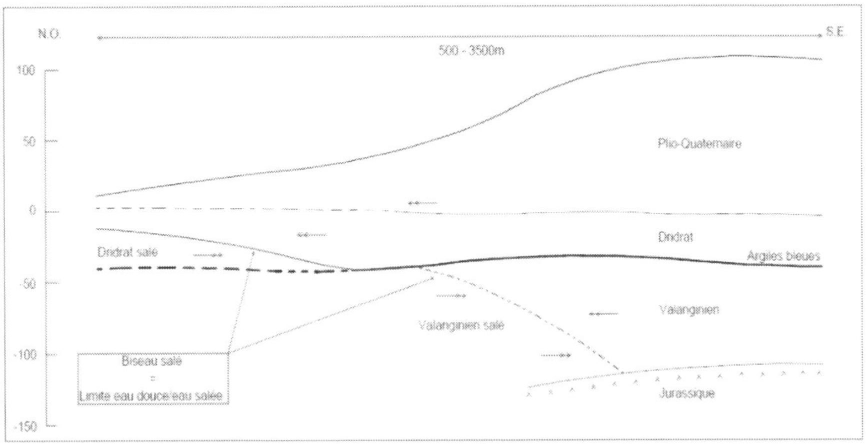

Fig.7. Schematic section of the marine invasion

IV-3 Presentation of interpretative maps

The results obtained by correlations of the geoelectric cuts made it possible to establish the following documents:

- Map of the isohypses of the limestone roof of Dridrat (Middle Hauterivian);
- Map of the isohypses of the marl roof of Safi (Lower Hauterivian);
- Map of the isohypses of the limestone roof of Valanginien and Upper Jurassic (Safi mare wall);
- Map of the salted bevel roof;
- Synthesis map.

IV-3-1 Map of the isohypses of the roof of the limestone of Dridrat (Fig. 8)

This document corresponds to the outline of the isohypses of the roof of the resistance set Rd, located below the deposits attributed to the Plio-Quaternary. It is, in fact, the roof of limestones of Dridrat, and as the case, the surface of erosion pre-quaternary.

This document shows:

- The presence, in the western part of the study area (near the shore), of a narrow, elongated depression along a SO-NE direction. At the bottom of this depression, the roof of the limestone of Dridrat is located with dimensions lower than 0 (z = 0 to -20 m);
- Towards the south-west, the roof of the Dridrat limestones shows a fairly regular morphology that rises to the surface, going up to grades greater than +70 m;
- On the south-east side of the depression, the roof surface of the Dridrat limestones narrows to the northeast due to the rise of the Jurassic formations.

 https://doi.org/10.1051/e3sconf/20183705003

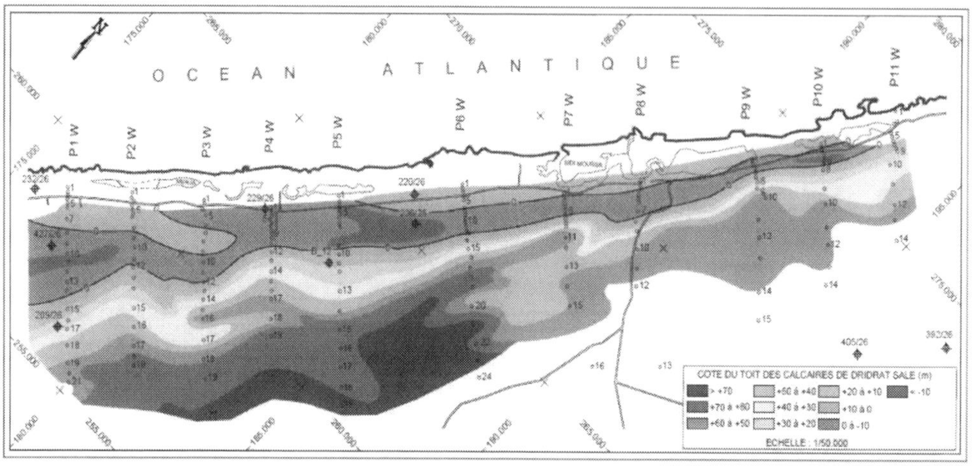

Fig.8. Map of the isohypses of the roof of the limestone of Dridrat (Middle Hauterivian)

IV-3-2 Map of the isohypses of the roof of the marls of Safi (Fig. 9)

This map gives an idea of the behavior of the marl roofs of Safi (Lower Hauterivian) which can play the role of an impermeable substratum at the base of Dridrat limestones. This substratum is intercalated between the Dridrat limestones of the Middle Hauterivian (resistant Rd) and the marly limestones of the resistant level Rv (Valanginian with fresh water) or the conducting level Cv (Valanginian salty).

This map shows:

- A depression can be pronounced, located between the contour level 0 and the coastal road Safi-El Jadida. This depression is extended in the direction SO-NE;
- In the southern part of the study area (south of the P6W profile), the roof of the marls of Safi draws an ascending trough structure towards the East
- The presence of a boundary, between the profiles P6W and P7W to the North-East of which, the roof of the marl becomes undetectable by the electrical soundings.

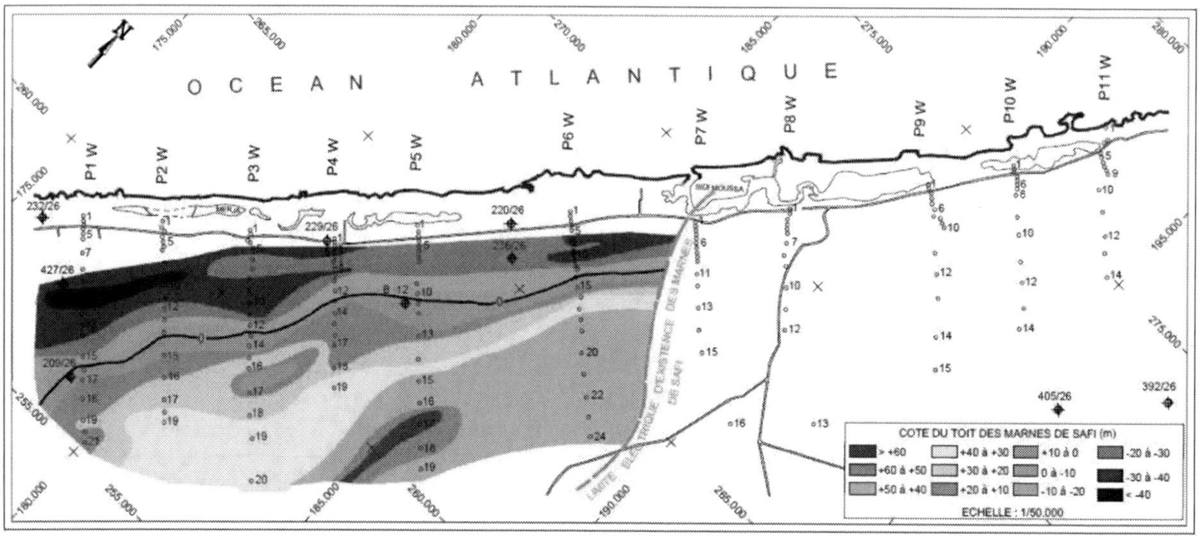

Fig.9. Map of the isohypses of the marl roof of Safi (Lower Hauterivien)

IV-3-3 Map of the isohypses of the limestone roof of the Valanginian and Upper Jurassic

This map (Fig. 10) corresponds to the marl wall of Safi. It gives an idea of the roof morphology of Valangian limestones (Rd) and resistant formations of the Upper Jurassic (Rp).

On this map, the roof of the Valanginien and Upper Jurassic has a structure, oriented SO-NE, which goes from the northwest (<-50 m) to the southeast (rating> +40 m). Northeast of the profile P6W, there is the limit beyond which the highlighting of the marls of Safi becomes very difficult.

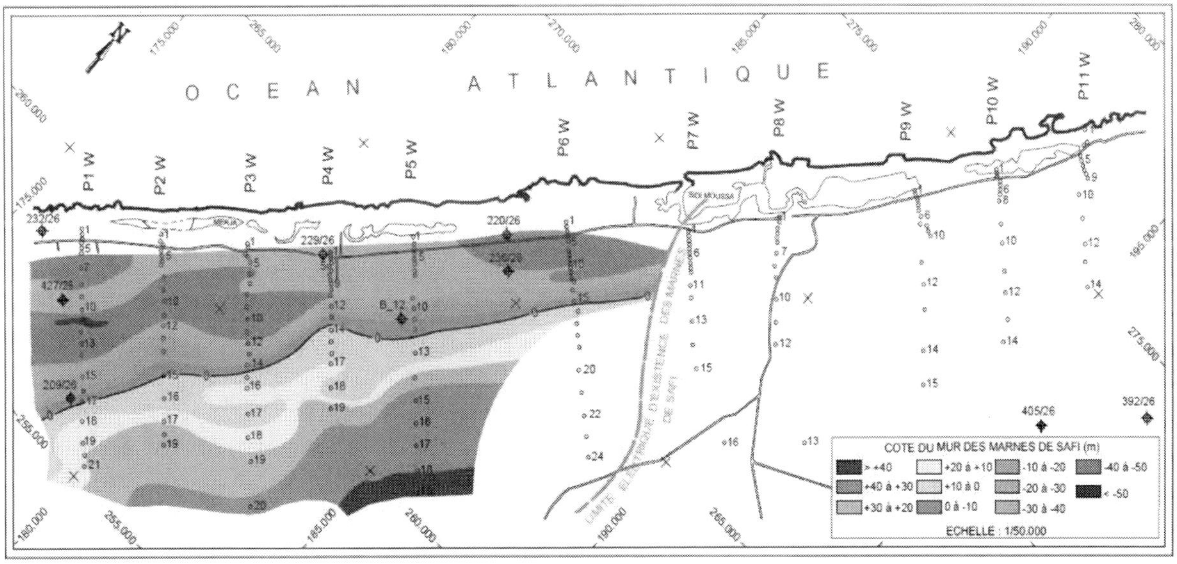

Fig.10. Map of the isohypses of the limestone roof of Valanginien and Upper Jurassic (Safi mare wall)

IV-3-4 Salted bevel roof map (Fig. 11)

This map, drawn from the elevations on the geo-electric sections, gives an idea of the shape of the roof surface of the salt wedge.

On this map we note the following observations:

- The presence of a salt wedge boundary along the shore. In the northeastern part, this limit is located at a short distance from the shore (900 m) whereas in the southwestern part this distance can exceed 3000 m;
- The roof of the salted bevel is characterized by:
  - ➢ A first rise, ranging from -110 to -20, which would be assimilated to the invasion of the carbonate formations of the Valanginien;
  - ➢ A second ascent, ranging from -20 to the sea coast for a distance of about 1000m. The latter, less clear than the first, would indicate the salt invasion in the higher levels, attributed to the limestones of Dridrate.

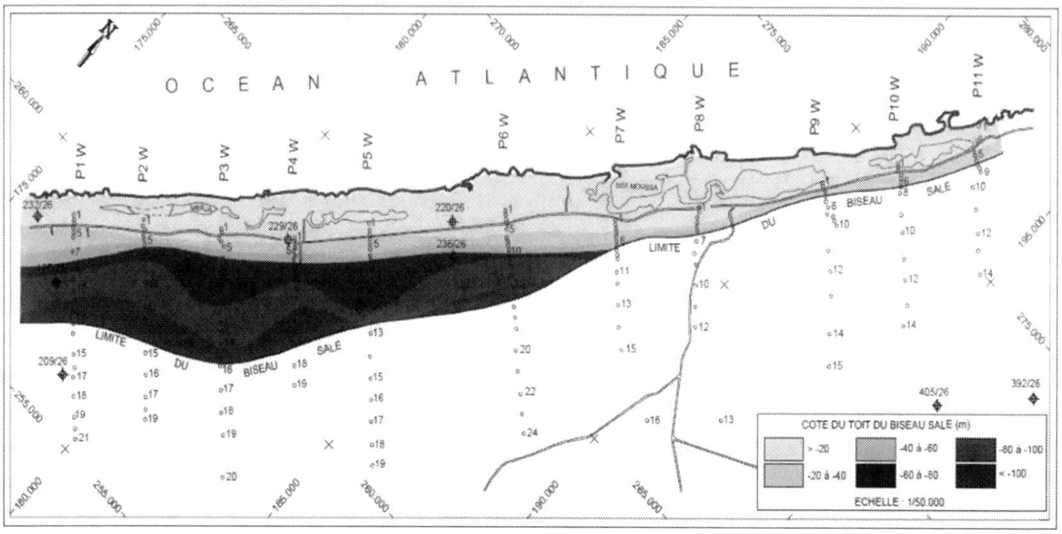

Fig.11. Map of the isohypses of the salted bevel roof (freshwater / saltwater interface)

IV-3-5 - Synthesis map (Fig. 12)

This map shows the hydrogeological limits of the Hauterivo-Valanginian aquifer system, deduced from the roof maps of Resistant Rd (or Rdv) and the Safi marl wall (roof of Rv or Cv), as well as electrical cuts. We observe successively downstream upstream:

- The marine invasion limit of the Dridrat limestone, located between the eastern borders of the Merjas and the Safi - El Jadida coastal road. A strip of approximately 300 to 600 m completely invaded is defined (zone I indicated on the document);
- The limit of invasion of valanginian limestones is behind the first zone: about 1000 m wide, the Valangin aquifer is saturated with salty water. Only Dridrat limestones would contain "soft" water in Zone II;
- Upstream in zone III, the two carbonate levels seem free of any salt contamination. This area represents about 65% of the area studied, which represents an uncontaminated area of nearly 110 km^2.

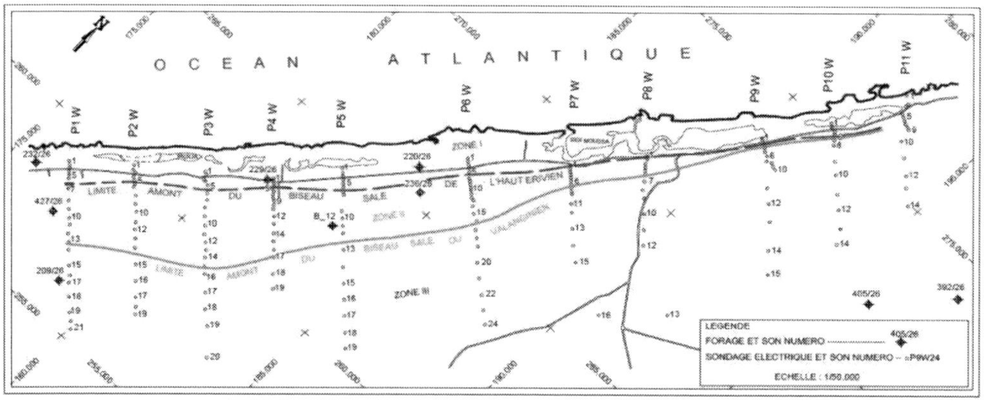

Fig.12. Synthesis map

V- CONCLUSION

The electric prospecting by electrical soundings and the hydrogeological synthesis made it possible to propose a diagram on the aquifer system of the Walidia sahel and its downstream limit of equilibrium with the oceanic waters :

1- The salted bevel :

- The salt wedge is followed along the entire ocean edge. The contaminated zone has a fringe with an average width of approximately 2300m (900 m with a P11 profile at 3800 m with a P3 profile);
- The differentiation of two distinct boundaries, materializing two salty bevels, storied and of variable importance:
  - ➢ The upper salty wedge of the Middle Hauterivian (Dridrat limestones) is the least important, remaining at a short distance from the shore (600-1400 m). this weak invasion of Dridrat limestones is evidence of their good hydrodynamic characteristics;
  - ➢ The lower saline bevel of the Valanginian penetrates further upstream into probably karstified limestones. The invasion limit is steeper, its upstream end extends up to 3800 m from the shore.

2- The identification of two aquifer systems :

- The aquifer, represented by Dridrat limestones of about 30 m of power (when it is not subject to erosion pre-quaternary). The sheet of these limestones is rarely in charge, the piezometric level does not exceed the zero rating;
- Below is a potential aquifer represented by past marly Valanginian limestones. It is separated from Dridrat limestones by Safi marls whose thickness is variable and most often reduced. Locally a significant decrease in the thickness of these marls or a total disappearance can allow the two aquifers to communicate.

3- To the East and North-East of Sidi Moussa, the highlighting of Cretaceous limestones due to a significant rise of the Jurassic formations. In this sector, only the upper Jurassic resistant horizon could present a potential objective. However the interest is limited, due to the presence of anhydrite and gypsum in these levels.

BIBLIOGRAPHY

1. ONI, 1963 : Etude par prospection électrique dans la plaine des Doukkala. Région de Sidi Bennour, 16 Août-20 Novembre (Compagnie Africaine de Géophysique).
2. M .Ferré and J. P. Ruhard, « Les Bassins des Abda Doukkala et du Sahel d'Azemmour à Safi. Plaines et bassins du Maroc Atlantique » (éditions du service géologique du Maroc), Ressources en Eau du Maroc, Tome II.
3. Aboumaria KH., 1993. Les formations quaternaires du Sahel des Doukkala (Meseta occidentale marocaine) : mise en place et évolution post–sédimentaire. D.E.S., Thèse de 3ème cycle Fac. Sci. Rabat , 8tab .,88 fig., 186p.
4. Akil M., 1990. Les dépôts quaternaires littoraux entre Casablanca et Cap Beddoza (Méseta côtière marocaine) : Etude géomorphologique et sédimentologiques. Thèse doc. Etat, Fac. Sci. Rabat. 417p.
5. Astier, J.L., 1971. Géophysique appliquée à l'hydrogéologie. Masson, Paris.
6. DRPE, 1992. Etude par prospection électrique : Région des Abda et des Doukkala, (21 mars 1990-10 juillet 1990, 08 octobre 1992-15 novembre 1992). GEOATLAS. Tome 1 : texte et planches.
7. Fakir Y., 1991- Caractérisation hydrogéologique et hydrochimique des aquifères côtiers du Sahel de Safi à Oualidia (Méséta côtière- Maroc). Thèse 3éme cycle. Univ. Cadi AYYAD. Marrakech.

8. Gigout M., 1952. Les ressources en eau souterraine des Doukkala et Abda (Maroc occidental). Rapp. Inéd. Second. Thèse.
9. Merrad S., 1997. Contribution à l'étude hydrogéologique et hydrochimique des aquifères côtiers du Sahel entre Sidi Abed et Oualidia (Maroc). Mém. C.E.A., Univ. Chouaib Doukkali. El Jadida.
10. Souhel A. et El Achheb A., 2000. Cadre géologique des principaux aquifères de la plaine des Doukkala. Actes de la 2 ème session de l'Université de Printemps des Doukkala Abda, Pub. Univ. Chouaib Doukkali-El Jadida, n° 3, pp. 71-73.

Solar system design for water pumping

HADIDI *ABDELKADER* [(1)*], YAICHI *MOHAMMED*[(1)]

[1] Unité de recherche en Energie Renouvelables en milieu saharien, URERMS, Centre de Développement des Energies Renouvelables, CDER, 01000, Adrar, Algeria

Abstract. In our days, it seems to us that nobody can suspect it on the importance of water and energy for the human needs. With technological advances, the energy need does not cease increasing. This problem of energy is even more sensitive in the isolated sites where the use of the traditional resources proves often very expensive. Indeed, several constraints, like the transport of fuel and the routine maintenances of the diesel engines, return the search for an essential alternative energy source for this type of sites. It summer necessary to seek other resources of energy of replacement. Renewable energies, like photovoltaic energy, wind or hydraulic, represent a replacement solution par excellence and they are used more and more in our days more especially as the national territory has one of the solar layers highest with the world. The duration of insolation can reach the 3900 hours/year on the Sahara. The energy acquired daily on a horizontal surface of 1m2 is about 5kWh, that is to say meadows of 2263kWh/m2/year in the south of the country. The photovoltaic energy utilization for pumping of water is well adapted for more the share of the arid and semi-arid areas because of the existence in these areas of an underground hydraulic potential not very major. Another very important coincidence supports the use of this type of energy for the water pumping is that the demand for water, especially in agriculture, reached its maximum in hot weather and dryness where it is precisely the moment when one has access to the maximum of solar energy. The goal to see an outline on the general composition of a photovoltaic system of pumping, as well as the theoretical elements making it possible to dimension the current pumping stations.

Keywords: Water, Energies Renewable, Photovoltaic Pumping, Requirement out of water, Agriculture.

*Corresponding author: hadidiabdelkader@gmail.com

1. Introduction

Small-scale pumping is one of the most interesting uses of solar energy. Indeed, the maximum intensity of solar radiation generally corresponds to the period of the most important pumping water requirements. On the other hand, the fact that this energy is available just at the point of use, the farmer is freed from problems related to the supply of fuel, or the existence of easily accessible transmission lines. At present, the main obstacles hampering the use of solar pumps on a larger scale are their high cost and the too-recent nature of this technology. The development of a reliable and reasonably priced solar pump - which would be very likely within a few years - could be a blow to agriculture in the third world.

Pumping in the Saharan regions is the only solution for the irrigation of agricultural land and drinking water. As a result of the field experience and the various studies carried out, irrigation by photovoltaic pumping systems is the best way for a rapid development of agriculture in the Saharan regions [1]. Particular attention should be given to this project considered strategic and this for several reasons among them: Richness of the water region at shallow depths (Pm=15m); very large area of land that can be exploited and stability of the population in isolated areas.

2. Pumping methods

To pump water with a photovoltaic system, two techniques are possible:
In the first technique, solar energy is consumed in "real time"; This is called "pumping in the sun". This solution requires the storage of water in a tank (water pumped during the day is stored for later use in the evening, for example). The second method is to use energy storage, this time, via batteries. The energy stored in the day can be used later to pump water [2].

3. Datas and Methodes

The pumping stations consist of one or more pumps which suck up volumes of water and relieve them under a certain pressure in the pipes of the network (see Figure 1).

3.1. Solar pump system

A solar pumping system usually consists of [3]:

The photovoltaic generator, the electric pump unit, control electronics and control and the storage part.

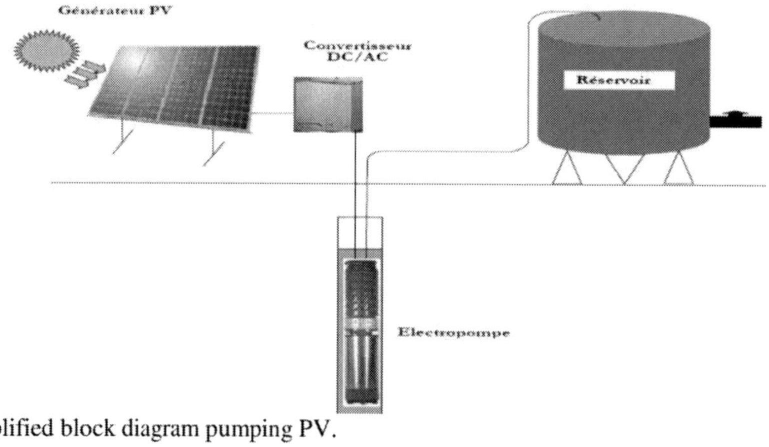

Fig.1. Simplified block diagram pumping PV.

3.1.1. The photovoltaic generator

For higher powers, it is necessary to combine serial and parallel multiple modules (Figure 2). In some applications, it is possible to use one or more modules of several tens of cells. For larger applications, the PV generators are grouped in a field of several modules (a few hundred).

The curve of I-V operating a seriated-parallel association of solar modules will have a similar evolution operating in the basic cell curve, but by changing the échèles on both axes. The performance of a PV generator are determined from these curves.

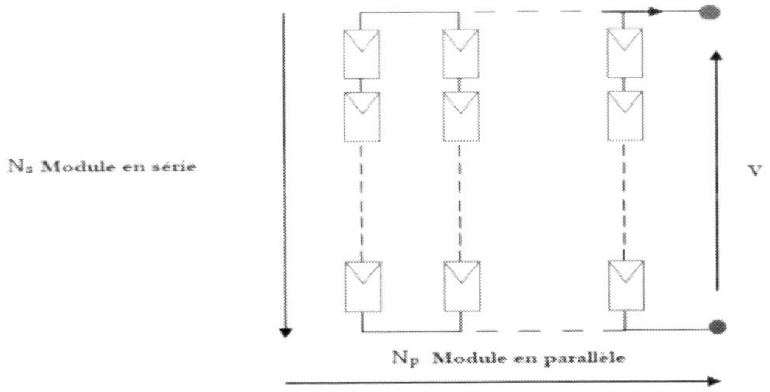

Fig.2. Module Ns put in series and Np in parallel.

Using this equation below for a mixed group Ns and Np:

$$I = N_p I_{Ph} - N_P I_0 \left[e\left(N_s V + (IN_s R_s)/\mathrm{N}_P \right) - 1/nN_s N_T \right) \right] - \frac{N_s V + I_s R_s / N_p}{\frac{NsR_p}{N_P}}$$

(1)

3.1.2. The electric pump unit

A pump is a device for sucking and discharging a fluid. There are two main types of pumps: centrifugal pumps and positive displacement pumps. These are suitable for raising low water flows at high pressures [2,4].

➢ The centrifugal pump

The centrifugal pump conveys the kinetic energy of the fluid motor by a rotational movement of rotary wheels or fins. The water returns to center of the pump will be pushed outward and upward through the centrifugal force of the blades.

➢ Characteristics of a centrifugal pump

- Centrifugal pumps are widely used for applications with photovoltaic energy because the couple drive of the pump is zero at start ;
- The pump runs with very low sunshine, the motor can provide fast speed nearly constant ;
- Use for high flows.

The parameters necessary to dimension the solar pump are the operating flow and the height at which the pump will have to pump. This is increased by pressure drops and pressure in the pipe.

The flow delivered by a pump is the amount of water it delivers during a given time interval. In solar pumping, the flow is often expressed in m3 per day.

3.2. The general pressure drop

The general pressure drop depends on:

• The pressure drop is logically directly proportional to the length of the pipe: it increases when the pipe length increases.

• When the diameter decreases, the pressure drop increases considerably. The liquid has more difficulty flowing so friction increases for an identical flow rate.

• The higher the flow rate (higher speed), the greater the frictional forces for the same diameter.

3.2.1. The total head gauge Hmt

This is the difference in pressure in meters of water column between the suction and discharge ports. This height can be calculated as follows [5]:

$$Hmt = Hg + \Delta p \qquad (2)$$

With:

• Hg: Geometric height between the pumped water table (dynamic level) and the utilization plane (see Figure 3). It is calculated by the following formula:

$$Hg = A + B + C \qquad (3)$$

• Δpc: Pressure losses produced by the friction of water on the walls of the pipes. They are expressed in meters of water and are a function of distance lines (D), their diameter and the pump capacity.

The calculation of the linear head loss, that corresponding to the general flow in a rectilinear duct, is given by the following this equation:

$$\Delta P = \rho g \Delta H = \lambda L \rho V^2/(2D) \qquad (4)$$

• Δp: linear pressure loss in Pa, λ: pressure loss coefficient (number without dimension), p: density of water in kg /m3, V: flow velocity in m/s, D: diameter Hydraulic tube in m, L: length of tube in m.

• Ns=C+B: The static level in a well or borehole is the distance between the ground and the pump before the water surface. [3 ,4]

• Nd=C+B+A: The dynamic level of a well or borehole is the distance the soil and the surface of the water for pumping to a given flow rate. For the calculation of the HMT, the dynamic level is calculated for an average flow rate.

• Folding=Nd-Ns: The difference between the dynamic level and the static level.

• Maximum Drawdown: is the maximum acceptable drawdown before stopping the pump.

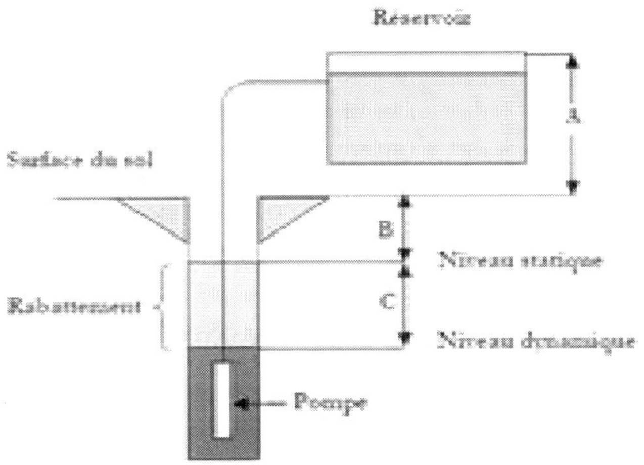

Fig.3. Hauteur gauge total.

4. Puming System Design

Two methods are used for sizing photovoltaic pumping systems: an analytical method or graphical method. These methods allow to design a photovoltaic pumping system to meet the water needs of a determined consumption.

4.1. Graphical method

These are graphic facilitating direct reading digital calculations. Graphics for spontaneously determine the result of calculations in a system of predefined lines and prepared beforehand, using the performance charts pumps supplied by the manufacturer that provides a function of irradiation overall Pc power required developed by the panels to operate the pump in this range of flow, and Hmt. These are quite handy (Figure 4) and summarize the performance of different pumps for the conditions of their use [5,6]. The graphs are operated by a direct reading without having to perform additional plots by reading data directly lying at the intersection of corresponding lines by reading the dot competing in relation to the needs of the speaker. The equipment manufacturers are developing such diagrams based on calculated or measured data. This kind of graphs shows the possible configuration of an electric pump.

https://doi.org/10.1051/e3sconf/20183706001

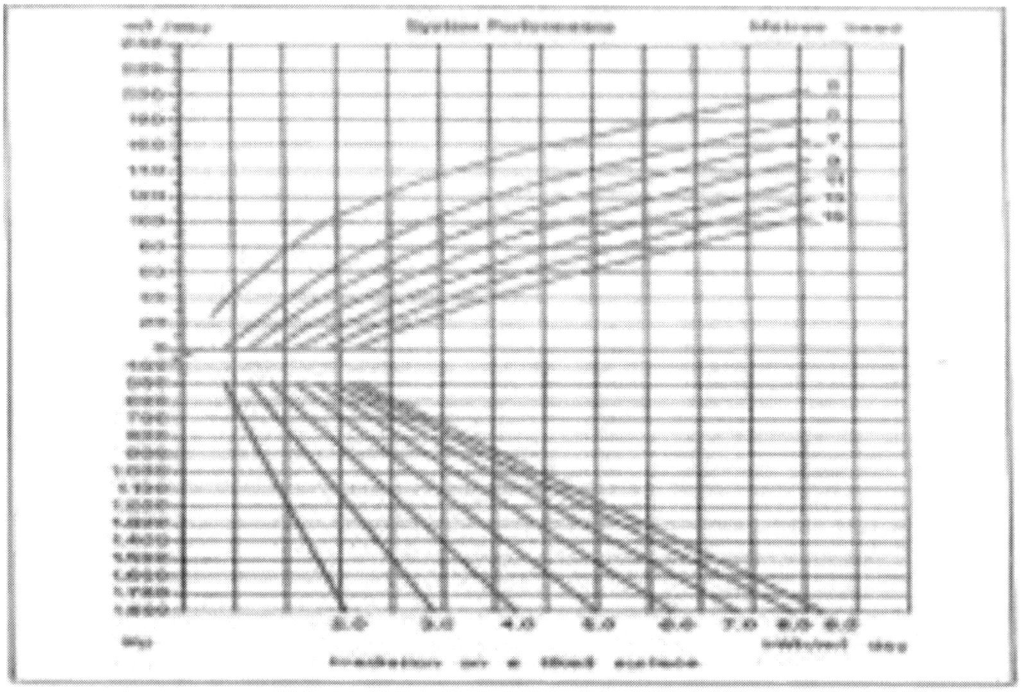

Fig.4. Typical performance curve of pump conditions of use.

4.2. Analytical method

The different steps for the design of a pumping system are [5,6]: Assessment of water needs, water power calculation necessary, determination of solar energy available and component selection. Once known the daily volume Q(m3/day), the total head Hmt and the incident average irradiation G in terms of generating the corresponding rated power of the photovoltaic generator, Pc is calculated by the following expression [7] :

4.2.1 Estimates of water requirements

For a tropical region, water requirements can be defined using the following values of the Table1.

Table 1. Hydraulic Region data

Humans: 1000hab, Person 5 to 10 liters/day minimum, normal living conditions 30 liters/day
Animals: 100 Sheeps and goats 5 liters/day, 8 Horses 40 liters/day, 2 Ane 20 liters/day, 4 Camel 20 liters/day
Irrigation: Agriculture maraichère 60m^3/hectare/day for 1 hectare, Rice 100 m^3/ha/day, cane sugar 65m^3/ ha/day, Cotton 55 m^3/ha/day

We summarize the calculated in Table 2

Table 2. Water requirements

Number N		daily hydraulic Q [m^3/day]	
N_1	1000	Q_1	30
N_2	100	Q_2	0.005
N_3	8	Q_3	0.04
N_4	2	Q_4	0.02
N_5	4	Q_5	0.02
N_6	0.25	Q_6	0.06
N_7	0.25	Q_7	100
N_8	0.25	Q_8	65
N_9	0.25	Q_9	55

$$Q=N_1Q_1+(N_2Q_2+N_3Q_3+N_4Q_4+)+(N_5Q_5+N_6Q_6+N_7Q_7+N_8Q_8+N_9Q_9)$$
$$=30+0.5+3.2+0.04+4.2+15+25+16.25+13.7=\mathbf{107.94m^3\ /day} \quad (5)$$

4.2.2. Calculation of daily hydraulic energy required

Once the necessary volume of water needs for each month of the year and the characteristics of the well defined, we can calculate hydropower daily and monthly average needed from the relationship [8]:

$$E_h = Ch \cdot Hmt \cdot Q \quad (6)$$

with:

Eh: Hydropower kwh/day ; Ch: Constant hydraulic [kg.s.h/m2]=g.ρ/3600, g: the gravity (9.81m/s2), Q: Water volume (m3/day), Hmt: Total height gauge (m) and ρ: Density of water (1000kg/m3).

$$Hmt= 20 +10\% \cdot 20=22m \quad (7)$$

Ch=9.81·1000/3600=2.725kg.s.h /m2, so $Eh=2.725 \cdot 22 \cdot 107.94=6471.001Wh/day$ (8)

4.2.3. Calculation of daily electrical energy required:

The energy required to raise a certain quantity of water to a certain height during a given day is calculated from the following equation:

$$Ee= Eh/(\eta mp \cdot \eta ond) \quad (9)$$

Where, Ee: Electric energy expressed in kWh/day, ηmp: The performance of the pump unit, generally between 30% and 60%, ηond: The inverter efficiency.

ηmp·ηond=0.264 (we use 0.3), So, Ee= 6471.001/0.3= 21750Wh/day (10)

4.2.4. Determination of the available solar energy

The design method used is based on the calculations of monthly daily average values of solar radiation Gd available to the inclination β of the photovoltaic (PV) relative to the horizontal plane. This must be done to maximize the conversion of sunlight into electrical energy [7,8].

In our Région (Adrar province) $Gd=5.8kW/m2/day$ (11)

4.2.5. The Size of the PV array

The power Pc provided by the PV generator in the measurement standard conditions CSM (Illuminance (1000W/m2) and the temperature 25°C) [8].

$Pc = \eta g \cdot A \cdot G$ (12)

with:

Pc: The power output of the generator (W) under MSC (peak power).
ηg: The efficiency of the generator to the reference temperature (25 °C).
A: The active area of the generator (m2).
G: The illumination in the MSC conditions.

The daily electrical energy is given by the equation:

$Ee = \eta pv \cdot A \cdot Gd(\beta)$ (13)

ηpv: The average daily output of the generator in the operating conditions ;
Gd(β): The average daily radiation incident on the level of modules β inclination (kWh/m2/day)
ηpv: The yield is calculated by the following formula:

$\eta pv = Fm(1 - \gamma (T-Tr)\, \eta g$ (14)

Or :

Fm: Coupling factor, defined as the ratio between the electrical energy generated under the conditions of operation and electric energy that would generate in the system working at maximum power point.
γ: Cell temperature coefficient. γ assumes values between 0.004 and 0.005/°C for modules in mono and poly crystalline silicon, and 0.001 and 0.002 for amorphous silicon modules.
T: average temperature of the cells during hours of ensoleillement.
The calculation of power in watts peak.
Substituting equations (9) (12) (13) (14), we obtain the peak power of the generator:

$$P_c = \frac{G}{F_m[1-\gamma(T-T_r)]G_d(\beta)} \cdot \frac{E_h}{\eta_{MP} x \eta_{Ond}} \quad (15)$$

We can put directly:

$Pc=Ee/(Gd)=21570/(0.7 \cdot 5.8)=5313Wc$ (16)

0.7 : average daily efficiency
We can choose panel PV with her peak power Ps =75Wc,
Module of total Number puted in series and in parallel Nt

$Nt=Pc/Ps=5313/75=71module$ (17)

So we must take the correct $Pc=71 \cdot 75=5325Wc$ (18)

4.2.6. Pump dimensioning

The choice of the pump is based on the following two factors: The discharge head and HMT The hourly flow rate Qh.

$$Q_h = \frac{Q\left(m^3/day\right)}{h} \quad (19)$$

h: is the number of hours maximum sunshine in (1000W/m2)

$Q_h = 107.94/5.8=18.61=19m^3/h$ (20)

5. Conclusion

In order to define the useful power to the operation of the pump developed by the solar panels, the minimum data are: The geographical location to determine the overall solar irradiation, the daily water requirements during the maximum need period to determine the flow, the total head to measure the static level, maximum drawdown, tank height and pressure drops due to the piping.
For define the monthly total daily irradiation kWh/m2/day with respect to the latitude of the working area, choose a pump according to the flow rate and the total head, (use the performance charts of the pumps supplied by the manufacturer), select the inverter suited to the pump (voltage and power), determine the peak power Pc required for the operation of the pump by the analytical or graphic method, select the type of solar panel (set their nominal power Pn), determine the number of panels by Pc/Pn relationship, check the rated voltage for inverter operation (depending on model), determine the number of serial/parallel modules (the connection form).

6. Bibliographie

[1] S. LABED, "Le pompage photovoltaïque et le développement des régions sahariennes", Colloque International sur les Ressources en Eau Souterraines dans le Sahara (CIRESS) Ouargla - 12 et 13 décembre 2005. Pompage Photovoltaïque "Rev. Energ. Ren. Vol. 8 (2005) 19-26.
[2] A. Labonne, "Alimentation d'une pompe à Burkina Faso", 2004.
[3] B. Molle, "Les stations de pompage individuelles pour l'irrigation", juin 1996.
[4] J. Royer, T. Djako, "Le pompage photovoltaïque", Manuel de cours à l'intention des ingénieurs et des techniciens", Université d'Ottawa, 2002.
[5] A. Hadidi, "Synthèse Bibliographie sur le pompage solaire et le dimensionnement de la station de pompage photovoltaïque", Rapport annuel, Mai 2016.
[6] A. Djafour, "Etude d'une station de pompage photovoltaïque dans une région saharienne", Thèse de magistère, université de Ouargla, 2000.
[7] A. Hadj Arab, M. Benghanem et A. Gharbi, "Dimensionnement de Systèmes de Pompage Photovoltaïque", Rev. Energ. Ren. Vol. 8 (2005) 19 – 26.
[8] B. Multon, H. B. Ahmed, N. Bernard, « Les moteurs électriques pour applications de grande série », Pierre-Emmanuel CAVAREC Antenne de Bretagne de l'École Normale Supérieure de Cachan, Revue 3EI juin 2000.

Modeling and simulation of the shading effect on the performance of a photovoltaic module in the presence of the bypass diode.

Mohamed Zebiri[1,*], *Mohamed* Mediouni[1], and *Hicham* Idadoub[1]

[1]LASIME, National School of Applied Sciences, BP 1136, Agadir, Morocco

Abstract. In photovoltaic renewable energy production systems where production is dependent on weather conditions, maintaining production at a suitable level is more than essential. The shading effect in photovoltaic panels affects the production of electrical energy by reducing it or even causing the destruction of some or all of the panels. To circumvent this problem, among the solutions proposed in the literature we find the use of by-pass diode and anti-return diode to minimize these consequences.In this paper we present a simulation under Matlab-Simulink of the shading effect and we compare the current voltages characteristics (I-V) and power voltage (P-V) of a photovoltaic system for different irradiations in the presence and absence of diode by -pass. For modeling, we will use the diode model and the Lambert W-function to solve the implicit equation of the output current. This method allows you to analyze the performance of a panel at different shading levels.

Key word*: Photovoltaics,* modeling *and simulation, shading, bypass diode, W-function of Lambert.*

1. Introduction

The power produced by a photovoltaic field is the sum of the powers of all the cells composing it [1], but in reality, it is lower there, and that depends on the meteorological conditions. Among the conditions affecting this production we encounter the shading effect that will be presented in this paper. In a field of photovoltaic panels, some modules or cells of a module can be under the shade caused by clouds, trees, buildings, bird droppings.... Moreover, in these conditions, the PV characteristics become more complex and nonlinear they vary with solar insolation and have multiple peaks.

However, it is important to understand and predict them in order to extract as much power as possible. For this reason, over the years, researchers have been studying the characteristics of PV modules and the meteorological factors affecting them. Herin Patel [2] compiled a list of the work done in the study of these characteristics during changes in insolation, temperature or load and their effect on available power.

* Corresponding author: m.zebiri@uiz.ac.ma

To study the shading effect we will simulate three modules in series that are subject to different irradiations and we will compare the characteristics obtained in the case of presence of diode bypass and in its absence.

2. Materials and methods

2.1. Photovoltaic equivalent model

An equivalent circuit can represent a photovoltaic cell. Several models used in the literature and vary in complexity and precision, operating in various conditions. The one-diode model is the most widely used to represent a PV cell, thanks to its simplicity and precision [1, 2, 3, 4] figures one, it consists of a photocurrent source (I_{ph}), a diode, an internal resistance (R_s) and a shunt resistor (R_h). The model with by-pass diode figure 2, [4, 5, 6] is used to study the effect of shading on the photovoltaic modules using this diode.

2.1.1. Model without bypass diode

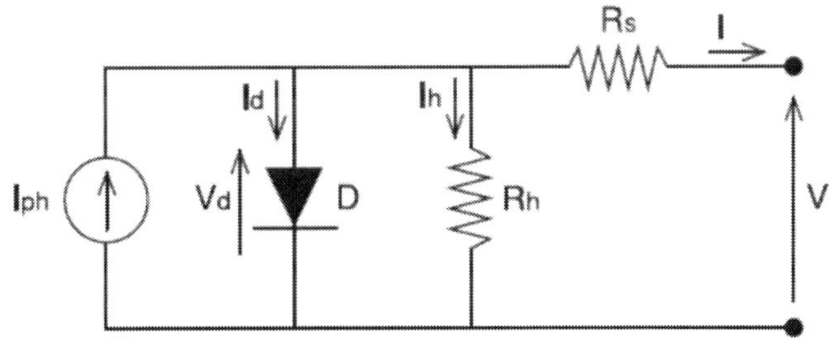

Fig. 1: Electrical model of a PV panel.

$$I = I_{ph} - I_d - I_h \tag{1}$$

$$I_h = (V+I_sR_s)/R_h \tag{2}$$

$$I_d = I_{sat,d}\left(e^{(\frac{V_d}{V_{t,d}})} - 1\right) \tag{3}$$

$$I = I_{p\square} - I_{sat,d}\left(e^{(\frac{V_d}{V_{t,d}})} - 1\right) - (V + I_sR_s)/R_{\square} \tag{4}$$

2.1.2 Model with bypass diode

This is the same model as in Figure 1 with a diode D_b in anti-parallel.

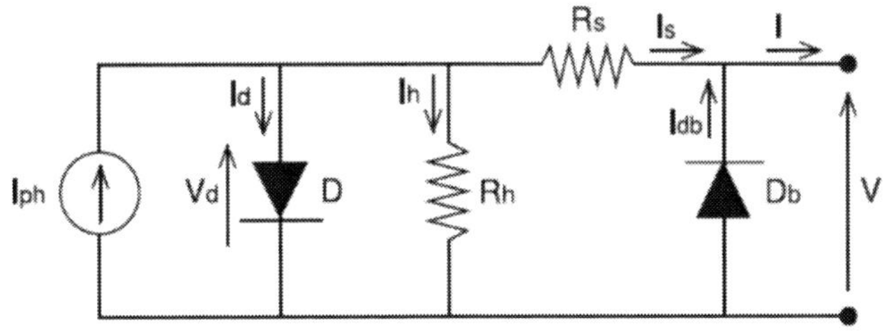

Fig. 2: Electrical model of a PV panel with bypass diode Db [2].

In the normal operating mode of the photovoltaic panel, the open circuit voltage is of the order of 0.6 V for the crystalline cells while in reverse bias operation, the voltages can reach more than -20V [4, 5, 6]

Kirchhoff's law gives the current generated by the PV cell:

$$I = I_{ph} - I_d - I_h + I_{db} \tag{5}$$

$$I_h = (V+I_sR_s) / R_h = (V+ R_s(I-I_{db})) / R_h \tag{6}$$

$$I_{db} = I_{sat,db}\left(e^{(\frac{-V}{V_{t,db}})} - 1\right) \tag{7}$$

$$I = I_{ph} - I_{sat,d}\left(e^{(\frac{V_d}{V_{t,d}})} - 1\right) - (V+ R_s(I-I_{db})) / R_h + I_{sat,db}\left(e^{(\frac{-V}{V_{t,db}})} - 1\right) \tag{8}$$

With I_{db}, and V_t, are respectively current db and thermal voltage of the bypass diode.

The system of equations (5, 6, 7, 8) clearly shows that the current I of the PV is a non-linear and implicit function of the voltage V, the irradiation level and the temperature [5]. Nevertheless, such a non-linear system can be solved in one of the computing environments, such as Matlab. In this way, we can obtain a nonlinear relation between the current I and the voltage V. Such a relation is reported in the equation (9) [5, 7] it makes use of the function W of Lambert of the term θ of which the value depends on the voltage V and is reported in equation (10)

$$I = \frac{R_{s\square}(I_{p\square}+I_{sat,d})-V}{R_{s\square}+R_s} + I_{sat,db}\left(e^{\frac{-V}{V_{T,db}}} - 1\right) - \frac{V_{T,d}}{R_s} LambertW(\theta) \tag{9}$$

With:

$$\theta = \frac{(R_h R_s/(R_h+R_s))I_{sat,d}e^{\left[R_h R_s(I_{ph}+I_{sat,d})+R_h V/V_{T,d}(R_h+R_s)\right]}}{V_{T,d}} \tag{10}$$

2.2. Partial shading and bypass diodes

In nature the uniform irradiation is not always satisfied because of the shadows of buildings or trees, the fluctuation of the atmosphere, clouds and variations of the solar angle and the loss of power occurs precisely by the effect of shading. The impact of this effect is reduced by several factors including the bypass diode. In practice, a module contains bypass diodes to prevent damage due to reverse bias on shaded cells. These diodes are placed in anti-parallel with a group of 18 cells in general.

The bypass diode will start driving as soon as a shaded cell is reverse biased. It allows the current from the non-shaded modules to shunt around the shaded cells through this diode thus protecting this cell group and limit the shading effect to only the neighboring group of cells protected by the same bypass diode. The effect of shading on the output of a PV module is non-linear because a small amount of shadow on part or the entire module can result in a significant reduction in output power.

If the generated current (I_{ph}) of a module decreases to a level lower than the current generated by all the modules in series, the bypass diode limits the reverse voltage to a value lower than the breakdown voltage of the PV cells. In other words, and as illustrated in figure 5, the third bypass diode begins to operate when equation (11) is satisfied.[8]

$$I_{pva} > I_{ph}\ (i) \tag{11}$$

2.3. Simulation

In our simulation we used three modules (Kyocera KC130GT figure 4) connected in series figures 3a and 3b. The modules receive different irradiations. At first, we conduct our simulation without bypass diode figure 3a, and then we use a bypass diode for each module figure 3b.

The numbered modules 1, 2, 3 are illuminated first by an irradiation of 1000 w / m² each in both cases, i.e. without shadow, then they will be illuminated respectively by following variants irradiations between (1000 w / m², 600 w / m² and 200 w / m²) in both configurations, see table 1.

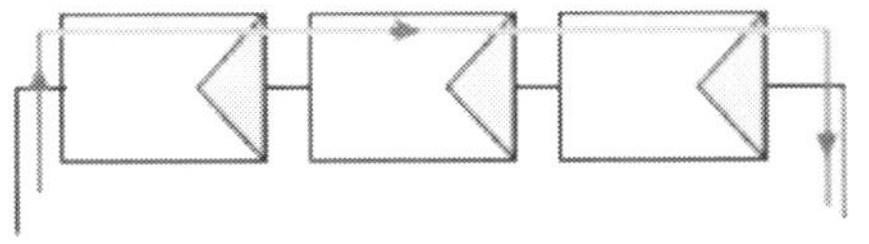

Fig. 3a: Three PV modules connected in series without bypass diode.

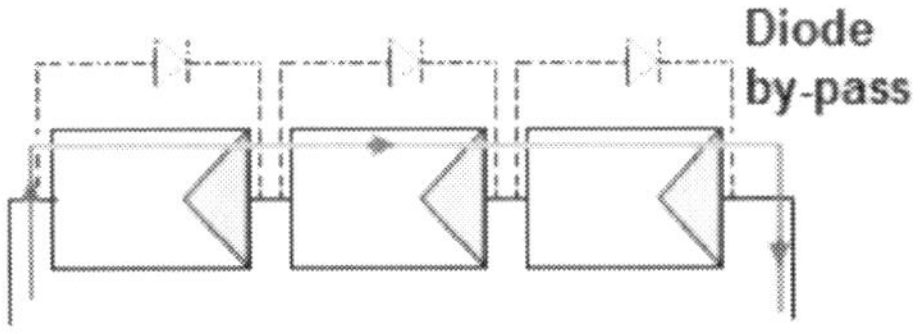

Fig. 3b: Three PV modules connected in series with bypass diode.

■ Electrical Performance under Standard Test Conditions (*STC)	
Maximum Power (Pmax)	130W (+10%/−5%)
Maximum Power Voltage (Vmpp)	17.6V
Maximum Power Current (Impp)	7.39A
Open Circuit Voltage (Voc)	21.9V
Short Circuit Current (Isc)	8.02A
Max System Voltage	600V
Temperature Coefficient of Voc	-8.21×10^{-2} V/℃
Temperature Coefficient of Isc	3.18×10^{-3} A/℃

*STC : Irradiance 1000W/m², AM1.5 spectrum, module temperture 25℃

Fig. 4: Kyocera KC130GT module specifications

Table 1 Different simulation scenarios in the presence and absence of bypass diode.

	Irradiation in w/m² of Module 1	Irradiation in w/m² of Module 2	Irradiation in w/m² of Module 3
1° scenario	1000	1000	1000
2° scenario	1000	1000	200
3° scenario	1000	600	200

 https://doi.org/10.1051/e3sconf/20183706002

3. Results and discussion

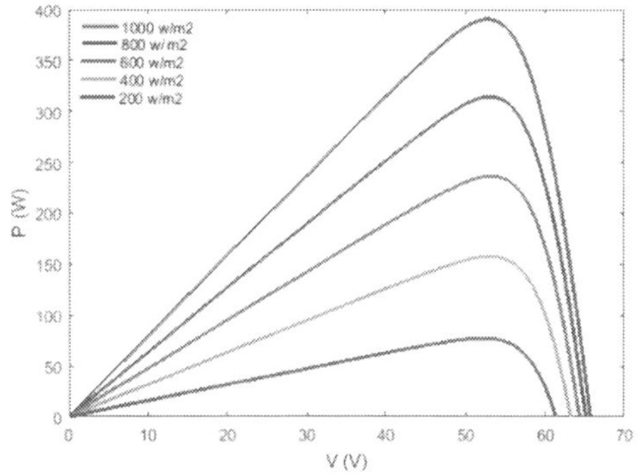

Fig. 5a: P-V Characteristics of three PV cells with different irradiance without bypass diode.

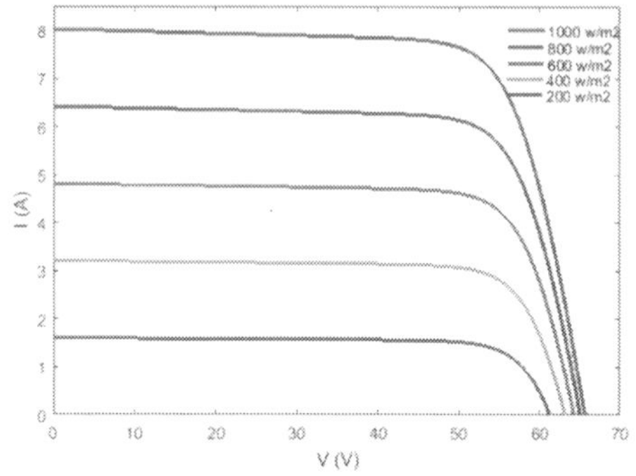

Fig. 5b: I-V Characteristics of three PV cells with different irradiance without bypass diode.

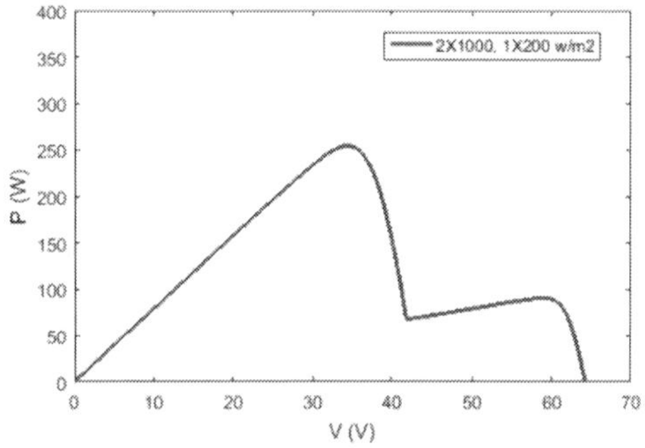

Fig. 6a: P-V Characteristics of three PV cells with different irradiance with bypass diode.

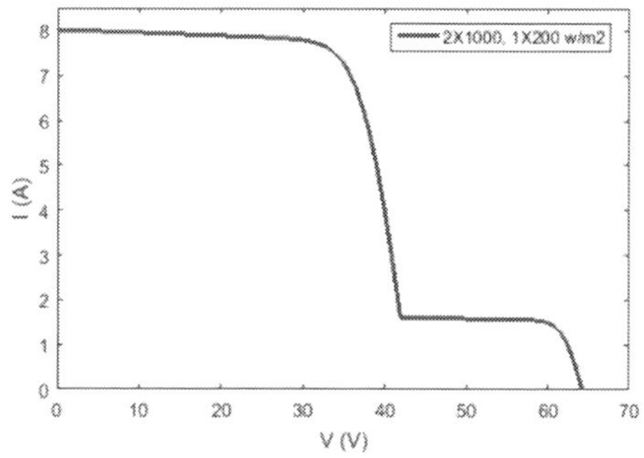

Fig. 6b: I-V Characteristics of three PV cells with different irradiance with bypass diode.

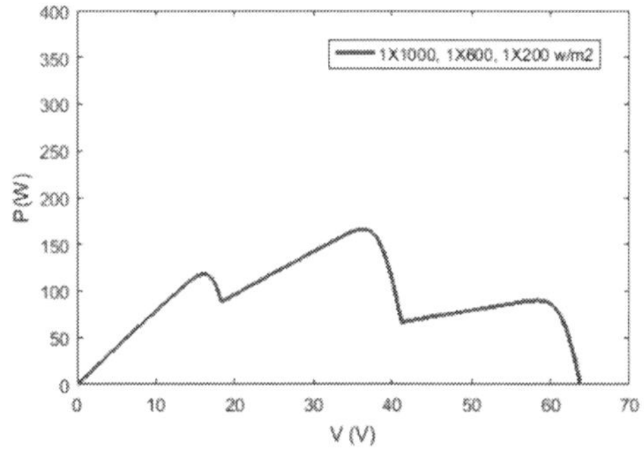

Fig. 6c: P-V Characteristics of three PV cells with different irradiance with bypass diode.

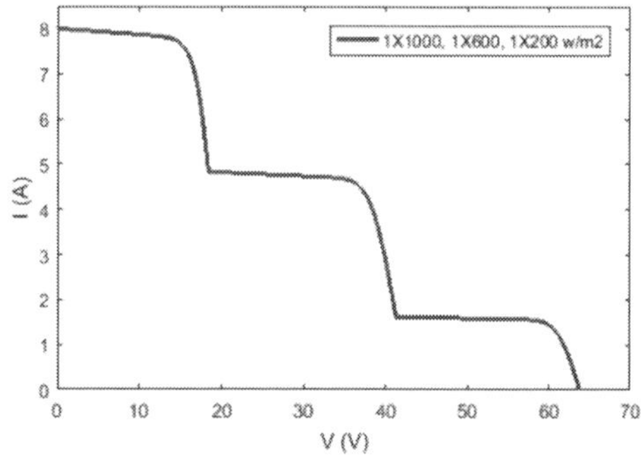

Fig. 6d: I-V Characteristics of three PV cells with different irradiance with bypass diode.

Figures 5a and 5b show the characteristics (P-V) and (I-V) of the PV modules with variable illumination but the same for all the modules, at constant temperature in the presence and absence of the bypass diode. From the graph we find that as the irradiation decreases, the current and the output voltage also decrease, which implies the decrease of the output power.

Without a bypass diode, the output power of the three modules is severely affected and is equivalent to the power of the least lighted module. We notice that the shading causes a significant degradation of the power output.

On the other hand, in the presence of a bypass diode the power is maintained at higher levels and we observe global and local maximum peaks present in the characteristic (PV) figure 6a and 6b and the power produced is greater figure 7a than in the absence of the bypass diode figure 7b. The figure 8 shows the path taken by the current produced by the modules in full sunlight. The diode becomes active and in this state, its impedance becomes lower than that of the shaded module and the current of the other modules shunts around the module that through this diode. This prevents the unshaded modules from forcing a current against the polarized state in inverse of the cells of the shaded module and therefore, overheating (hot spot) is avoided for the module under shading [9].

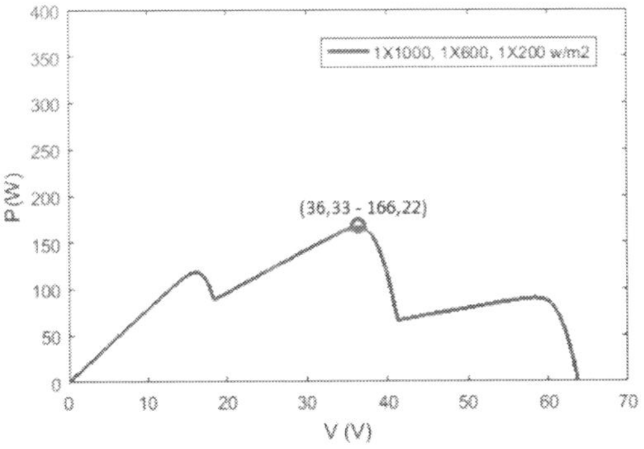

Fig. 7a: Characteristic (P-V) at different illuminations in the presence of bypass diode.

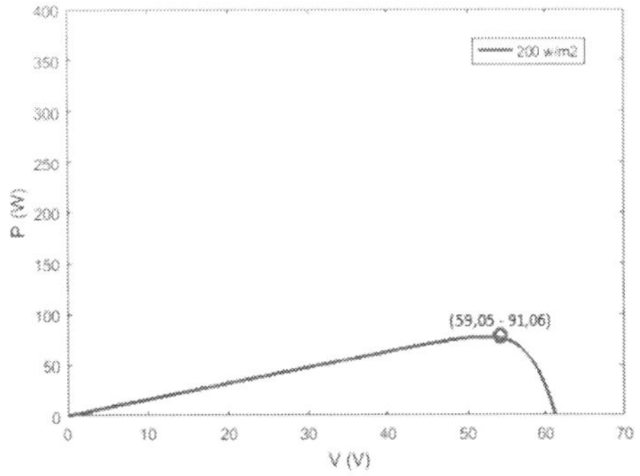

Fig. 7b: Characteristic (P-V) at the same illuminations as in 7a but without a bypass diode.

In figure 7 we compare the maximum power point of the characteristic (P-V) with bypass diode figure 7a and without bypass diode figure 7b. We note that with a bypass diode the maximum power is 166.22 w while without a bypass diode it is only 91.06 w. Because of the presence of a bypass diode, the non shaded module can still produce full power and PV system gets higher power output.

We can say that a bypass diode :

1 Shunte the current of the module around the shaded cells figure 8 ;

2 Protects the solar cells in the modules against the formation of hot spots ;

3 Reduces the reverse voltage through the shaded cells in the modules

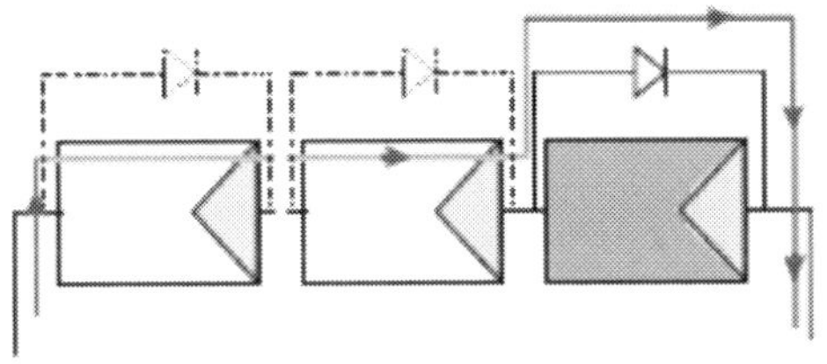

Fig. 8: The path taken by the current in case of shading.

The results obtained by our simulations are the same obtained by works in the literature Table 3 gives a list of some works.

Table 3: List of some works on the shading effect and by-pass diode.

Auteur	Modèle
K. Matter [3]	A diode with Simulink
G. Petrone [5]	Analytical with Lambert W-function
J. D. Bastidas-Rodríguez [7]	A diode with Lambert W-function
R. Ramabadran [10]	A Matlab Analytical Diode
E. Díaz-Dorado [11]	Overlapped and non-overlapped bypass diodes
A. Mermoud [12]	Bypass diode analysis

Conclusion

In this work, we studied the effect of different shading conditions on the output power of PV modules. We used Matlab / Simulink to perform our simulations and used the Kyocera KC130GT module specifications. The results obtained under partial shading conditions clearly show that the performance of the panels varies and strongly depends on these conditions. So note that in the presence of the bypass diode the characteristic (P-V) has several peaks, the number of which depends on the illumination values received by the modules under shading. This feature contains a global spike and other locales. While in the absence of this diode, we have one and the only whatever the insolation of the modules and this peak is the maximum power produced by the least lighted module. We conclude that the existence of these diodes improves the value of the output power in cases where one of the modules is in partial or full shading conditions. The study of the characteristics of a photovoltaic field in the event of shading and creating a database of these characteristics will be used to make adequate decisions and in time to prevent probable failures and thus to ensure a better maintenance intervention.

References

1 L. Bun, *Détection et localisation de défauts pour un système PV*, tel-00647189v1 (2011)

2 H. Patel and V. Agarwal, Senior Member IEEE, IEEE **23**, (march 2008)

3 K. Matter, H. J. El-Khozondar, R. J. El- Khozondar, T. Suntio, IRJET, **02** Issue 02, (2015)

4 G. Notton, I. Caluianu, I. Colda et S. Caluianu,. R E R **13** 49 – 62, (2010)

5 G. Petrone, G. Spagnuolo, M. Vitelli. S E M & S C **9,** 1652–1657, (2007)

6 M.L. Orozco-Gutierrez, J.M. Ramirez-Scarpetta, G. Spagnuolo, C.A. Ramos-Paja. R E **55**, 417 – 427, (2013)

7 J. D. Bastidas-Rodríguez, L. A. Trejos-Grisales, D. González-Montoya, C. A. Ramos-Paja, G. Petrone, G. Spagnuolo, E P S R **155** 67–79, (2018)

8 M. Seyedmahmoudian, S. Mekhilef, R. Rahmani, Y. Rubiyah, E. T. Renani, Energies**,** **6**, 128-144, (2013)

9 P. Bulanyi, R. Zhang, *The 52nd Annual Conference of the Australian Solar Council* (Solar2014)

10 R. Ramabadran, R. Gandhi Salai, B. Mathur, J S D, **2**, No. 3, (11/2009)

11 E. Díaz-Dorado, A. Suárez-García, C. Carrillo, and J. Cidrás, (SPEEDAM 2010)

12 A. Mermoud, T. Lejeune, (25th European Photovoltaic Solar Energy Conference. 2010)

AUTHOR INDEX

Abdelhamid Bouzidi 81
Abdelkader Hadidi 321
Achiban Hassan....242
Aderoju Olaide....198
Afilal Tribak A. 119
Aghzzaf Brahim 285
Ait Taleb A. 93
Ait Yacine Zehor....73
Akhssas Ahmed....285
Alili L. 1, 11, 29
Allouch M. 187
Amechrouq Ali 242
Arifi Karim....64
Arioua Abdelkrim....233
Baite W. 268
Baite Wissal 46
Belbahloul Mounir....110
Belghyti Driss 40
Ben Said H.....268
Benaabidate Lahcen....222
Benbrahim Hafsa 162
Ben-Said E. 1, 11
Bouaakkaz Brahim 257
Bouchaou Lhoussaine....257
Bouignane Aziz....222
Boukdir A. 1, 11, 29, 46, 268, 307
Boukdir Ahmed....198, 233, 285
Boukhalef Laila....129
Bouqbis Laila 129
Boutirame Fatima....285
Boutirame Ikram 285
Chentouf M....187
Cherif Harrouni Moulay 129
Cherkaoui M. 102
Dahbi S. D....46
Darwesh Nabil....40
Dekayir Abdelilah....58
Diouri Mohamed....212
Echakraoui Z.....268
Echakraoui Zhour....198
Ekouele Mbaki V. R....268
El Abid Abdallah 64
El Abidine El Morjani Zine....257
El Alaoui Lamiae 58
El Amrani Idrissi Abdelkhalek....233
El Bakkali Mohammed....40
El Baraka N. 93
El Faleh E. M....180
El Fjiji H. 268
El Hammadi A....93
El Harti Abderrazak....73
El Kharrim Khadija 40
El Maslouhi M. R. 268
El Maslouhi Rachid....198
El Mderssa Mohamed....242
El Qacimi N....93
Elblidi Souad....64
Elfjiji H....307
Elhimri Hamza 257
Elissami A. 46, 268
El-Khantoury Ismail 242
Elmaghnougi I. 119
Essaadaoui Y....102, 138
Fatni A....93
Fekhaoui Mohammed 64
Foutlane Mohamed 250
Guerner Dias António....198
Habib Akka....81
Haida Souad....250
Hamad H. 138
Hamzaoui Najia El....153
Hasnaoui Mustapha 20
Hassan Ben – Saïd El 198
Hassan Ouallal....153
Hefiane Fatima Zahra....64
Idadoub Hicham....330
Ikhmerdi H. 1, 11, 29, 46
Jammoukh Mustapha 171
Kadiri L....102, 138
Kammann Claudia....129
Karaoui Ismail....233
Kossir A....11
Kritihi Assia....20
Lafreme C. 102
Lahcen Messaoudi....153
Laknifli A. 93
Lebkiri A.....102, 138
Lebkiri I. 102, 138
Maatouk M. 119
Mahboub A. 1
Mamouni R. 93
Manar Ahmed....285
Mansouri Khalifa 171
Maslouhi M. R. 29
Maychal Abdelaziz....20
Mediouni Mohamed....330
Mesmoudi H....46
Messlouhi Rachid 233
Mohamed Azrour....153
Mohammed Machkor 153
Mohammed Sbiti....153
Mohammed Yaichi....321
Mouatassime Sabri El....233
Moujahid Y. 180
Mouradi Abdellah 73
Msaad Asmaa 110
Msatef Karim....222
Naser Ramzy....40
Ouaissa Khadija 20
Ouargaga F. 180

AUTHOR INDEX

Ouass A. 102, 138
Oumessoud Youness 20
Rifi E. H. 102, 138
Sabri E. 46
Saffaj N. 93
Said Housni 81
Salhi Bachir 171
Sibari Hayat 250
Tafoughalti K. 180
Tahiri Abdelouahid 212
Tahri Latifa 64
Taous Ali 242
Werner Koyro Hans 129
Yahyaoui Ahmed 64
Younsi A. 1
Zebiri Mohamed 330
Zitouni A. 1, 46, 268, 307
Zitouni Abdelhamid 198
Zohra Ainlhout Lalla Fatima 129
Zouhri Abdeljalil 110